Modelling and Simulation in Science

THE SCIENCE AND CULTURE SERIES – ASTROPHYSICS

Series Editor: A. Zichichi

6th International Workshop on Data Analysis in Astronomy

Modelling and Simulation in Science

Erice, Italy 15 – 22 April 2007

Edited by

Vito Di Gesù
Università degli Studi di Palermo, Italy

Giosuè Lo Bosco
Università degli Studi di Palermo, Italy

Maria Concetta Maccarone
IASF-Pa/INAF, Italy

World Scientific

NEW JERSEY • LONDON • SINGAPORE • BEIJING • SHANGHAI • HONG KONG • TAIPEI • CHENNAI

Published by

World Scientific Publishing Co. Pte. Ltd.

5 Toh Tuck Link, Singapore 596224

USA office: 27 Warren Street, Suite 401-402, Hackensack, NJ 07601

UK office: 57 Shelton Street, Covent Garden, London WC2H 9HE

British Library Cataloguing-in-Publication Data
A catalogue record for this book is available from the British Library.

MODELLING AND SIMULATION IN SCIENCE
Proceedings of the 6th International Workshop on Data Analysis in Astronomy << Livio Scarsi >>

ISBN-13 978-981-277-944-1
ISBN-10 981-277-944-2

Printed in Singapore by World Scientific Printers (S) Pte Ltd

"Modelling and Simulation in Science"
Sixth International Workshop of the
"Data Analysis in Astronomy ≪Livio Scarsi≫" Series
EMFCSC, Erice, Italy
15-22 April 2007

ORGANIZING COMMITTEES

DIRECTOR OF THE WORKSHOP

V. Di Gesú – CITC, Università di Palermo, Palermo, Italy

INTERNATIONAL STEERING COMMITTEE

G. Fiocco	– Università "La Sapienza", Rome, Italy
S. Fornili	– Università di Milano, Crema, Italy
J. Knapp	– University of Leeds, Leeds, UK
M.C. Maccarone (Chair)	– IASF-Pa/INAF, Palermo, Italy
F. Murtagh	– University of London, Egham, Surrey, UK
S. K. Pal	– ISI, Kolkata, India
M. Parrinello	– ETH, Zurich, CH
B. Sacco	– IASF-Pa/INAF, Palermo, Italy
M. Scarsi	– Biozentrum, Basel, CH
A.A. Watson	– University of Leeds, Leeds, UK
B. Zavidovique	– Université Paris 11, Orsay, France

LOCAL SECRETARIAT

G. Lo Bosco	– Università di Palermo, Palermo, Italy
EMFCSC Staff	– Ettore Majorana Foundation and Center for Scientific Culture, Erice, Italy

PROCEEDINGS EDITORS

V. Di Gesú	– CITC, Università di Palermo, Palermo, Italy
G. Lo Bosco	– Università di Palermo, Palermo, Italy
M.C. Maccarone	– IASF-Pa/INAF, Palermo, Italy

PREFACE

The Data Analysis in Astronomy Workshop series, started more than 20 years ago, is aimed at providing an updated overview of advanced methods and related applications to data analysis issues in astronomy and astrophysics. The series, with its previous five sessions, strongly contributed to stimulate and enforce the scientific interaction between astrophysicists and data analysis community, which discussed, debated and compared methods and results, theories and experiments.

The first edition (Erice 1984) was mainly devoted to the presentation of emerging Systems for Data Analysis (MIDAS, AIPS, RIAIP, SAIA). New methodologies for image and signal analysis were also presented with emphasis on cluster and multivariate analysis, bootstrap methods, time analysis, periodicity, 2D photometry, spectrometry, and data compression. A session was dedicated to Parallel Processing and Machine Vision.

The second workshop (Erice 1986) reviewed data handling systems planned for large major satellites and on-ground experiments (CGRO, HST, ROSAT, VLA). Data analysis methods applied to physical interpretation were considered. New parallel machine vision architectures were presented (PAPIA, MPP), as well as contributions in the field of artificial intelligence and planned applications to astronomy (expert systems, pictorial databases).

The third edition (Erice 1988) was dedicated to emerging topics (chaotic processes, search for galaxy chains via clustering, search of burst with adaptive growing) for solutions in frontiers of astrophysics (γ-ray astronomy, neutrino astronomy, gravitational waves, background radiation, extreme cosmic ray energy spectrum).

The fourth workshop (Erice 1991) provided a review of large working experiments at different energy spectra (HST, ROSAT, CGRO); goals, problems, solutions, and results of data analysis methods to experimental data were discussed. The Italian/Dutch X-ray satellite SAX was also presented. A compared review of the surviving data-analysis systems from the Erice 1984 workshop was presented (MIDAS, ESIS, EXSAS, COMPASS).

The fitfh edition (Erice 1996) mainly referred to the data analysis problems present in all the fields from radio to gamma-ray astronomy, and to the multiwavelength approach, taking into account the currently advanced methods of data fusion, information retrieval, high-speed computing. A special session was devoted to the successful launch of the X-ray astronomy satellite BeppoSAX and to its early scientific results.

All proceeding of the Data Analysis workshops were published in the Ettore Majorana International Science Series.

The sixth edition (Erice 2007) held at the Ettore Majorana Foundation and Center for Scientific Culture, Erice, Italy, was the gateway to other scientific areas. The Workshop with the subtitle of "Modelling and Simulation in Science" addressed the basic approach to the world of simulation and modelling in three branches of Science — Astrophysics, Biology and Climatology. The present status of art and adopted research lines were reported and future developments anticipated. The impact of new technologies in the design of novel data analysis systems, the interrelation among different fields such as e.g. Cosmology, Bioinformatics, Earth environment, represented the logical fallout of the Workshop. The job of putting together outstanding people from different scientific areas was hard, but today more than ever, it seems appropriate to cite the phrase, quoted by many authors, *"A mind is like a parachute. It doesn't work if it not open"*.

This proceedings includes all papers presented during the Workshop and it is organized in three main sections:

- *Astrophysics, Cosmology, and Earth Physics*;
- *Biology, Biochemistry, and Bioinformatics*;
- *Methods and Techniques.*

The success of the Workshop was the result of the coordinated effort of a number of people, from the entire Scientific Committee (Giorgio Fiocco, Sandro Fornili, Johannes Knapp, Maria Concetta Maccarone, Fionn Murtagh, Sankar Pal, Michele Parrinello, Bruno Sacco, Marco Scarsi, Alan Watson, and Bertrand Zavidovique) to the Local Secretary (Giosuè Lo Bosco), and all participants who presented contributions and/or took part in the discussions.

We wish to thank the National Institute for Astrophysics INAF, and the Università degli Studi di Palermo for their support and for including the Workshop in cultural events of the Bicentennial of the University of Palermo. Finally, we thank the entire staff of the Ettore Majorana Foundation and Centre for Scientific Culture for their support and invaluable help in organizing a successful Workshop.

Vito Di Gesù
Giosuè Lo Bosco
Maria Concetta Maccarone

On the behalf of Prof. A. Zichichi (President of the EMFCSC) the "DAA - Data Analysis in Astronomy" Workshops are from now dedicated to "Prof. Livio Scarsi" who was the enthusiastic inspirer of the series.

Memory of Livio Scarsi (25 May 1927 - 16 March 2006)

Livio Scarsi was one of the major protagonist of the physics, astrophysics and space research of the 20th century. Born 25 May 1927 in Rocca Grimalda, Italy, his reach and exemplary scientific career is substantiated by the huge number of responsibilities, assignments, collaborations, academic and honorary positions and awards. Chairman of international research programs and space missions, Livio Scarsi has carried out functions of management and scientific advisor in many institutions, such as the Italian Consiglio Nazionale delle Ricerche, the Servizio Attivitá Spaziali (now Agenzia Spaziale Italiana), the European Space Agency and the Russian Academy of Science. Member of the Accademia dei Lincei, the Academia Europea and the International Astronautics Academy, he was awarded the "*Bruno Rossi Prize*" of the American Astronomical Society and received the Laurea Honoris Causa of the *Université de Paris 7 Denis Diderot.*

Graduated in physics at the University of Genoa, Italy, he began his scientific activity in the field of elementary particles and cosmic rays. First as a student and then collaborator of Giuseppe Occhialini, he became a Physics Professor at the University of Milan and a collaborator of the Saclay Center of Nuclear Studies, France, pursuing his interests in the field of "new particles" of cosmic radiation using the technology of nuclear emulsions flown in the upper atmosphere with stratospheric balloons.

At the end of the 50's he moved to the United States. At the Massachusetts Institute of Technology, with the Bruno Rossi Group; together with John Linsley he realized, in the desert of Volcano Ranch, New Mexico, the first giant array for Extensive Air Showers. Thanks to John and Livio they discovered the existence of cosmic particles of very high energy ($> 10^{19}$eV).

Back in Italy, after a short parenthesis at the University of Rome, Livio Scarsi became in 1967 Full Professor of Advanced Physics at the Sciences Faculty of the University of Palermo, where he activated a new field of research: the High Energy Astrophysics. He continued pursuing his interests in the research on rare components of the Cosmic Radiation with detectors on board of stratospheric balloons and rockets. One of the most relevant scientific results was the detection of pulsed emission of Gamma Radiation above several GeV from the Crab Nebula Pulsar PSR0531+21.

This activity continues with COS-B, the first European survey satellite to explore the gamma-ray sky. COS-B provided the first complete map of the γ-ray emission in the Galaxy above 50 MeV, together with the identification of galactic and extragalactic sources and the first catalogue of γ-ray sources above 50 MeV, promoting the gamma-ray astronomy to an adult and recognized branch of Astronomy.

In the course of the 70's, the research activities of the group led by Scarsi grew to such an extent as to necessitate the establishment in Palermo of the Istituto di Fisica Cosmica ed Applicazioni all'Informatica, IFCAI (now IASF Palermo), of the National Research Council, specifically dedicated to the realization of great projects of Space research. Livio Scarsi was appointed Director of the new Institute.

The inclusion of the name "Informatica" reflects his deep understanding and intuition of the fundamental role played by information science methods for a better understanding of complex experimental data. Following this idea, Livio Scarsi promoted, in the middle of the 80's, the "Data Analysis in Astronomy" Workshop series at the "Ettore Majorana Foundation and Centre for Scientific Culture" in Erice, Italy. This marked the beginning of similar symposia worldwide. With its five editions, the series has provided an updated overview of advanced methods and related applications to astronomy and astrophysics, allowing astrophysicists and computer scientists to discuss, debate and compare results and methods, both in theory and in experiments. The sixth edition followed the spirit and the indications Livio provided until few months ago.

The most remarkable success of Livio Scarsi has surely been the realization of the satellite for X-astronomy BeppoSAX, launched in 1996 and named in honor of Giuseppe (Beppo) Occhialini. BeppoSAX has been a space venture of extraordinary success and a landmark in X-ray astronomy. It has promoted a fundamental progress in the various branches of galactic and extragalactic high-energy astrophysics, documented by more than 2000 scientific articles and reports. The highlight is represented by the discovery of the source counterpart of the Gamma Ray Burst (GRB), solving a mystery remained untouched for about 30 years following the first detection of GRBs. For this, Livio Scarsi, as leader of the BeppoSAX Team, was awarded the *1988 Bruno Rossi Prize* of the American Astronomical Society.

Livio Scarsi entered in the new millennium with the proposal of a new and ambitious space mission. The project, named EUSO (Extreme Universe Space Observatory), concerns the realization of a sophisticated instrument for the detection of cosmic rays of highest energy. More than 100 researchers from scientific institutions in Europe, USA and Japan responded to this challenge. Livio will be remembered by his numerous colleagues and friends as the leader of great international collaborations. They will never forget Livio's juvenile enthusiasm and great humanity.

Antonino Zichichi

CONTENTS

PART A

Astrophysics, Cosmology and Earth Physics

SIMULATIONS FOR UHE COSMIC RAY EXPERIMENTS

JOHANNES KNAPP

School of Physics and Astronomy, University of Leeds, Leeds LS2 9JT, UK
E-mail: j.knapp@leeds.ac.uk
www.ast.leeds.ac.uk/~knapp

The simulation of air showers in the atmosphere is indispensable for experiments aiming at cosmic rays and gamma-rays well above 100 GeV. Simulations are equally important for the data interpretation, the optimization of reconstruction algorithms and the design of new experiments. Over the last 15 years the quality of simulations has greatly improved, mainly due to better hadronic interaction models and the vast increase in computing power. This article reviews the current status.

Keywords: UHE Cosmic Rays, Air Showers, Monte Carlo Simulations.

1. Cosmic Rays and Air Showers

The cosmic ray energy spectrum extends over many orders of magnitude, to energies energies above 10^{20} eV. Energies above 10^{18} eV are called ultra-high energies (UHE). UHE cosmic rays (CR) are, therefore, by far the most relativistic particles known. The spectrum follows nearly a power law which falls with rising energy about like $\propto E^{-3}$. At the highest energies, the flux is smaller than 1 particle per km^2 and century. So far it is unclear where these particle come from. As they are likely charged, they are deflected in galactic and intergalactic magnetic fields such that their arrival direction does not point back to their origin. No anisotropy has yet been observed. For the very highest energies this is expected to change. For CRs above about 6×10^{19} eV the universe is opaque due to their interactions with the cosmic microwave background, thus, the most energetic CRs should come from nearby sources (< 100 Mpc), and the intergalactic magnetic fields are believed not to be strong enough to bend the CR trajectories significantly. Therefore, the detection of an anisotropy and CR sources seems likely, provided one can collect a sufficient number of events. These rare cosmic rays can only be detected via their interaction in the atmosphere which produces billions of secondaries scattered over tens of square kilometers on the ground. While this particle multiplication helps greatly in detecting the showers, it means that energy and mass of the primary particle have to be deduced from the properties of the air shower. The development of an air shower depends not only on the primary mass and energy, but also on the details of the electromagnetic and hadronic interactions in the atmosphere, the particle transport and decay, and on statistical fluctuations in the individual processes.

2. The Pierre Auger Observatory

The Pierre Auger Observatory [1] has been conceived to answer the main questions related to the highest energy cosmic rays: Where do they come from? What are they? How are they accelerated? The Pierre Auger Observatory is located in Argentina and consists of an array (SD) of 1600 water-Cherenkov detectors which are arranged on a hexagonal grid with 1.5 km distance, covering a total of 3000 km^2. In addition, 24 fluorescence telescope (FD) survey the atmosphere over the array to record the nitrogen fluorescence light induced by the numerous secondaries in a shower. While the SD operates 100% of the time, the FD can only work in dark, clear nights which amounts to a duty cycle of 10% only. The FD, however, provides a calorimetric energy measurement, which is model independent, whereas the energy reconstruction of SD events relies on simulations, which are uncertain. On the other hand, the aperture is constant and easy to determine with the SD, but highly variable with energy and uncertain with the FD. Thus, the two techniques complement each other and allow valuable cross-calibration and systematics checks. Despite the low flux of cosmic rays, with Auger about 10 events per year with energies $> 10^{20}$ eV can be detected. Fig. 1 shows schematically an air shower and the two experimental techniques. In Fig. 2, a water-Cherenkov detector is shown in the field. It contains 12 t of water in which relativistic shower particles produce Cherenkov light. This is recorded by 3 photomultipliers (PMTs) which are read out

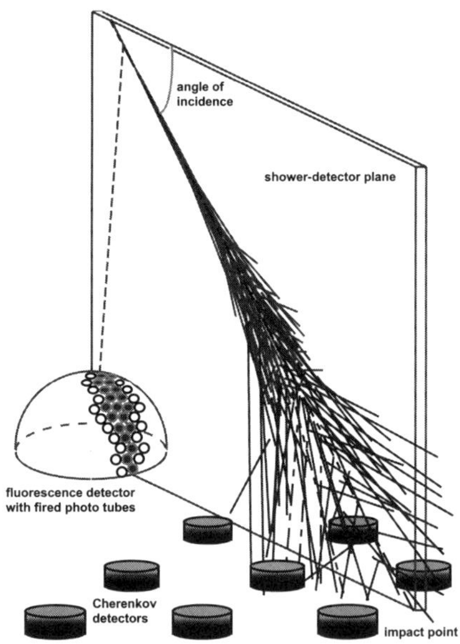

Fig. 1. Hybrid detection of air showers with the Pierre Auger Observatory: an array of surface detectors combined with fluorescence telescopes.

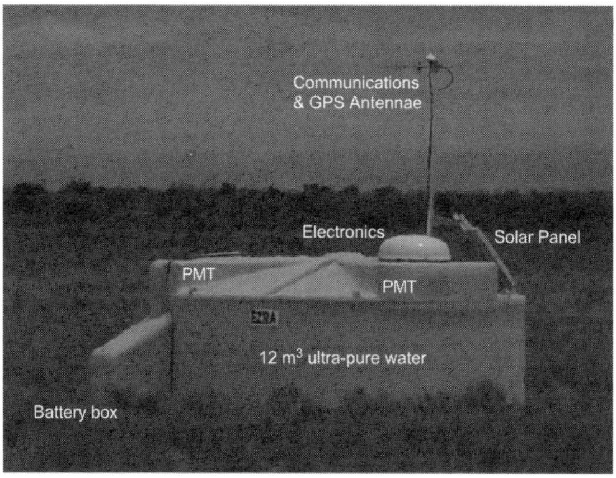

Fig. 2. One of 1600 Auger water-Cherenkov detectors in the field in Argentina.

all 25 ns with Flash ADCs. The data are transfered to the central data acquisition via wireless radio links. Fig. 3 shows a fluorescence telescope with the aperture and the ring of corrector lenses (Schmidt optics), the focusing aluminum mirrors and the 400 pixel PMT camera.

Fig. 3. An Auger Fluorescence Telescope. Left: focusing mirror. Right: aperture with filter and corrector ring and 440 pixel PMT camera.

3. Simulations versus Models

The following definitions capture well the two main types of numerical tools applied in astroparticle physics: **Simulation** is the imitation of the behavior of some situation or process by means of a suitably analogous situation (e.g. on a computer). A **Model** is a simplified or idealized description or conception of a particular system, situation, or process, that is put forward as a basis for theoretical or empirical understanding. It is a conceptual or mental representation of something.

Large and complex problems can usually be dissected in smaller and simpler, but inter-dependent, sub-problems. The simulation provides then the *numerical convolution* of these individual parts to a greater and more complex whole. Actually, this is how nature works, and the analysis of ever more fundamental sub-processes is one of the most successful strategies of science to advance our understanding. If the sub-processes are known in all details, then the numerical simulation produces the *correct* result, with all correlations, biases and selection effects, even with new features emerging from the complex interplay of the sub-processes. If not all details are known or it is impractical to do a full simulation (which is often the case) then models of reality are used employing simplifications, assumptions, or approximations. But the more simplification are made, the more care is needed to ensure that the model is good enough for the specific purpose and that the simplifications do not affect the results. Therefore, the simulation of elementary processes is the method of choice for a complex problem, such as the formation of air showers in the atmosphere.

4. Air Shower Simulations and the CORSIKA Program

Air shower physics aims at the a priori unknown energy and mass of the primary particles. These have to be reconstructed from the properties of the showers. Monte Carlo methods, using random numbers, naturally account for the statistical nature of the particle production and tracking processes, and give automatically the correct fluctuations of air shower observables. The great challenge of air shower simulations is that interactions of nuclei, nucleons, pions and all other particles that can be produced in interactions with nuclei in the atmosphere need to be simulated for energies from 1 MeV all the way up to $> 10^{20}$ eV, and that nuclear and hadronic, diffractive and non-diffractive, and low and high energy interactions are all modeled in a consistent way. While many of the processes relevant to the shower development (e.g. electromagnetic particle production, decays, particle transport, ...) are well known and thus easy to simulate, the details of the high-energy nuclear and hadronic interactions are uncertain. There is no fundamental theory underpinning theses reactions and the energies of interest are orders of magnitude beyond what is reached by man-made accelerators. Moreover, the huge number of secondaries ($> 10^{12}$ secondaries in a 10^{20} eV shower) requires statistical subsampling (thinning), where only about 1 in 10^5 particles is followed, to reduce computing time and disk space. Thus, applications of models cannot be entirely avoided and it has to be checked what the effects of the largely extrapolated hadronic interaction models and the drastically reduced particle sample, with its artificially enhanced fluctuations, on the properties of the full shower are.

The most popular and general tool for air shower simulations is the CORSIKA program [2, 3]. CORSIKA simulates the fully 4-dimensional development of air showers by following each individual particle to its interaction or decay. The mass, position, energy and direction of all particles arriving at the observation levels are stored for subsequent analysis. CORSIKA is composed of a framework that treats particle tracking, decays and in- and output and a series of interaction modules that have been developed independently and are applied within CORSIKA. These are the well-proven EGS4 package, for the simulation of all known electromagnetic processes, the packages GHEISHA, FLUKA, UrQMD for low-energy hadronic interactions and QGSJET, DPMJET, SIBYLL, neXus, models at low and high energies (for references on the interaction models used see the CORSIKA documentation [3]). The hadronic interaction packages have a considerable complexity of their own and runtime and code size vary by factors 2-40. Currently, the recommended modules are FLUKA and QGSJET II, as FLUKA describes the low-energy interaction in greatest detail, and QGSJET seems to agree best (at the 20-30% level) with a variety of astroparticle experiments from a wide range of energies. The availability of several modules within CORSIKA allows the assessment of systematical errors due to the choice of the model.

CORSIKA is used from GeV up to energies beyond 10^{20} eV, by experiments measuring cosmic rays in emulsions, by ground-based gamma-ray experiments, cosmic ray arrays of all sized and space experiments. Even underground experiments use CORSIKA to investigate their background of atmospheric muons. Overall, good agreement between experiments and simulations is found, indicating that cosmic rays are nuclei of mixed composition (as for CR at lower energies) and not photons or neutrinos. The differences between various hadronic primaries, i.e. proton to Fe, are more difficult to measure. Differences in mass-sensitive observables, such as the height of the shower maximum X_{max} from the FD, the signal risetime or the shower front curvature, both seen with the SD, are not much larger than the systematic errors and the fluctuations, such that results on mass composition remain quite uncertain. However, today a mixed nuclear composition seems consistent with measurements over most of the energy range. This was not the case 15 years ago. Then the state-of-the-art models predicted totally different showers than the measured ones and could not be used to interpret the data. The agreement between models and data is illustrated in Fig. 4 where predictions of the position of the shower maximum X_{max} for p, Fe and gamma-ray primaries are compared with experimental data. While the data clearly rule out a dominant fraction of gamma-rays at high energies, they are well between the prediction for p and Fe. In view of the fact that hadronic interaction models are not based on a fundamental theory, and that they are tuned at energies below 10^{12} eV and then extrapolated by 8 orders of magnitude in energy, this agreement is remarkable.

CORSIKA has options for very inclined and upward going showers, Cherenkov and fluorescence light production, modified atmospheric profiles, detailed interaction tests, electromagnetic pre-showering in the Earth magnetic field, and many other special applications. There are also options to visualize the shower development. Fig. 5 shows a simulated proton shower of 10^{15} eV at 45°. A rich structure is visible and in the periphery of the shower individual particles and their reactions can be seen. A collection of images and shower movies are available from Ref. 5. They illustrate the complexity of the shower development and allow identification of many physics processes implemented in the program.

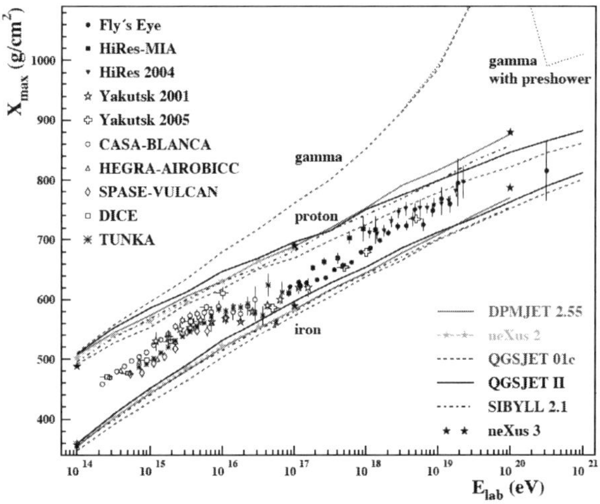

Fig. 4. X_{max} as function of energy. Models are compared to experimental data. A mixed composition is consistent with the data, a large fraction of primary photons is not.

5. Some Selected Details

Decay versus interaction

A simple example to illustrate the workings of the Monte Carlo method is the calculation of the point of the next interaction for a particle where both, inelastic collision and decay compete (e.g. for the $\pi^{\pm}$). The distribution of decay times is given by $\frac{dN}{dt} \propto e^{-t/(\tau_0\gamma)}$, where τ_0 is the lifetime of the particle at rest and γ is its Lorentz factor. From this distribution a specific t is picked at random which can be directly translated into a decay length s_{d} (in units of cm) via $s_{\mathrm{d}} = \beta ct$. Then, an interaction path length x is picked at random from the interaction length distribution $\frac{dN}{dx} \propto e^{-x/\Lambda_0}$ where Λ_0 is the interaction mean free path (in units of g/cm^2). x is then converted into an interaction distance s_{i} (in cm) via $s_{\mathrm{i}} = x/\rho$, with ρ being the height-dependent density of the atmosphere. As decay and interaction

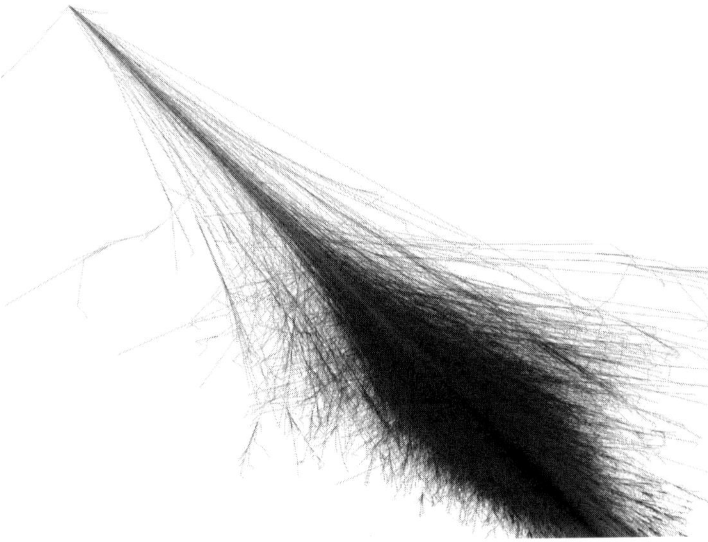

Fig. 5. A simulated shower of a 10^{15} eV proton at 45° zenith angle. The full height of the image is 30 km.

are independent of each other, the process with the shorter distance will happen. If $s_\mathrm{i} < s_\mathrm{d}$ the particle is tracked for s_i cm and then an interaction is performed, otherwise it is tracked for s_d cm and a decay is simulated. This is a very simple and fast algorithm that mimics exactly the process in nature, including the energy and density dependent ratio between interaction and decay and the fluctuations in path length.

Random Numbers

The Monte Carlo method relies on the use of random numbers to select specific values from an allowed range with a given distribution. While random numbers could be produced from truly random processes, such as radioactive decays or electronic noise, this is not practical for computer programs as they have to run in a reproducible way in order to search for programming errors. Therefore, algorithms are used to create pseudo-random numbers in a deterministic way, which behave in all respects like true random numbers, i.e. all digits and all combinations of digits appear with equal probability and there are no correlations within the sequence. But pseudo-random number sequences are not really random and have by construction a finite sequence length (period). For a good overview of random number generation see Ref. 4. Random number generators usually create numbers which are uniformly distributed between 0 and 1. From these it is then easy to create numbers with any other distribution. A very simple uniform generator is the linear congruent generator. It starts with a seed R_0 and creates random numbers U_j recursively via the simple formulae $R_j = a(R_{j-1} + b) \mod m$, and $U_j = R_j/m$. It has 3 free parameters: a, b and m, which are all integers. R_j are then integers between 0 and $m - 1$ and U_j are the real pseudo-random numbers between 0 and 1. The maximum period is m, but the actual period depends on choice of a and b. It is not clear a priori which choice of a and b gives a good performance. Typical period lengths are 10^{11} to 10^{15}, which is not enough for serious applications. Moreover, random numbers from this generator are correlated: k-tupels of random numbers usually lie on (k-1)-dim hyper planes and the less significant bits are usually less random. More complex generators are designed by combining two or more simpler methods. For instance, one can combine two random numbers from different generators with "+", "-", or "exclusive or" bit operations, or use the random number sequence of one generator as address to pick the random number from a second sequence. These methods produce much better randomness and far larger sequence lengths. In CORSIKA the generator RANMAR (from the CERN software library) is used. It produces 32-bit floating point numbers uniformly distributed between 0 and 1, and allows for 900.000.000 independent sequences of $\approx 2^{144} = 10^{43}$ period length each. There are even better generators, but the better the random number generator, the more computing time it requires. CORSIKA needs about 5×10^9 random numbers per hour of shower simulations. This amounts already to about 30% of the computing time.

Thinning

The runtime and disk space needed for one shower scales roughly with energy: it is about 1 h $\times$ E/10^{15} eV (on a modern workstation) and 300 MB $\times$ E/10^{15} eV, respectively. This makes the simulation of showers above about 10^{16} eV very difficult. A shower of 10^{20} eV, which contains about 10^{12} secondary particles, would run for 11 years and produce 30 TB output. Therefore, a speed-up mechanism is introduced, whereby only a statistical subset of particles is followed. The particles that are followed acquire a weight to account for the discarded particles. This procedure is called *statistical thinning*. It is very similar to election polls, where the final election result is forecast from questioning a small, but representative, sub-sample of voters. Thinning accelerates the simulation typically by a factor of 10^5, leaving the energy conserved, and the average particle numbers and distributions unchanged in regions where still enough particles are present. However, by thinning the fluctuations are artificially enhanced, which becomes noticeable in sparsely populated tails of the distribution. The statistical weights of the particles at observation level have to be removed by a suitable re-sampling procedure before the detector response can be simulated.

6. Outlook

The current uncertainty for most of the observables in an air shower is below 30%. Data from the new accelerators RHIC and LHC on nuclear and hadronic interaction cross-sections and particle production, at higher energies than available before, will constrain hadronic interaction models and thereby improve the extrapolation. Therefore, the residual systematic uncertainties may become soon smaller. However, there are many uncertainties on the 10% level in various part of the simulation programs, such that an overall precision of 10% seems very difficult to achieve.

Acknowledgement

The author is very grateful for the invitation to the 6th International Workshop on Data Analysis in Astronomy: *Modelling and Simulation in Science.* This article is dedicated to the memory of Livio Scarsi, who was a great scientist, a friendly and inspiring colleague and a deeply human person.

References

[1] For information on the Pierre Auger Observatory and first publications see `http://www.auger.org` and follow the link to "Scientific and Technical Information".
[2] D. Heck, J. Knapp, J.N. Capdevielle, G. Schatz, T. Thouw Forschungszentrum Karlsruhe Report FZKA 6019 (1998)
[3] `http://www-ik.fzk.de/corsika`
[4] F. James, 'A review of pseudorandom number generators', Computer Physics Communications 60 (1990) 329-344
F. James, 'Monte Carlo theory and practice', Rep. Prog. Phys. vol. 43 (1980) p. 1145-1189
[5] CORSIKA shower images and movies:
`http://www.ast.leeds.ac.uk/~fs`
`http://www.ast.leeds.ac.uk/~knapp/movies/EASmovies.html`
`http://www-ik.fzk.de/corsika/movies/Movies.htm`

DETECTOR MODELING IN ASTROPARTICLE PHYSICS

SERGIO PETRERA

INFN and Dipartimento di Fisica, Università di L'Aquila,
L'Aquila, via Vetoio, 67010, Italy
E-mail: sergio.petrera@aquila.infn.it

Detector modeling is an important step for the interpretation of experimental data in astroparticle physics. In this paper the most specific features of such process are shown, making use of two remarkable examples: the atmospheric neutrinos in MACRO and the Ultra High Energy cosmic rays in the Pierre Auger experiment.

Keywords: Astroparticle Physics, Simulation

1. Introduction

Detector modeling is a crucial step for the interpretation of experimental data in astroparticle physics. This modeling is usually done using the same simulation tools as in experiments at particle accelerators. Among these the GEANT4 simulation toolkit [1] is widely used. It is the last generation of a very successful family of simulation codes (GEANT [2]), initially developed at CERN for particle physics experiments, aiming to provide simulation tools of the passage of particles through matter. More recently GEANT has been rewritten as GEANT4 adopting the software engineering methodologies and the Object Oriented technologies. Furthermore its application has been extended to more fields, such as medical physics, astrophysics, space applications, background radiation studies, etc. For more specific purposes experiment specific simulation codes are developed as well.

Despite several common points with particle physics, there are features that are specific of detector simulation in astroparticle physics:

- apart few cases (e.g. space searches), the observation is always *indirect.*
- This means that one has to infer primary physics parameters from *secondary particles.*
- Consequently, the detector simulation has to be naturally extended to the *surrounding environment*, where such particles are produced and develop.

In order to make these points clear and possibly more evident, I will proceed through examples. Next section will show the detector simulation done for the atmospheric neutrino observation in MACRO. In this case the rock surrounding the detector becomes part of the detector simulation. In the following section the case of UHE cosmic rays observed in the Pierre Auger Observatory is shown. Here the Earth atmosphere above the detection apparatus plays the same role in the detector simulation.

2. MACRO as a ν detector

The MACRO apparatus was located in the Gran Sasso underground laboratory, with a minimum rock overburden of 3150 hg/cm^2. It was a large rectangular box divided longitudinally into 6 supermodules and vertically into a lower and an upper part, called the *attico*. For a full description of the apparatus see [3]. The active elements were liquid scintillation counters for time measurement and streamer tubes (ST) for tracking. The lower half of the detector was filled with trays of crushed rock absorber alternated with ST planes, while the *attico* was hollow and contained the electronics racks and work areas. The rock absorbers set a minimum energy threshold for vertical muons of 1 GeV.

The tracking system allowed the reconstruction of the particle trajectory in different views [3]. The intrinsic angular resolution for muons typically ranged from $0.2°$ to $1°$ depending on track length; the angular spread due to multiple Coulomb scattering in the rock and to kinematical angle for neutrino-induced muon was larger than this resolution.

The scintillator system consisted of horizontal and vertical layers of counters. Time and longitudinal position resolution for single muons in a counter were about 0.5 ns and 10 cm, respectively. Two different thresholds were used for the timing of these two outputs and the redundancy of the time measurement helped to eliminate spurious effects.

Thanks to its large area, fine tracking granularity and electronics symmetry with respect to upgoing and downgoing flight directions, the MACRO detector was a proper tool for the study of upward traveling muons. Further, it was sufficiently massive (5.3 $kton$) that it also collected a statistically significant sample of neutrino events induced by internal interactions.

2.1. *Atmospheric neutrinos and their oscillation*

Bruno Pontecorvo was the first, already in 50's, to mention the possibility of neutrino oscillations, more precisely of neutrino $\leftrightarrow$ antineutrino transitions in vacuum [4]. Since then, many experiments on neutrino oscillations have been made with solar, reactor, accelerator and atmospheric neutrinos.

Assuming mixing of two neutrino flavors (for example, ν_μ and ν_τ) and two mass eigenstates (ν_2 and ν_3), the survival probability for muon neutrinos is

$$P(\nu_\mu \rightarrow \nu_\mu) = 1 - \sin^2 2\theta_m \sin^2\left(1.27\Delta m^2 L_\nu / E_\nu\right) \tag{1}$$

where θ_m is the mixing angle, $\Delta m^2 = m_3^2 - m_2^2$ is the difference of the squares of the eigenstate masses (in eV^2), L_ν is the neutrino path–length (in km) and E_ν is the neutrino energy (in GeV).

Atmospheric neutrinos are a unique "beam" to investigate neutrino oscillations since they cover a region of parameter space until now unexplored by man-made neutrino beams. The energies extend from fractions of GeV up to several tens of TeV, and the baseline varies from about 20 km at the zenith to about 13000 km at the nadir.

In MACRO the streamer tubes and the scintillation counters made possible the identification of neutrino events on the basis of time-of-flight (ToF) measurements, as well as by topological criteria. Four different classes of neutrino events were detected (see Fig. 1):

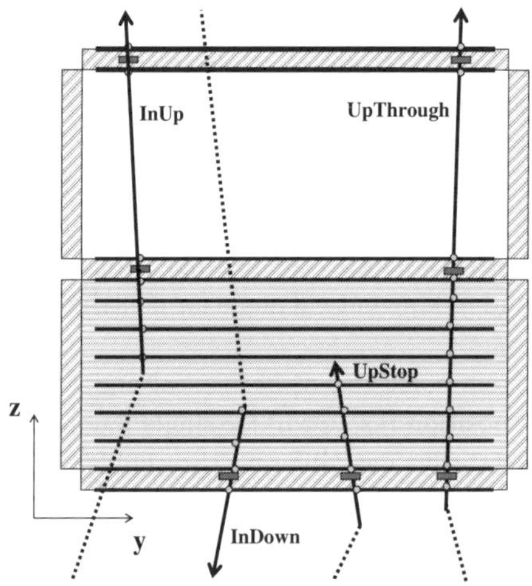

Fig. 1. Topologies of events induced by muon neutrino interactions inside or below the detector. The circles indicate the ST hits and the boxes, the scintillator hits.

-*UpThrough*. This is the largest neutrino event sample for MACRO. It is characterized by upward-going muons crossing the detector; they originated in charged current (CC) ν_μ–interactions in the rock below the apparatus. The direction of flight is determined by measuring the ToF given by

$$\frac{1}{\beta} = \frac{c}{v} = \frac{c\,(t_2 - t_1)}{l_{sci}} \tag{2}$$

where t_1 and t_2 are the times measured in higher and lower scintillator planes, respectively, and l_{sci} is the path–length between the scintillators. Therefore $1/\beta$ is $+1$ for downgoing tracks and -1 for upgoing tracks to within measuring errors. For about 50% of the events the ToF is redundantly measured by more than two scintillator layers.

-*InUp*. This class includes events with an upward-going track starting inside the lower part of the detector due to a ν-interaction there. As for the *UpThrough* events, the ToF is also measured for this class.

-*UpStop.* These events are caused by ν-interactions in the rock below the apparatus producing an upgoing track which stops inside the detector. Only the lower scintillator layer is fired and so a ToF measurement cannot be made.
-*InDown.* Events due to an internal interaction associated with a downward-going track. In this case also, only the lower layer of scintillators is crossed.

The last two classes of events are recognized by means of topological criteria. Both have a track with one end in the bottom layer of scintillators.

2.2. *Physics and detector simulation*

The simulation of atmospheric neutrino events required the development of physics generators based on atmospheric neutrino fluxes and neutrino cross sections.

The FLUKA [5] atmospheric neutrino flux was used at low energy. This choice was made because of the completeness of the code and of the agreement between the new measurements of the low energy primary CR spectrum. For the *UpThrough* events, we used the Bartol flux [7]. We have checked that, to within 5%, FLUKA and Bartol calculations give the same predictions for the ratios quoted above.

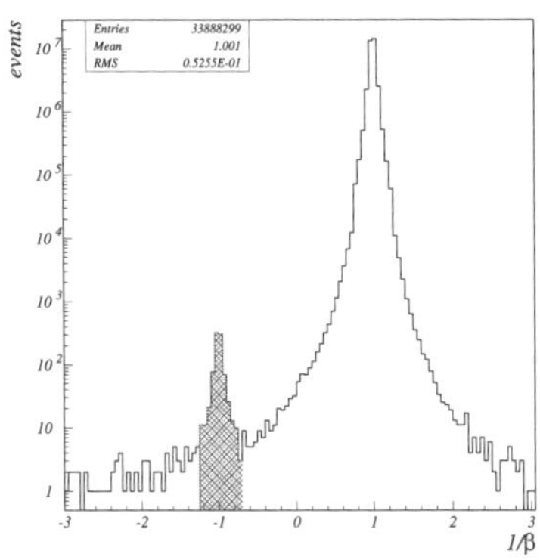

Fig. 2. $1/\beta$ distribution for long tracks. The shaded area refers to *UpThrough* events.

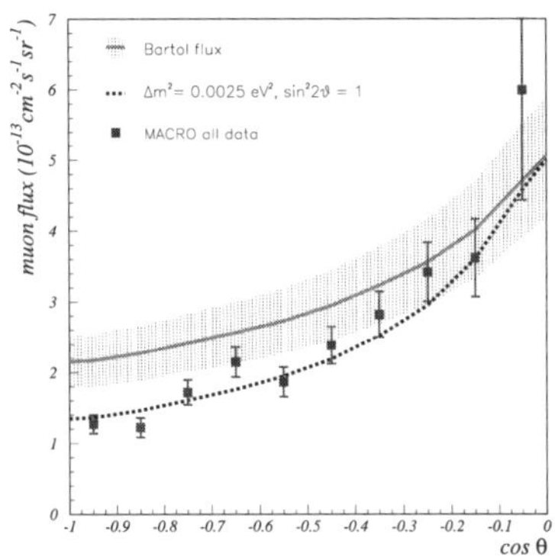

Fig. 3. Comparison between the measured angular distribution of the *Up-Through* muon flux and the MC prediction assuming ν-oscillations with the MACRO parameters. The shaded area and the included line shows the non-oscillated flux with its uncertainty.

Different neutrino event generators were used and the interactions were simulated both inside the detector and in the surrounding rock. The energy loss for muons propagating through rock was taken from [8], adjusting the energy loss for the chemical composition of the Gran Sasso rock.

A MC program (GMACRO) based on the GEANT package was used to simulate the response of the detector. The ST response was simulated by introducing the processes which affect the distributions of cluster widths (induced charge distribution on the pick-up strips, electronics performance, delta-ray production and so on). The signal from a particle traversing a scintillation counter was approximated by assuming that the energy loss occurred in the middle position between entry and exit points at the average of entry and exit times.

It has be pointed out that in the simulation the detector is extended to the rock surrounding the apparatus. This medium has an important role in three different processes:

- the production of neutrino induced muons;
- the muon transport to the apparatus;
- the background evaluation of fake upward muons.

2.3. *Data interpretation*

The *UpThrough* topology is very clear in MACRO and their detection is robust. This can be easily recognized from Fig. 2. The tracks with $1/\beta$ in the signal range $[-1.25, -0.75]$ are accepted as upward-going at the end of the analysis chain. The up-down symmetry of the detector and of the analysis chain allows to keep the downward-going muons together with the upward-going sample, which is smaller by about a factor of 10^{-6}.

Fig. 3 shows the measured angular distribution of the *UpThrough* muon flux. The comparison with the (non-oscillated) MC prediction shows remarkable differences, even including its uncertainties. A fit to the flux assuming ν-oscillation gives better results and allows to estimate the oscillation parameters in (1).

The discussion of this physics result is beyond the aim of this paper and can be found in [9], including many details of the analysis.

3. Cosmic Rays studies with the Pierre Auger experiment

The Pierre Auger Observatory [10] is an international experiment with the goal of exploring with unprecedented statistics the cosmic ray spectrum above 10^{19} eV. Of particular interest are cosmic ray particles with energy exceeding 10^{20} eV. At these energies they interact with cosmic microwave background radiation thus generating a spectrum cutoff, known as GZK effect [11]. This effect attenuates the particle flux except if their sources are in our cosmological neighborhood (< 100 Mpc). Furthermore protons of these energies may point back to the source and open a new kind of astronomy with charged particles.

The extremely low rate, a few particles per $\mathrm{km}^2 \cdot \mathrm{sr} \cdot$ century, of cosmic rays above the GZK cutoff requires a large area detector. The Auger Southern Observatory, in advanced stage of construction close to the town of Malargüe, Province of Mendoza, Argentina, covers an area of 3000 km^2 (see Figure 4).

Cosmic rays are detected by the Auger Observatory with two different experimental techniques. The Surface Detector (SD), a giant array of 1600 water

Cherenkov tanks, placed over the Observatory area with a spacing of 1.5 km, measures the shower particle density and arrival times at ground level. Presently about 1200 SD tanks are taking data.

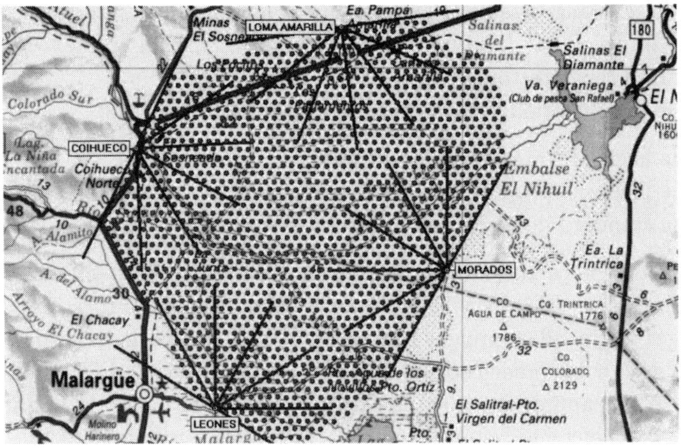

Fig. 4. A map of the Auger Southern Observatory. Dots represent SD tanks. FD Eyes are shown with their fields of view.

The Fluorescence Detector (FD), composed by a set of 24 telescopes, measures the longitudinal development of the cosmic ray shower in the atmosphere above ground. The telescopes are arranged in four peripheral buildings (*Eyes*), each housing 6 of them, overlooking the SD array. All the FD buildings are completed, with their FD telescopes taking data.

3.1. *The detector simulation*

With the purpose of studying the primary spectrum and composition from ground, data simulation requires an intermediate step. This deals with the interaction of primaries in the atmosphere and their further development into showers. This step is usually done as a separate process. Sometimes the shower generation is a preliminary step of the overall simulation process. More frequently, since it is strongly time consuming, showers are generated with pre-set features and stored in data libraries for further use. The output of this simulation contains: the particle content and the energy released along the shower development (the "longitudinal profile"); the list and the properties of the particles reaching ground level (the "ground particles"). There are several codes (e.g. CORSIKA, AIRES, CONEX, etc.) performing this task. They do not use standard simulation tools (e.g. GEANT), but consist of standalone code. They are interfaced with codes modeling the hadronic interaction at the relevant energies (e.g. QGSJet, Sibyll, Nexus, etc.). Details and references about this simulation process are given in the paper by J. Knapp [12].

The detection simulation follows the shower generation process and accesses shower data. In the case of Fluorescence Detection the longitudinal profile (more precisely the energy released along the shower development) has to be converted into photons, modeling the fluorescence and the Cherenkov emissions. These photons have to be transported to the telescope windows, taking into account possible scattering (Rayleigh and Mie) processes. This interface, from profile to telescope, is embedded into the detector simulation framework.

The role of simulation is then crucial in three different steps:

- the hadronic interaction and the shower development;
- the photon transport to the FD telescopes;
- the standard detector simulation (in the SD array, in the FD telescopes).

In the first two steps the atmosphere is modeled and is part of the simulation. Therefore in Auger the atmosphere plays the same role as the rock in MACRO.

In Auger a graded approach has been adopted for each basic detector: fast simulators with home made code, GEANT4 fast simulation and GEANT4 full simulation. The last approach, which is the most time consuming, is commonly used as a reference, preferring faster codes when high statistics is required.

3.2. *The Surface Detector*

The Surface Detector (SD) is made of water Cherenkov tanks. The tanks have 3.6 m diameter and 1.2 m height to contain 12 m^3 of clean water viewed by three 9" photomultiplier tubes (PMT). A solar panel and a buffer battery provide electric power for the local intelligent electronics, GPS synchronization system and wireless LAN communication. A picture of one tank in the field is shown in Figure 1.

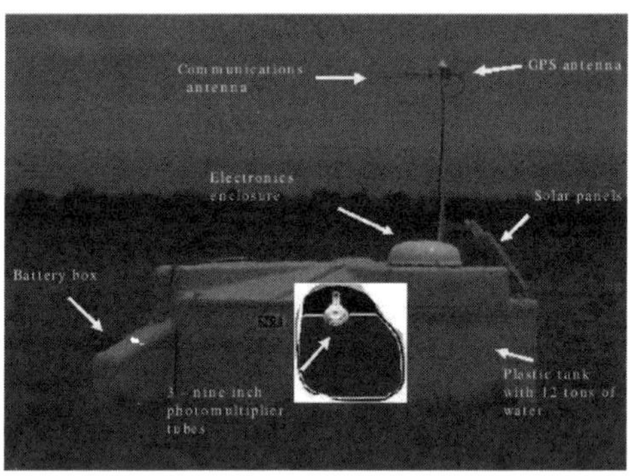

Fig. 5. A SD tank in the field. The main components of the detector are sketched in the figure.

The signals are continuously digitized with 16 bit dynamic range at 40 MHz sampling rate and temporarily stored in local memory. The trigger conditions include a *threshold trigger* (one or more FADC counts above 3.2 Vertical Equivalent Muon [*VEM*] in each of 4 or more tanks) and a *time over threshold trigger* (12 FADC bins exceeding 0.2 VEM in sliding window of 3 μs in each of 3 or more tanks).

Fig. 6 shows the implementation of the SD simulation in the Auger offline framework. The simulation is split into sequences of self contained processing steps, called *modules.* This modular design allows collaborators to easily exchange code, compare algorithms and build up a wide variety of applications by combining modules in various sequences. More details about this framework can be found in [13].

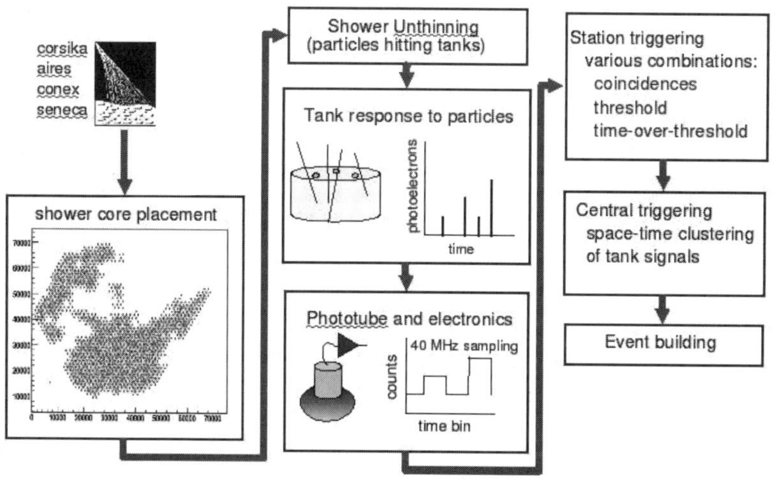

Fig. 6. Steps for simulating the surface array. Each simulation step is encapsulated in a software module.

3.3. *The Fluorescence Detector*

The Fluorescence Detector (FD) consists of 24 wide-angle Schmidt telescopes grouped in four stations. Each telescope has a 30° field of view in azimuth and vertical angle. The four stations at the perimeter of the surface array consist of six telescopes each for a 180° field of view inward over the array.

Each telescope is formed by hexagon shaped segments to obtain a total surface of 12 m^2 on a radius of curvature of 3.40 m. The aperture has a diameter of 2.2 m and is equipped with optical filters and a corrector lens. In the focal surface a photomultiplier camera detects the light on 20×22 pixels, each covering 1.5°×1.5°. The total number of photomultipliers in the FD system is 13,200. PMT signals are continuously digitized at 10 MHz sampling rate with 15 bit dynamic range. The FPGA-based trigger system is designed to filter out shower traces from the random background of 100 Hz per PMT. The atmosphere parameters are monitored, making use of laser beams, LIDAR's, calibrated light sources and continuous recording of weather conditions.

Fig. 7 shows the implementation of the FD simulation in the Auger offline framework. In particular the two modules in the up left boxes, namely the *ShowerLight-Simulator* and *LightAtDiahragmSimulator*, exploit the photon emission and their transport from the shower to the telescope in the atmosphere. The other modules handle the optical and electronic response of the detector.

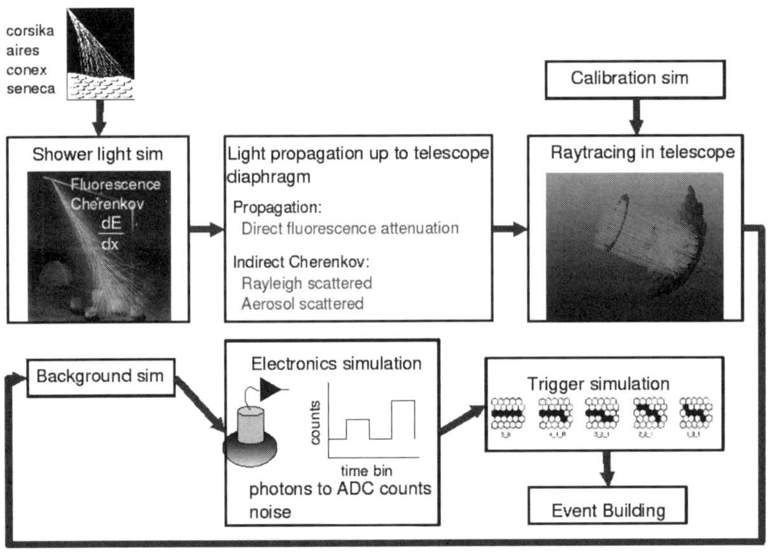

Fig. 7. Steps for simulating the fluorescence telescopes. Each simulation step is encapsulated in a software module.

3.4. *Hybrid events*

The events simulated in the Auger offline framework are written in the standard data acquisition format. Therefore they can be reconstructed as real events. Furthermore, in order to follow accurately the evolution of the detector and then reproduce its aperture at any time, for each event simulated at a given time the actual configuration of both SD and FD systems is retrieved from the relevant databases. The atmospheric data are retrieved as well from the atmospheric monitoring database. The analysis of these events provides information very useful to relate the measured parameters to the actual shower parameters and is then a fundamental tool for the interpretation of real data in terms of the underlying physics.

A case of particular interest in Auger is the class of hybrid events. These events are simultaneously detected by both systems. In this case the simulation makes use of both simulation sequences shown in Figs. 6 and 7.

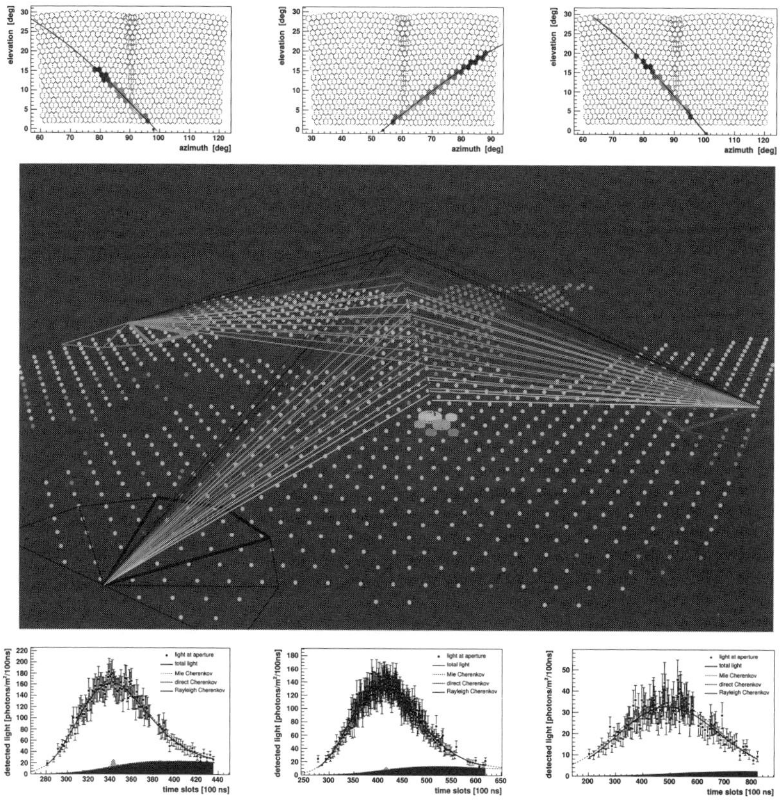

Fig. 8. A simulated hybrid event and its reconstruction. The upper pads show the pixels with first level trigger. The central figure shows the 3D pictorial view of the event as seen by the SD and FD systems. The lower pads show the reconstructed longitudinal profiles. The multiple pads refer, from left to right, to the Los Leones, Los Morados and Coihueco buildings.

Figure 8 shows one of the simulated hybrid events after reconstruction. This event hits the ground at about the same distance (ranging from 21 to 28 km) from the three Eyes active at the event time (Los Leones, Los Morados and Coihueco). The SD tanks active at the same time are also visible in the 3D picture. The event has an energy of 25.4 EeV and is reconstructed with energies of 25.0, 25.4 and 24.5 EeV respectively from Los Leones, Los Morados and Coihueco respectively. The error on these energies, including the fit procedure, but not the systematic uncertainties, are for all eyes below 10%. [a]

---

[a]This paper is dedicated to the memory of Livio Scarsi, whom I had the pleasure to meet for the first time as a student, at a balloon launch in Trapani Birgi in the mid 70's and who ever touched me for his warm humanity.

References

[1] Geant4 Collaboration, "Geant4 - a simulation toolkit", Nucl. Instr. Meth. A506 (2003) 250
[2] R. Brun, M. Hansroul and J.C. Lassalle, GEANT User's Guide, CERN DD/EE/82 edition, 1982
[3] M. Ambrosio *et al*, Nucl. Instrum. Methods A486 (2002) 663
[4] B. Pontecorvo, J. Exp. Theor. Phys. 33 (1957) 549 and J. Exp. Theor. Phys. 34 (1958) 247
[5] G. Battistoni *et al*, Astropart. Phys. 19 (2003) 269, Erratum 291
[6] T.K. Gaisser *et al*, Proceedings of 27th International Cosmic Ray Conference, Hamburg (2001) 1643
[7] V. Agrawal *et al*, Phys. Rev. D 53 (1996) 1314
[8] W. Lohmann *et al*, CERN-EP/85-03 (1985)
[9] M. Ambrosio *et al*, Eur. Phys. J. C36 (2004) 323
[10] The Pierre Auger Collaboration, "Properties and performances of the prototype instrument for the Pierre Auger Observatory", Nucl. Instr. Meth. A523 (2004) 50-95
[11] K. Greisen, Phys. Rev. Lett. 16 (1966) 748;
G.T. Zatsepin, V.A. Kuzmin, Sov. Phys. JETP Lett. 4 (1966) 78
[12] J. Knapp, "Simulations for Ultra High energy Cosmic ray Experiments", these proceedings
[13] S. Argirò, *et al*, "The Offline framework of the Pierre Auger Observatory", submitted to Com. Phys. Comm. (2007)

SIMULATING A LARGE COSMIC RAY EXPERIMENT: THE PIERRE AUGER OBSERVATORY

TOM PAUL

Northeastern University
Boston, MA 02115, USA
**E-mail: t.paul@neu.edu*

In this article we describe some of the techniques employed to simulate the response of the Pierre Auger Observatory to the extensive air showers produced by ultra high energy cosmic rays. This observatory is designed to unveil the nature and the origins of cosmic rays with energies in excess of 10^{19} eV, and comprises one nearly completed site in Argentina as well as a planned sister site in the northern hemisphere.

Complimentary air shower detection methods are employed; at the southern site, water Cherenkov detectors sample portions of the shower arriving at the ground, while a system of telescopes observes the nitrogen fluorescence which air showers induce in the atmosphere. We explain how the detector response to air showers is simulated for these different experimental approaches. Further, we elaborate on the more general software requirements imposed by the disparate simulation and reconstruction tasks taken on by the large, geographically dispersed collaboration which operates the observatory. We provide an overview of the framework which was devised to accommodate these requirements and motivate the choices of underpinning technologies.

Keywords: Cosmic rays; Simulations; Software Framework

1. Introduction

Ultra-high energy cosmic ray observatories aim to discover the origins and composition of the highest energy particles ever observed. The flux of cosmic rays with energies below about 10 TeV is sufficiently large to allow direct observation by detectors carried aboard balloons or satellites. Above this energy, however, the flux becomes so low that direct observation is impractical. Fortunately, at very high energies it is possible to use the Earth's atmosphere as a detection medium. Ground-based experiments with large apertures and exposure times can then observe the particle cascades generated when primary cosmic radiation interacts with atomic nuclei high in the atmosphere. Such cascades are known as extensive air showers (EAS), and can spread out over a large area by the time they arrive at the Earth's surface. Several techniques have been employed to study these EAS, including use of particle detectors to sample portions of the particle cascade arriving at the ground as well as fluorescence telescopes to observe the cascade as it develops in the atmosphere.

Computer models provide an essential aid in establishing the relationship between the energy, flux and chemical composition of primary cosmic ray particles and the EAS observables to which they give rise. The first step in understanding the details of this involves simulation of the physics processes leading from the first interaction of the primary cosmic ray to the shower of particles which is ultimately observed. A significant effort has been devoted to this problem [1], and the issue is discussed in other articles in these proceedings [2]. In this note we consider the second step in the process: modeling the response of the detector apparatus to the EAS. Simulation of the detector allows researchers to relate the signals generated by the detector readout to the observables of interest, and to quantify the effects of detector idiosyncrasies and backgrounds. Simulations are also important for computations of the instrument's aperture, which is used to convert the number of observed events at a particular energy to the cosmic ray flux. Such issues are discussed in other contributions to these proceedings [3]. Here we will consider the specific case of the Pierre Auger Observatory [4]. This case study is an interesting one to consider, as it illustrates some of the challenges posed by assembling simulation codes for a complex cosmic ray experiment operated by a large, geographically dispersed collaboration.

The article is organized as follows. Sec. 2 provides a brief description of the Pierre Auger Observatory. Sec. 3 describes the main features of the software framework which allows collaborators to work together to build up the various pieces required for a full simulation the observatory. Sec. 4 then provides an brief overview of how simulations proceed from an EAS model up through generation of the signals registered by the observatory instruments.

2. The Pierre Auger Observatory

The Pierre Auger Observatory exploits two complimentary methods to measure the properties air showers. Firstly, a collection of telescopes is used to measure the fluorescence light produced by excitation of nitrogen induced by the cascade of particles in the atmosphere. Fluorescence light enters a telescope through a 1.1 m diaphragm, and is focused by a spherical mirror on a set of 440 photomultiplier (PMT) tubes. Twenty-four telescopes are in operation, distributed at four sites of 6 telescopes each. The second shower detection method employs an array of detectors on the ground to detect particles as the air shower arrives at the Earth's surface. Each of these surface detectors consists of a cylindrical tank containing 12 tons of purified water instrumented with three photomultiplier tubes to detect the Cherenkov light produced when particles pass through it. When deployment is completed, there will be a total of 1600 surface detectors spaced 1.5 km apart on a hexagonal grid. A schematic depiction of the surface array and fluorescence telescope layout is shown in Fig. 4. In order to observe potential cosmic ray sources across the full sky, the baseline design of the observatory calls for two sites, one in the southern hemisphere and one in the northern. The southern site is located in Mendoza, Argentina, and construction there is nearing completion. The state of Colorado in the USA has been selected as the location for the proposed northern site.

Each detection technique has its own strengths and limitations. The fluorescence telescopes can only operate effectively when the sky is moonless and clear, and thus runs with roughly a 10% duty cycle. Furthermore, the fluorescence detection aperture is somewhat challenging to compute, as it depends on the shower energy as well as atmospheric conditions, background light, and the performance of the instrument. Simulations are invaluable for taking such details into account. In contrast, the surface detectors operate continuously and have a fixed aperture for showers with energies sufficiently above threshold.

On the other hand, the fluorescence telescopes measure quantities which tend to be more directly related to properties of the primary cosmic ray than the surface array does. For instance, the fluorescence measurement is more-or-less calorimetric, with shower brightness related to shower energy. In addition, the depth at which the shower reaches maximum brightness can be directly observed by the telescopes, and this quantity provides clues about the chemical composition of the primary particle. In contrast, the surface array does not directly observe the shower as it evolves in the atmosphere. Instead, the Cherenkov detectors measure signals related to track length of particles traversing the water volume for tanks at different distances from the shower axis, as well as the shape and overall time structure of the shower front. Interpreting these measurements in terms of primary energy and composition relies on computer modeling of both the shower properties and the detector response. Showers which are detected by both instruments are called hybrid events, and provide an invaluable tool for cross-calibration and for understanding the particular systematics associated with both instruments.

3. Software Framework

Simulation of large experiments requires a common software framework which is flexible enough to accommodate the collaborative effort of many physicists developing a variety of applications over a long time period[a]. The offline software framework [5] of the Pierre Auger Observatory was designed to provide an infrastructure to support the distinct computational tasks necessary not only to simulate, but to reconstruct and analyze data gathered by the observatory. It features mechanisms to retrieve data from many sources, deal with multiple file formats, support "plug-in" modules for event simulation, reconstruction and analysis, and manage the abundance of configuration data needed to direct a variety of applications. An important design goal was to ensure that all physics code is "exposed" in the sense that any collaboration member can replace existing algorithms with his or her own in a straightforward manner.

The offline framework comprises three principal parts: a collection of processing *modules* which can be assembled and sequenced through instructions provided in an XML [6] file, an *event* structure through which modules can relay data to one

[a]The Pierre Auger Observatory will operate for about 20 years, for example.

another and which accumulates all simulation and reconstruction information, and a *detector description* which provides a gateway to data describing the configuration and performance of the observatory as well as atmospheric conditions as a function of time. The principal ingredients are depicted in Fig. 1.

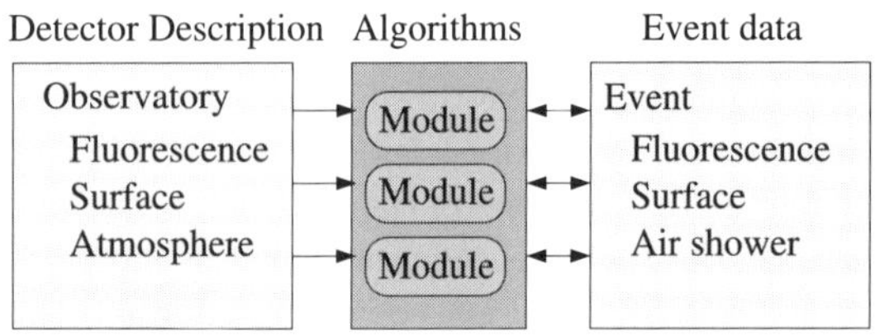

Fig. 1. General structure of the offline framework. Simulation and reconstruction tasks are broken down into modules. Each module is able to read information from the detector description and/or the event, process the information, and write the results back into the event. The Detector description provides a single gateway to information about he observatory's performance as a function of time. Such data typically reside either in XML files or in MySQL [7] databases.

This approach of pipelining processing modules which communicate through an event serves to separate data from the algorithms which operate on these data. Though this approach is not particularly characteristic of object-oriented design, it was used nonetheless since it better suits the requirements of collaborating physicists who wish to develop and refine simulation and reconstruction algorithms.

These principal components mentioned above are complemented by a set of foundation classes and utilities for error logging, physics and mathematical manipulation, as well as a unique package supporting abstract manipulation of geometrical objects.

4. Detector Simulation

Dedicated air shower simulation packages [1] generate output files containing either a description of the longitudinal development of the shower as it descends through the atmosphere, a list of particles hitting the ground, or both. Simulation of the observatory response takes such files as input, and proceeds through a succession of stages to simulate the behavior of all detector components, and ultimately produces a data file mimicking that produced by the actual detector. In the offline framework described in Sec. 3, each of these simulation and reconstruction stages is encapsulated in a module.

We now describe some of the physics modules used for simulating the surface and fluorescence detectors, though due to space constraints it is impossible to present all the details. For many steps in the simulation procedure there may exist a number of plausible approaches, ranging from fast parametrizations to extremely detailed modeling of detector behavior. Owing to the modular nature of the offline framework, different approaches to a particular simulation step can be plugged into the full simulation chain and easily compared with one another and with real data.

4.1. *Surface array*

The surface array simulation starts from a list of ground particle positions, momenta and particle types produced by an EAS simulation program. The shower core is placed within a model of the surface array layout, and a list of particles to inject into each tank is then determined. EAS simulations generally employ a thinning procedure [8] to reduce computational load, and consequently output lists of weighted particles. The algorithms used to inject particles into surface detector stations must therefore convert distributions of weighted particles to distributions of particles with unity weight. Next, simulation of the response of each of the tanks to the injected particles is performed, ultimately resulting in histograms of the times when photoelectrons are released from the PMT photocathodes. These photoelectron distributions are then passed through a simulation of the front-end electronics, culminating in digitization of the signal and application of per-tank triggering algorithms which distinguish potentially interesting signals from background noise. Finally, tanks which generate local triggers are passed to a central triggering module, which searches for space-time clusters of tanks consistent with an EAS. Events passing the central trigger can then be reconstructed and analyzed with the same software used for real data.

To provide a bit more detail, we consider the case of the module which simulates an individual water tank.

4.1.1. *Water tank simulation*

As mentioned in Sec. 2, each surface detector is is composed of a tank of purified water instrumented with three photomultiplier tubes. The water is contained within a Tyvek[b] bag which serves as a diffusive light reflector. At ground level, an EAS is composed primarily of electrons, gamma rays and muons with typical energies below about 10 MeV for electrons and photons and 1 GeV for muons. Electrons and muons traversing the water emit Cherenkov photons which can be reflected from the Tyvek bag until they either strike one of the photocathodes or are absorbed by the water or by the bag. Gamma rays entering the tank will almost always Compton scatter or pair produce, with the resulting charged particles generating Cherenkov light. δ-rays produced by particles traversing the water also contribute to the number of Cherenkov photons at the level of about 10%. Charged particles will radiate Cherenkov light until they either exit the tank or lose enough energy to drop below threshold for Cherenkov production.

The software framework described in Sec. 3 supports implementation of alternative approaches to modeling these processes, allowing both fast parametrizations

[b]Tyvek is a trademark of the DuPont company.

that simply sum up the expected contribution to the total signal for each particle based on its type and trajectory, to very detailed simulations in which each particle and each Cherenkov photon is individually tracked, with all physics processes taken into account. A detailed simulation may be desirable for studies of composition, while a faster but less detailed simulation may be sufficient for aperture studies of the observatory operating in hybrid mode.

A detailed tank simulation module has been prepared using the the `Geant4` [9, 10] toolkit. An illustration of a simulated tank is shown in Fig 2. It is important to

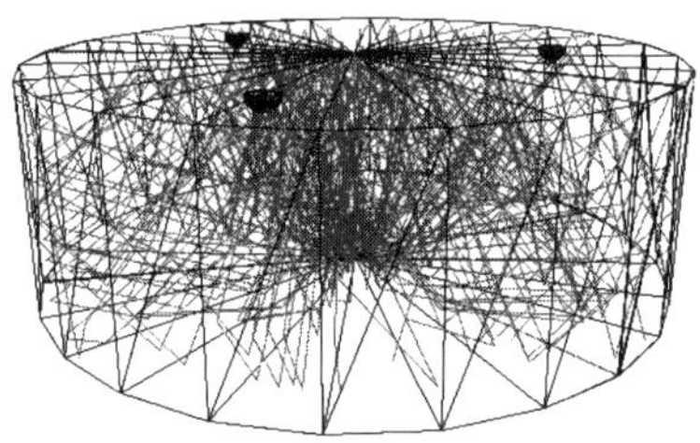

Fig. 2. Simulation of a particle entering an Auger water tank from above and radiating Cherenkov light. Housings for the three photomultiplier tubes are visible at the top of the tank. The cylindrical water volume has a 10^2 m surface and is 1.2 m high.

compare the predictions of simulations to measurements of the behavior of a single tank [11, 12]. We consider one example as an illustration. The response to single muons has been measured using a hodoscope to select through-going atmospheric muons with various trajectories through a tank [11]. Simulations mimicking these measurements have been performed, with some results shown in Fig. 3. In the figure one observes that both data and simulation exhibit a linear relation between the muon track length in the tank and the size of the signal, except for the case of long tracks for which both simulation and data show an enhancement in the signal size. As explained in the figure, this enhancement is a consequence of Cherenkov light directly entering the PMT without first being diffused by the Tyvek bag. Correct simulation of such details is particularly important for the study of highly inclined air showers.

4.2. *Fluorescence telescopes*

Simulation of the fluorescence telescopes begins from a description of air shower longitudinal development written by an EAS simulation program. Data from this file are used to compute the fluorescence and Cherenkov light emitted by the shower during its development. The propagation of this light up to the telescope entrance is then simulated, taking account of attenuation and scattering in the atmosphere using either parametrizations or measurements taken at the experimental site [14]. A ray-tracing module is used to follow photons as they enter the telescope, reflect

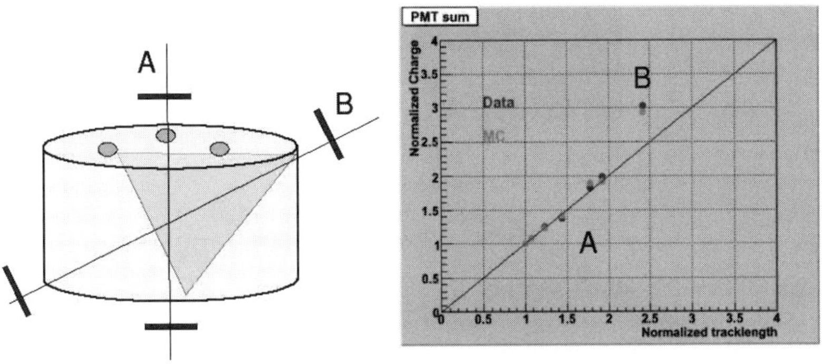

Fig. 3. To study water tank behavior and verify simulations, scintillator paddles are used to select throughgoing muons with different trajectories and track lengths (left). Trajectories with short track lengths (A) generate signals proportional the the track length. Due to the tank geometry, trajectories with a long track lengths (B) tend to be inclined enough to emit Cherenkov light directly into the PMTs, as indicated schematically by the Cherenkov cone shown around trajectory B. On the right, we plot the simulated (MC) and measured (Data) relation between track length and signal size for such an experiment. For short track lengths, a linear relation is seen (the line is meant to guide the eye). The enhancement from direct Cherenkov light is observed for longer track lengths in both simulations and measurements.

off the mirror and hit the array of PMTs. The PMT signals are then fed to a simulation of the readout electronics and triggering algorithms. Fluorescence detector simulation procedures are discussed elsewhere in more detail [13].

Figure 4 shows a fully simulated and reconstructed hybrid event, in which the steps described in Sec. 4.1 and 4.2 have been employed to produce a data file in the same format used by the data acquisition systems. This event was subsequently passed through the same reconstruction codes used to process real data.

5. Conclusions

Discovering the origins and composition of the highest energy cosmic rays involves relating quantities of interest, such as cosmic ray energies and chemical composition, to the properties of the extensive air shower they generate. The behavior of detector instruments used to measure such quantities must be taken into account in order to convert the raw data to a physics result. Though back-of-the envelope estimation is useful to provide guidance, the details can become so intricate that computer simulations become indispensable. In large collaborative experiments, different researchers may favor different approaches to a particular aspect of the detector response simulation. In order to encourage a variety of ideas to flourish, a common software framework is essential. Such frameworks eliminate the need to duplicate code required for common tasks, provide easy access to the work of other collaborators, and support quality control. As such, they constitute an integral component of the experiment. The simulation, reconstruction and analysis software of the Pierre

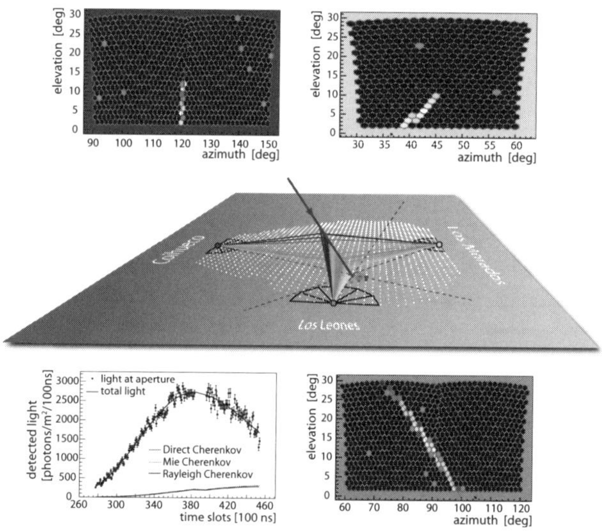

Fig. 4. Result of the a sequence of modules which first simulate then reconstruct a hybrid EAS event. The figure shows the detector including the grid of 1600 surface detectors and the three (of four) fluorescence detectors which triggered on the event. The three camera images show the image of the shower recorded on the telescope pixels. The plot on the lower left shows the light profile arriving at the Los Leones telescope, indicating contributions from different light sources.

Auger Observatory aims to fulfill these requirements, and provide a foundation for the computer modeling needed to shed light on the origins and nature of the highest energy cosmic rays.

6. Acknowledgments

This work was supported in part by the U.S. National Science Foundation.

References

[1] D. Heck, J. Knapp, J.N. Capdevielle, G. Schatz, T. Thuow, Report FZKA 6019 (1998).
S.J. Sciutto, "AIRES: A system for air shower simulations (version 2.2.0)," arXiv:astro-ph/9911331.
T. Bergmann *et al.*, "One-dimensional hybrid approach to extensive air shower simulation", Astropart. Phys. **26**, 420 (2007).
H.J. Drescher and G.R. Farrar "Air shower simulations in a hybrid approach using cascade equations" Phys. Rev. **D67**, 116001 (2003).

[2] J. Knapp, "Simulation of Ultra High Energy Cosmic Ray Experiments"
P. Necesal, "Testing of Cosmic Ray Interaction Models at the LHC Collider", these proceedings.

[3] S. Petrera, "Detector Modelling in Astroparticle Physics", these proceedings.

[4] J. Abraham *et al.* [Pierre Auger Collaboration], "Properties and performance of the prototype instrument for the Pierre Auger Observatory", Nucl. Instrum. Meth. A **523**, 50 (2004).

[5] S. Argirò *et al.* [Pierre Auger Collaboration], "The offline software framework of the Pierre Auger Observatory", *Presented at 2005 IEEE Nuclear Science Symposium and Medical Imaging Conference, El Conquistador Resort, Puerto Rico, 23-29 Oct 2005.* arXiv:astro-ph/0601016

[6] `http://www.w3.org/XML/`.

[7] `http://dev.mysql.com`.

[8] A.M. Hillas,"Shower simulations, Lessons from MOCCA" Nucl. Phys. Proc. Suppl. **52B**, 29 (1997).
A.M. Hillas, Proc. of the Paris Workshop on Cascade simulations, J. Linsley and A.M Hillas (eds.), 39 (1981).

[9] S. Agostinelli *et al.*, "G4, a simulation toolkit", Nucl. Instrum. Meth. A **506**, 250 (2003);

[10] L. Anchordoqui, T. McCauley, T. Paul, S. Reucroft, J.Swain and L. Taylor, "Simulation of water Cherenkov detectors using Geant4" Nucl. Phys. Proc. Suppl. **97**, 196 (2001).

[11] A. Creusot *et al.* "Response of the Pierre Auger Observatory water Cherenkov detectors to muons", *Presented at 29th International Cosmic Ray Conference (ICRC 2005), Pune, India, 3-11 Aug 2005.* FERMILAB-CONF-05-282-E-TD.

[12] P. Allision *et al.* "Observing muon decays in water Cherenkov detectors at the Pierre Auger Observatory" *Presented at ICRC 2005.* arXiv:astro-ph/0509238
A. Etchegoyen *et al.* "Muon-track studies in a water Cherenkov detector" Nucl. Instrum. Meth. A **545**, 602 (2005).
M. Aglietta *et al.* "Calibration of the Surface Array of the Pierre Auger Observatory" Nucl. Instrum. Meth. A **568**, 839 (2005).

[13] L. Prado Jr. *et al.*, "Simulation of the fluorescence detector of the Pierre Auger Observatory", Nuc. Instrum. Meth. A **545**, 632 (2005).

[14] B. Keilhauer *et al.* "Atmospheric profiles at the southern Pierre Auger Observatory and their relevance to air shower measurements." *Presented ICRC 2005.* arXiv:astro-ph/astro-ph/0507275.
R. Cester *et al.* "Atmospheric aerosol monitoring at the Pierre Auger Observatory" *Presented ICRC 2005.* FERMILAB-CONF-05-293-E-TD.

TESTING OF COSMIC RAY INTERACTION MODELS AT LHC COLLIDER

PETR NEČESAL* and JAN ŘÍDKÝ (Supervisor)

Institute of Physics, AS CR,
Prague, Czech Republic
**E-mail: necesal@fzu.cz*
www.fzu.cz

Hadronic interaction models which are the most important components of Monte Carlo generators of extensive air showers are compared. Simulations of some types of nuclei collisions in LHC conditions are presented and comparisons between hadronic models are done.

Keywords: Hadronic interaction models; Monte Carlo generators; LHC; ATLAS.

1. Introduction

The interpretation of extensive air shower (EAS) measurements depends on Monte Carlo (MC) generators. The most important problem and the biggest challenge is to extrapolate accelerator data to ultra-high energies ($\simeq 10^{20}$ eV) encountered in Cosmic Ray (CR) interactions. The energy reached at present day colliders is much smaller than energy of EAS. The energy spectrum of cosmic rays is depicted in Fig. 1 [1, 2]. The description of interactions of hadronic particles in the showers with the nuclei in the atmosphere is the most crucial aspect of simulations. Therefore the reliability is indispensable. However, hadronic interactions are still inaccurately described in EAS energy range. Interaction models have to provide results in accordance with data acquired at colliders.

New opportunities arise with Large Hadron Collider (LHC) at CERN. The aim of the presented work is to study particle production in conditions of the ATLAS detector at LHC. It offers better conditions to study event generators as the energy 14 TeV in the center of mass system of proton-proton collisions corresponds already to the region above the 'knee' of the CR spectrum. Capabilities of ATLAS detector are enormous and therefore they could be utilized in nucleus-nucleus and proton-nucleus interactions relevant to cosmic rays studies [3].

A plan of this work is to compare PYTHIA (*6.221*) [4], HIJING (*hijing1.383* and *hipyset1.35*) [5], QGSJET (*qgsjet01c*) [6] and QGSJET-II (*qgsjet-II-03*) [7] at LHC energy region. QGSJET is one of the generators of high energy collisions (for particle with energy $E > 80$ GeV) included in CORSIKA (version *6.2040*) [8].

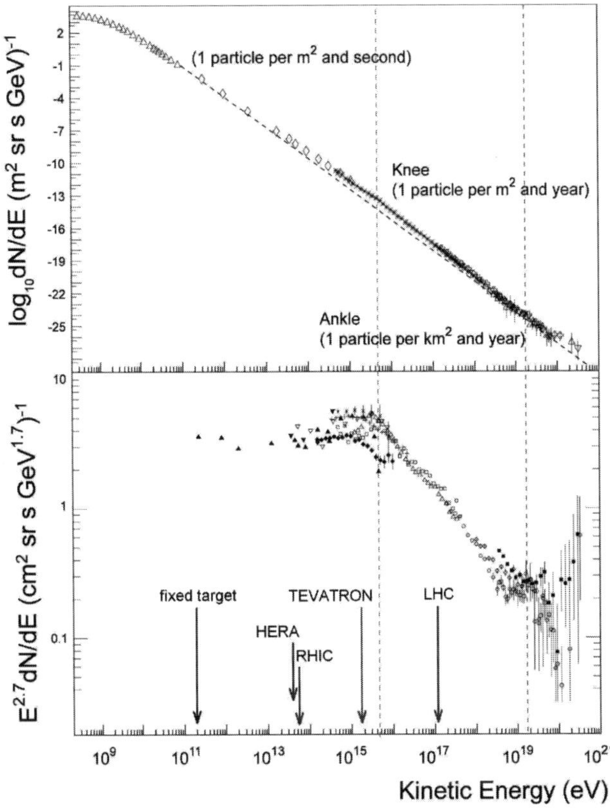

Fig. 1. The cosmic ray energy spectrum.

2. Comparison of Models

It is important to keep consistency between events generated by different simulators. Decays of particles were turned on (in HIJING and PYTHIA) or special subroutines were used in order to have the same particles in the final state, i.e. $\pi^{\pm}$, K_L^0, $K^{\pm}$, γ, p, $\bar{p}$, n, $\bar{n}$, $\mu^{\pm}$, $e^{\pm}$ and *neutrinos*. QGSJET does not contain decays of unstable particles and therefore subroutines from CORSIKA were used for decay particles. The production of charmed particles was switched on. Decays of charmed particles were carried out according to decay branching ratios [9]. QGSJET-II does not contain the production of charmed particles at all and subroutines from CORSIKA of version *6.2040* were used for decays of other unstable particles.

Generators differ from each other also in treatment of diffractive dissociation. Diffractive interaction is characterized by low-$p_{\perp}$ parton scatterings which are not calculable in QCD and must be described by phenomenological models of soft processes. In real experiment diffractive events can be selected by kinematical cuts. We can use e.g. so called *rapidity gap*, which is the rapidity interval with no particle production in the central region. It is useful to test generators with all relevant processes switched on, but with additional conditions which simulate the function of the detector trigger. Our condition was that at least one charged particle has to be in the pseudorapidity interval of $0 < \eta < 3$ and one charged particle in the region characterized by $-3 < \eta < 0$. Each of them has to have energy bigger than 2 GeV. All events which did not satisfy these conditions were excluded as diffractive ones.

To compare all four programs, we generated proton-proton collisions. The center of mass system energy of pp collision is $\sqrt{s} = 14\,\text{TeV}$. All generators were used in conditions producing events as close to minimum bias events as possible and $5 \cdot 10^5$ events were generated. The histograms shown below are normalized to 1000 events. Diffractive dissociation was switched on in HIJING and PYTHIA and of course diffractive events were not omitted in QGSJET and QGSJET-II. HIJING is not able to use double diffractive events in contrast to other used generators.

Figure 2 shows pion distributions which are in fact identical with distributions of charged particles. It is obvious, that a large number of PYTHIA events did not satisfy the additional conditions on charged particles and therefore PYTHIA produced the smallest number of pions. Except for QGSJET-II, the pion production of other generators starts at multiplicity above 10. QGSJET-II still produces non-negligible amount of events with small number of pions. All interactions have to satisfy the cuts on charged particles mentioned above. It means that even in collisions characterized by small number of secondary particles high $p_\perp$ has to be transferred. This type of interaction is called *hard diffraction*. QGSJET-II and HIJING produce similar pseudorapidity distributions. PYTHIA has a little peak in multiplicity which is caused by single diffractive dissociation. PYTHIA has very different shape of the pseudorapidity weighted by energy.

The atmosphere consists of N_2, O_2 and Ar with the volume fractions of 78.1 % , 21.0 % and 0.9 % and protons represent the vast majority of comic ray particles. Therefore pN collisions are very frequent and important for CR physics. PYTHIA can not be employed, because it generates only interactions between elementary particles. At LHC conditions nitrogen would have momentum of $p_N = 49\,\text{TeV}$ and proton $p_p = 7\,\text{TeV}$. This corresponds to *CMS* energy $\sqrt{s} \approx 37\,\text{TeV}$. After interaction four-momenta of secondary particles were transformed from particular generator frame to the frame, in which proton has momentum 7 TeV and nitrogen 49 TeV ('detector frame'). QGSJET and QGSJET-II are designed to generate $p \to A$ (or $A \to B$) collisions. Proton was the projectile and nitrogen was the target, this type of collision is natural for CR physics. $350 \cdot 10^3$ of minimum bias events were generated and histograms were rescaled to 1000 events for lucidity.

In order to demonstrate sensitivity of individual subdetectors of the ATLAS to the differences between generators we show the pseudorapidity coverage by particular subdetectors. Cuts on overall pseudorapidity imposed by the coverage of hadronic calorimeter (HDC), electromagnetic calorimeter (EMC), muon detectors (MD) and inner detector (ID) are applied to simulations.

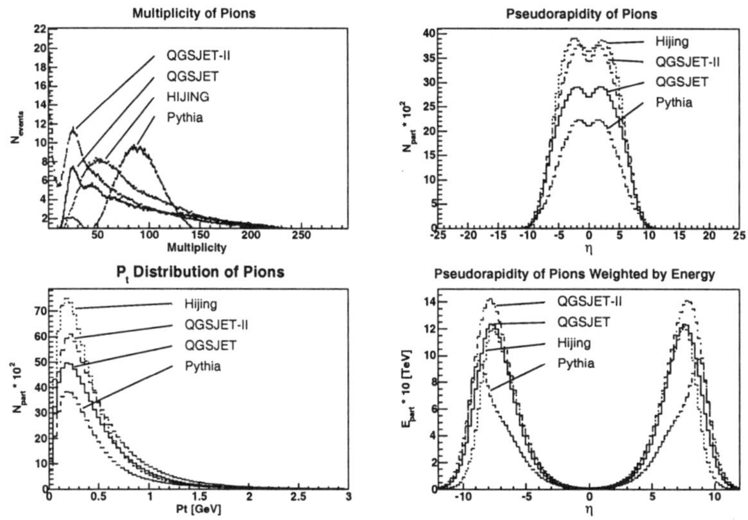

Fig. 2. Resulting pion spectra from pp collisions at $\sqrt{s} = 14\,\text{TeV}$ with trigger condition excluding diffraction. Histograms are normalized to 1000 events.

Simulations of pN collisions show little differences in pion distributions and therefore even in charged particle distribution among generators. The differences are visible in pseudorapidity distribution of antiprotons, kaons, protons and muons. In the latter case energy was used as weight (see Fig. 3). Protons and antiprotons produced by HIJING and QGSJET turn out to be distributed very similarly in the central region of hadronic calorimeter, however QGSJET-II gives approximately 70 % of the multiplicity of the other simulators. $\mu^{\pm}$ production together with muon pseudorapidity distribution weighted by energy seem to be the best indicator to differentiate between models. Significant differences can be found also in pseudorapidity distribution of $K^{\pm}$, but total $K^{\pm}$ multiplicity is nearly indeterminable due to total number of particles and their energy produced in interaction.

In addition to proton-proton and proton-nucleus collisions also nucleus-nucleus interactions are interesting from the point of view of CR physics. Nitrogen $^{14}_{7}\text{N}$ and iron $^{56}_{26}\text{Fe}$ are possible participants in CR interactions. Iron and nitrogen nuclei can be accelerated at LHC to momenta $p_{Fe} = 182\,\text{TeV}$ and $p_N = 49\,\text{TeV}$, respectively. The center of mass energy corresponding to collision of nitrogen and iron with momenta p_N and p_{Fe} is $\sqrt{s} \approx 189\,\text{TeV}$. Similarly to the tests described above, resulting distributions of secondary particles were compared in the 'detector frame', in which incident nuclei have mentioned momenta p_{Fe} and p_N. Because of the vast number of produced secondary particles it was sufficient to generate 10^5 of minimum bias events. In simulations of nucleus-nucleus interactions the diffractive dissociation was switched off in generator. No additional trigger conditions were set.

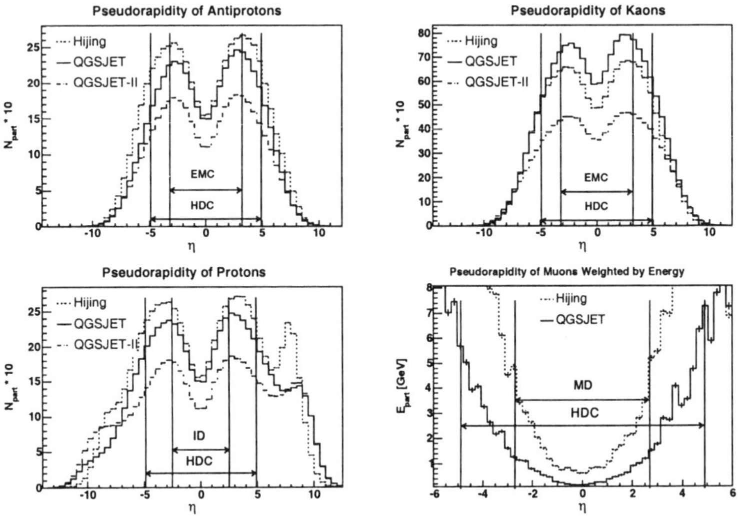

Fig. 3. Resulting histograms from pN collision at energy $\sqrt{s} = 37\,\text{TeV}$, pseudorapidity coverage of detectors are drawn by dashed lines. *HDC*, *EMC*, *MD* and *ID* denote hadronic and electromagnetic calorimeter, muon detectors and inner detector, respectively. Histograms are normalized to 1000 events.

Pseudorapidity cuts shown in histograms of variables from NFe collisions are drawn in Fig. 4. Differences among all generators are observable in charge particle pseudorapidity distribution (see the upper left histogram). Charged particle production by QGSJET in the central region represents only 60 % of charged particles produced by QGSJET-II. In contrast to previous case, QGSJET and QGSJET-II have very similar pseudorapidity distribution of kaons in the range covered by calorimeters. HIJING produces more kaons. Detection of protons would distinguish all generators. Production of muons and their pseudorapidity distribution is best suited for comparison of real data with simulations. Muon production in HIJING differs from that of QGSJET by a factor ≈ 2 or more in the whole range covered by hadronic calorimeter (see the bottom right histogram).

3. Conclusions

Four Monte Carlo generators: PYTHIA, HIJING, QGSJET and QGSJET-II were tested in conditions of the ATLAS experiment at LHC. Collisions with both diffraction dissociation and hard processes switched on were generated with additional conditions playing the role of the detector trigger. This approach was adopted because different generators treat diffraction in different ways. The aim was to test models at their full potential. As detectors are sensitive mostly to non-diffractive events we selected these events by means of simulated detector trigger. Switching off

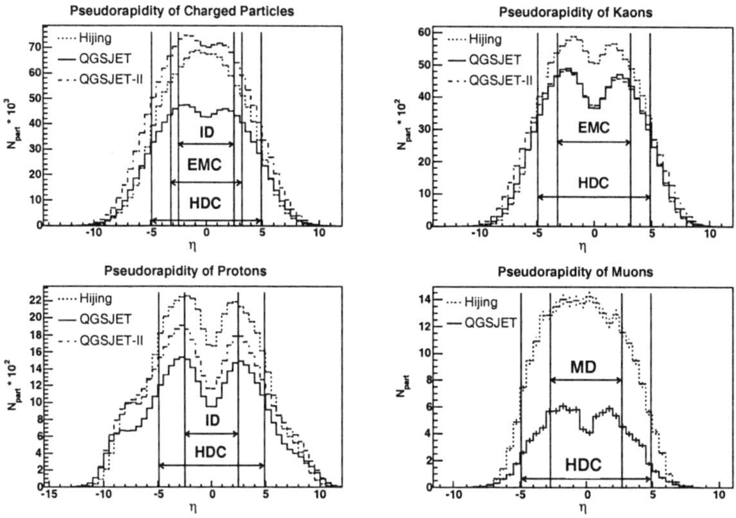

Fig. 4. Resulting distributions from NFe collisions at energy $\sqrt{s} = 189\,\mathrm{TeV}$, pseudorapidity coverages of detectors are drawn by dashed lines. *HDC*, *EMC*, *MD* and *ID* denote hadronic and electromagnetic calorimeter, muon detectors and inner detector, respectively. Histograms are normalized to 1000 events.

diffraction dissociation especially in pp collisions would represent for each generator different cut.

Simulations at LHC energies show differences between charged particle production. Particular generators are based on different theoretical approaches and philosophy and it is therefore very interesting and quite remarkable, that predictions are so similar in gross features (Tab. 1).

Table 1. The mean values of $\pi^{\pm}$ multiplicities, energy and $p_{\perp}$ for studied pp collisions. Energies and momenta are given in GeV units.

Collision	Quantity	PYTHIA	HIJING	QGSJET	QGSJET-II
pp	$\langle n \rangle_{\pi}$	90.6	105.2	91.7	91.0
	$\langle E \rangle_{\pi}$	51.27	41.60	57.12	50.87
	$\langle p_{\perp} \rangle_{\pi}$	0.47	0.42	0.47	0.52

Generators seem to differ mostly in "heavier" flavor production (not only charm, but already strangeness). This can be seen from large discrepancies in $K^{\pm}$ and $\mu^{\pm}$. Unstable particles decayed in simulations and therefore sources of $\mu^{\pm}$ were mainly charmed particles and B-mesons. The production of $K^{\pm}$ is important for CR physics as kaons contribute to muon production in extended air showers in a different way than pions. Particles with charm are significant sources of $\mu^{\pm}$ and muons seem to be the most convenient for testing the generators.

MC generator HIJING does not produce as much energetic particles in very forward regions ($|\eta| > 6$) as other generators. This can be seen in pseudorapidity histograms weighted by energy. Due to this fact, HIJING can not be used for CR shower simulations. On the other hand it is well tested generator in nucleus-nucleus interactions and it describes collective phenomena. Applicability of HIJING in astroparticle physics depends on improvement of diffractive dissociation treatment. HIJING is not able to generate double diffraction.

Charged particle production is quite in accordance among all generators in proton-nucleus collisions (in processes without diffractive dissociation), but differences increase in nucleus-nucleus collisions (e.g. nitrogen-iron), which gives another opportunity to test generators.

References

[1] Milke J. *et al.* (2004): Test of Hadronic Interaction Models with Kascade. *Acta Phys. Pol. B* **35**, 341–348.

[2] Gaisser T. K. (1990): Cosmic Rays and Particle Physics. Cambridge University Press, Cambridge.

[3] Aronson S. *et al.* (2002): A Nuclear Physics Program at the ATLAS Experiment at the CERN Large Hadron Collider. *Letter of Intent*, Brookhaven National Laboratory.

[4] Sjostrand S., Eden P., Friberg C., Lonnblad L., Miu G., Mrenna S., Norrbin E. (2001): High-energy physics event generation with PYTHIA 6.1. *Comput. Phys. Commun.* **135**, 238–259.

[5] Wang X. N., Gyulassy M. (1991): HIJING: A Monte Carlo model for multiple jet production in p p, p A and A A collisions. *Phys. Rev.* D **44**, 3501.

[6] Kalmykov N. N., Ostapchenko S. S., Pavlov A. I. (1997): *Nucl. Phys. B (Proc. Suppl.)* **52**, 17.

[7] Ostapchenko S. (2006): *Nucl. Phys. Proc. Suppl.* **B 151**, 143.

[8] Heck D., Knapp J., Capdevielle J. N., Schatz G. and Thouw T. (1998): CORSIKA: A Monte Carlo Code to Simulate Extensive Air Showers. *Report* FZKA 6029 Forschungszentrum Karlsruhe, http://www-ik3.fzk.de/~heck/corsika

[9] Particle Data Group, http://pdg.lbl.gov/.

OBSERVATIONS, SIMULATIONS, AND MODELING OF SPACE PLASMA WAVES: A PERSPECTIVE ON SPACE WEATHER

VIKAS S. SONWALKAR

Electrical and Computer Engineering Department, University of Alaska Fairbanks, Fairbanks, Alaska 99775, USA

E-mail: ffvss@uaf.edu

www.uaf.edu

Changes in solar activity lead to adverse conditions within the upper atmosphere which may cause disruption of satellite operations, communications, navigation, and electric power grid. The term space weather is used to refer to changes in the Earth's space environment. This paper reviews plasma waves, found in all parts of the ionosphere and magnetosphere, in the context of space weather. Generated by energetic particles within the magnetosphere, these waves in turn cause particle accelerations, heating, and precipitation, taking an active part in determining space weather. Terrestrial lightning is an important source of plasma waves and forms a link between the lower and the upper atmosphere. Though many aspects of plasma waves such as their morphology and association with energetic particles and geomagnetic phenomena are well established, their generation mechanisms in most cases remain elusive. Current research involving active and passive, ground and space borne experiments, modeling, and simulation is providing better understanding of plasma wave generation mechanisms and the relation of plasma waves to other space weather parameters such as variations in geomagnetic field and energetic particle fluxes resulting from solar storms. Potentially, plasma waves could serve as one of the key indicators of space weather.

Keywords: Plasma waves; Space weather; Magnetosphere; Lightning.

1. Introduction

We are familiar with weather on the Earth. We can feel the wind and rain, ice and hail. We can see the clouds and lightning and we can hear thunder. The effects of a great storm such as Katrina are devastating. We are less familiar with space weather. The storms occurring in the upper atmosphere, comprising of the ionosphere and the magnetosphere, cannot be felt by our senses, but its effects on the Earth could be enormous. A large space storm in 1989 caused failure of Hydro Quebec power grid leading to a blackout that lasted nine hours and cost more than a billion dollars.

The main players in the Earth's upper atmosphere are the cold plasma, high energy particles, geomagnetic field, and plasma waves. Sun is the main source of energy for the processes taking place in the ionosphere and magnetosphere. Violent and drastic changes on the Sun lead to geomagnetic storms, which often lead to adverse conditions in the near-Earth space environment. After briefly describing the Earth's atmosphere-ionosphere-magnetosphere system, solar radiation and its impact on the upper atmosphere, this paper provides a review of plasma waves including its role in determining, specifying, and forecasting space weather.

2. Atmosphere-Ionosphere-Magnetosphere System and its Solar Drivers

The Earth's atmosphere-ionosphere-magnetosphere is a complex and highly coupled system powered by the Sun (Figure 1). Above the thin layer of the Earth's neutral atmosphere lies the ionosphere and magnetosphere, a vast region containing cold and hot plasma (energetic particles), fluctuating magnetic fields, and a large variety of plasma waves.

The ionosphere begins at a height of about 70 km and contains enough electrons and ions to affect propagation of radio waves. It has been found convenient to think of ionosphere consisting of three regions: the D region (~70-90 km), the E region (~90-150 km), and the F region (~140-300 km). The electron density (N_e) increases more or less uniformly with altitude from the D region ($N_e \sim 10^4$ el/cc) reaching a maximum ($N_e \sim 10^6$ el/cc) in the F region (F_2 peak). Extending upward from the F layer is the magnetosphere, wherein the Earth's magnetic field largely controls the movement of ions and electrons. The magnetosphere contains distinct plasma regions and large scale current systems. The magnetosphere extends outward from the Earth about 60,000 km toward the Sun and has a tail that extends many times that distance in the direction away from the Sun. The magnetosphere contains a cold plasma, consisting mainly of electrons, H^+, He^+, and O^+ ions, and a few other ions in small numbers. The inner magnetosphere, called plasmasphere, is a high density (~100-1000 el/cc) cold plasma region that corotates with the Earth. The boundary of the plasmasphere is called plasmapause. The magnetosphere contains two zones of energetic particles, called the Van Allen radiation belts, . The inner belt, containing protons up to several hundred MeV, extends from roughly 1,000 to 5,000 km above the Earth's surface and the outer belt, dominated by electrons up to tens of MeV, from some 15,000 to 25,000 km. Ring currents of electrons and low energy protons (< 50 keV) reside within the outer radiation belt. Other large scale currents in the magnetosphere include field aligned currents at high latitude and tail current in the magnetotail region. The magnetosphere ends at a boundary known as the magnetopause, beyond which is the domain of the solar wind.

The solar wind, a constant outward plasma flow from solar corona, and the embedded interplanetary magnetic field (IMF) provides the energy, momentum and most of the mass that fills and powers the Earth's magnetosphere. Changes in solar activity leading to coronal mass ejections (CME), large solar flares, and high speed solar wind streams can severely influence the behavior of magnetospheric plasma and cause great variations in the motion and quantity of the energetic particles within the magnetosphere. Large enhancements in the particle number and motion

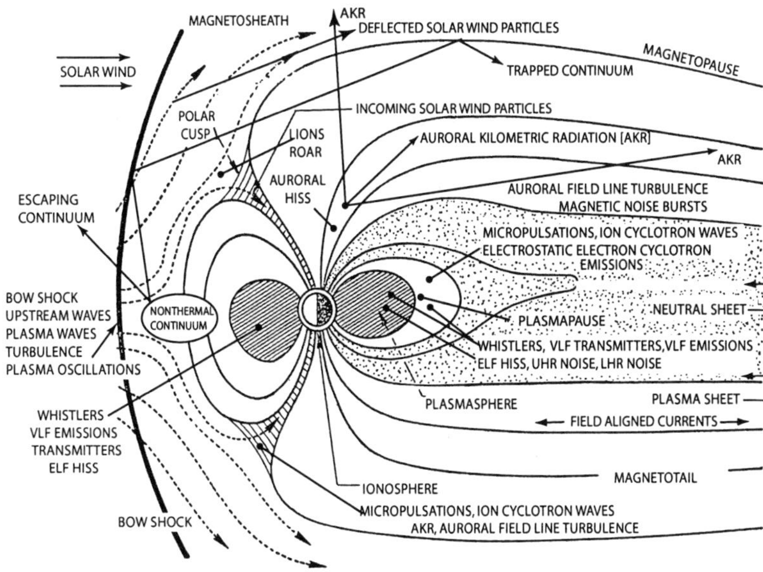

Fig. 1. Schematic showing various regions and features of the magnetosphere in the noon-midnight meridian. Also shown are the locations where plasma waves of various types are observed. (Adapted from S. D. Shawhan, *Rev. Geophys. Space Phys., 17, 4*, 705 (1979), and from V. S. Sonwalkar, *Lect. Notes Phys.*, 687, 141 (2006). With permission.)

can alter the magnetospheric configuration giving rise to geomagnetic storms and substorms, which are associated with changes in magnetic field, enhanced fluxes of energetic particles, increased magnetospheric and ionospheric currents, and increased auroral activity. In general, the term space weather is used to refer to conditions on the sun, in the solar wind and in the upper atmosphere that can influence the performance and reliability of space-borne and ground-based technological systems and can endanger human life or health. Space weather monitoring and forecasting is an important scientific and technological challenge facing the global community of space scientists [1].

3. Plasma Waves

3.1. *Observations of plasma waves*

Magnetospheric plasma consisting of electrons and ions of finite temperature and permeated by a magnetic field can support a large variety of electromagnetic, electrostatic and magnetosonic wave modes that can not exist in free space. A wave mode is characterized by a distinctive polarization, refractive index, and a range of frequency within which the mode can propagate. The allowed modes of propagation depend on the characteristic medium frequencies: electron and ion plasma and cyclotron frequencies. The plasma frequency is the natural frequency of oscillations of electrons or ions and the cyclotron frequency is the gyration frequency of electrons and ions. These frequencies depend on plasma density, composition, and the strength of the geomagnetic field. Because these parameters vary widely in different regions of the magnetosphere, distinctive wave activity is found in different parts of the magnetosphere. In the magnetosphere the observed range of plasma waves varies between a fraction of Hz (e.g. ion cyclotron waves) to a few MHz (e.g. auroral hiss, Auroral Kilometric Radiation).

Plasma waves are found in all parts of the magnetosphere (Figure 1). The names given to these diverse wave phenomena are generally indicative of one or more properties or features that each kind exhibits: frequency range, spectral characteristics, region, and local time of occurrence, and plasma wave modes. For example, ELF hiss, also called plasmaspheric hiss, occurs in the extremely low frequency (300 Hz-3 kHz, ELF) range inside the plasmasphere. Its spectrum resembles that of a band-limited thermal or fluctuation noise and can be identified aurally by a hissing sound. The information on plasma waves has been obtained from a large number of passive and active, and ground and space borne experiments [2–8]. Past observations have identified and measured plasma wave properties (wave mode, polarization, and intensity), their dependence on cold plasma parameters, and association with energetic particles and geomagnetic activity.

Figure 2 dynamic spectra shows an example of plasma waves observed by the Plasma Wave Experiment (PWE) instrument on the Combined Release and Radiation Effects Satellite (CRRES) satellite near the equatorial magnetosphere. The CRRES satellite (Perigee: 322 km; Apogee: 33,745 km; Inclination: 17.9°) was launched in 1990 into a geosynchronous transfer orbit for a nominal three-year mission to investigate fields, plasmas, and energetic particles inside the Earth's magnetosphere. A large variety of plasma waves, such as lightning-generated whistlers, chorus, and auroral hiss (VLF hiss), are also observed on the ground. Figure 3 (b and c) shows examples of auroral hiss observed at the South Pole Station, Antarctica. The auroral hiss is generally found in close association with geomagnetic field variations, visible aurora, and radar aurora (HF radar echoes scattered from ionospheric irregularities generated by precipitating auroral electrons).

3.2. *Generation and propagation of plasma waves*

Energetic particles are the main sources of plasma waves found in the magnetosphere [3, 6, 9]. A number of processes (e.g. plasma diffusion and convection across the magnetic field, pitch angle scattering, plasma drifts, particle acceleration) operating in the magnetosphere disturb the particle distribution function from its equilibrium state (generally a Maxwellian distribution), leading to new particle distributions which are unstable to the growth of plasma waves. Resonant conversion of kinetic energy of particles to wave energy or vice versa can take place by two dif-

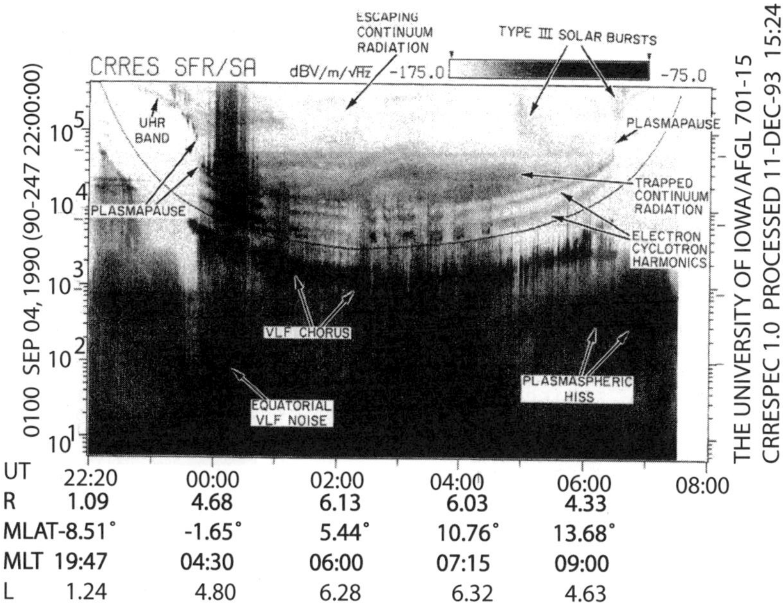

Fig. 2. A 10-hour CRRES PWE electric field spectrogram for orbit 100 showing various plasma waves observed in the equatorial region. The geomagnetic field was becoming disturbed following 2-3 very quiet days. The dark dashed line plotted on the spectrogram shows the electron cyclotron frequency (f_{ce}) calculated from the CRRES fluxgate magnetometer experiment. (Adapted from Anderson, R. R., *Dusty plasma, Noise, and Chaos in Space and Laboratory*, ed. H. Kicuchi, Plenum Publishing Corporation, New York, 1994. With permission.)

ferent mechanisms depending on whether the particle motion along the geomagnetic field (longitudinal motion) or the particle motion transverse to the magnetic field is the controlling factor. The former mechanism leads to flow or beam instabilities (or damping) and the latter to gyroresonance (or cyclotron resonance) instabilities (or damping). Lightning, very low frequency (VLF), and low frequency (LF) transmitters are other important sources of energy for plasma waves found in the magnetosphere.

Plasma waves can propagate long distances from their generation regions. [2, 5, 6, 10, 11] Wave propagation in a magnetoplasma is both anisotropic and dispersive. This leads to complex propagation paths that undergo multiple reflections within the magnetosphere. Excepting Auroral Kilometric Radiation (AKR) and escaping continuum radiation, most other plasma waves remain trapped in the magnetosphere. During their propagation waves may undergo amplification or damping as a result of wave particle interactions.

A small fraction of whistler mode ($f \leq f_{pe}, f_{ce}$) waves, including lightning-generated whistlers, ground transmitter signals, chorus, and auroral hiss, propagate

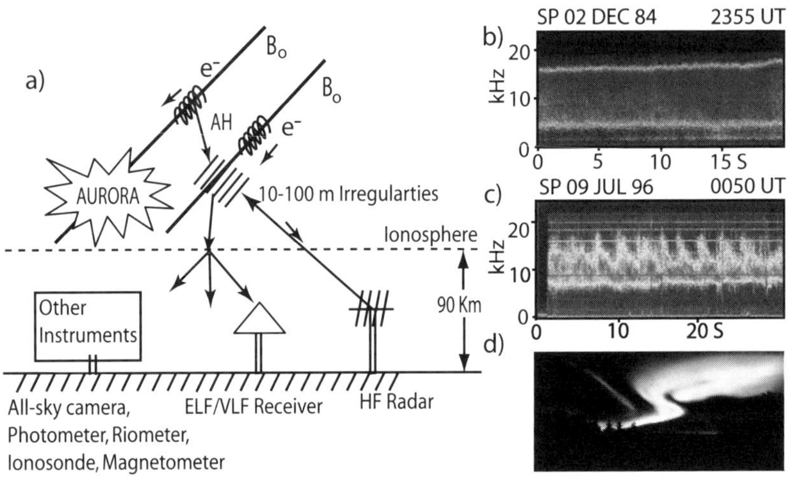

Fig. 3. (a) Schematic illustrating the propagation of auroral hiss (AH) from its source region on auroral field lines to the ground. The figure also shows complementary instrumentation often found at a high latitude ground station. (b), and (c) are examples of continuous and impulsive auroral hiss, respectively, observed at the South Pole Station, Antarctica. (d) Aurora observed at North Pole, Alaska. ((a), (b), and (c) are adapted from Sonwalkar et al., (2000). (d) Photo taken by Jan Curtis; courtesy of the Geophysical Institute, University of Alaska Fairbanks.)

down to the Earth [2]. Various aspects of the ionospheric plasma can profoundly affect the propagation of whistler mode energy from the magnetosphere to the ground and vice versa. Plasma density irregularities present in all parts of the magnetosphere and ionosphere reflect, refract, and scatter waves. [7–9, 11] Highly collisional D-region plasma can lead to a 10 to 20 dB reduction in wave energy that passes through this region (upward or downward). A consequence of D-region absorption in the daytime is that many magnetospheric wave phenomena are better observed at nighttime when the D-region absorption is minimum [2].

3.3. *Modeling and simulations of plasma waves*

Modeling and simulations play an indispensable role in our understanding of plasma waves. [7, 10–12] A typical plasma wave is observed far away from its generation region and thus the modeling and simulation scenario requires a two fold approach: (1) simulate generation process, typically nonlinear, using particle simulations, and (2) simulate propagation, typically linear, using ray tracing approach. The propagation of plasma waves from their source region to other locations in a smooth magnetosphere is well understood with the help of ray tracing simulations. However, the propagation of waves in an irregular magnetosphere containing large (1-100 km) and small scale (1-100 m) irregularities is not well understood. Similarly, self-consistent simulation of wave-particle interactions involved in wave generation or particle precipitation have proved difficult. Despite years of research, our understanding of the generation mechanisms of many of the commonly observed VLF emissions remain poorly understood. [6, 9]

3.4. *Contribution of plasma waves to space weather: wave-particle interactions*

Whistler mode waves and their interactions with energetic particles have been a subject of interest since the discovery of the radiation belts. These interactions establish high levels of ELF/VLF waves and play an important role in the acceleration, heating, transport, and loss of energetic particles in the magnetosphere via cyclotron and Landau resonances. [13] For example, wave-particle interactions have been found responsible for the decay of the ring current, the precipitation of electrons and ions to form diffuse aurora, loss of electrons to form the slot region between the radiation belts, energy transfer and heating at the collisionless bow shock and at the field aligned current regions. [3] Wave-particle interactions and the role they play in determining space weather and their importance relative to other processes taking place in the magnetosphere remains an active research area.

3.5. *Monitoring space weather using plasma waves*

Plasma wave observations provide information that is complementary to that obtained from other instruments (Figure 3). Ground and space based passive and active (wave injection) experiments have demonstrated that plasma wave can be used to remotely measure cold and hot plasma parameters. [2, 4, 7, 8] Remote sensing of cold plasma with wave technique is well established. Remote hot plasma diagnostics using waves has not been as successful, mainly due to our inability to quantitatively understand wave-particle interactions that lead to wave generation. However, as the computational techniques, particularly those involving parallel processing, improve, we may expect that ground based hot plasma diagnostics will become possible. Observations of chorus and auroral hiss will provide information on the ring current electrons and auroral electrons. Powerful ground based techniques have been developed in last 15 years to measure the high energy (MeV) electron precipitation in the lower ionosphere. [14] In these experiments, the perturbations in the amplitude and phase of a VLF transmitter signal propagating in the Earth-ionosphere waveguide are used to determine the modification to the lower ionosphere resulting from particle precipitation.

4. Concluding Remarks

A great advance has been made in the last 50 years in measuring the characteristics of a large variety of plasma waves in all parts of the magnetosphere. Both in situ spacecraft and remote ground observations have provided much complementary information on magnetospheric wave activity as a function of various geophysical parameters. To first order, free energy sources for the generation of these waves have been identified and in most cases they are found to be the energetic particle population of the magnetosphere. Details of the mechanisms of generation for most waves, however, remain unknown or controversial. For example, it is still being debated if the plasmaspheric hiss is generated by energetic electrons within the plasmasphere or is generated by lightning [15–17]. It is becoming increasingly clear that plasma waves play a fundamental role in the dynamics of magnetosphere via wave particle interactions and contribute to particle diffusion, precipitation, acceleration, and heating.

We conclude by pointing out the potentially important role that lightning may be playing in the physics of magnetospheric plasma waves. It has been widely assumed that most of the plasma waves are generated from background noise levels within the magnetosphere via wave particle interactions with energetic particles supplying the free energy. Past research [6] shows that lightning can be an important source of (1) plasmaspheric hiss believed to be responsible for the slot region in the radiation belts (2) lower hybrid waves that can heat and accelerate protons to suprathermal temperatures (3) ULF magnetic fields that can influence the generation and amplification of geomagnetic pulsations. In addition, lightning induced electron precipitation (LEP) events regularly occur throughout the plasmasphere and are important on a global scale as a loss process for the radiation belt electrons. Approximately 2000 thunderstorms are active near the Earth's surface at any given time, and on the average, lightning strikes the Earth ~100 times per second. The average lightning discharge radiates an intense pulse of ~20 Gigawatts peak power which propagates through the lower atmosphere, and into the ionospheric and magnetospheric plasmas, generating new waves, heating, accelerating and precipitating components of the charged particles comprising these plasmas. Thus future investigations should consider electromagnetic energy released in the thunderstorms as a potentially major source of free energy for the generation of magnetospheric plasma waves and for precipitation of particles, and should determine its implications for the atmosphere-ionosphere-magnetosphere coupling.

Acknowledgments

This work was supported by NASA under contract NNG04GI67G. The author thanks Amani Reddy for her assistance in preparing figures.

References

[1] P. Song, H. Singer, and G. Siscoe, The U. S. National Space Weather Program: A Retrospective, in *Space Weather* (American Geophysical Union, Washington, DC, 2001).

[2] R. A. Helliwell, *Whistlers and Related Ionospheric Phenomena*, (Stanford University Press, 1965).

[3] S. D. Shawhan, Magnetospheric plasma wave research 1975-1978, *Rev. Geophys. Space Phys.* **17**, 705 (1979).

[4] R. A. Helliwell, VLF wave simulation experiments in the magnetosphere from Siple Station, Antarctica, *Rev. Geophys. Space Phys.,* **26**, 551(1988).

[5] D. A. Gurnett and U. S. Inan, Plasma wave observations with the Dynamics Explorer 1 spacecraft, *Rev. Geophys. Space Phys.,* **26**, 285 (1988).

[6] V. S. Sonwalkar, Magnetospheric LF- VLF-, and ELF-waves, in *Handbook of Atmospheric Electrodynamics*, Ed. H. Volland, (BocaRaton, Fla., CRC Press, 1995).

[7] V. S. Sonwalkar, D. L. Carpenter, T. F. Bell, M. A. Spasojevic, U. S. Inan, X. Chen, J. Li, J. Harikumar, A. Venkatasubramanian, R. F. Benson, B. W. Reinisch, Diagnostics of magnetospheric electron density and irregularities at altitudes <5000 km using whistler and Z mode echoes from radio sounding on the IMAGE satellite, *J. Geophys. Res.*, **109**, A11212, (2004).

[8] R. F. Benson, P. A. Webb, J. L. Green, D. L. Carpenter, V. S. Sonwalkar, H. G. James and B. W. Reinisch, Active wave experiments in space: The Z Mode, *Lect. Notes Phys.,* (Springer-verlag, Berlin Heidelberg, 2006).

[9] J. W. LaBelle, and R. A. Truemann, Auroral radio emissions, *Space Sci. Rev.,* **101**, 295 (2002).

[10] I. Kimura, Ray paths of electromagnetic and electrostatic waves in the earth and planetary magnetospheres, *Plasma waves and Instabilities at comets and in Magnetospheres*, eds. B. T. Tsurutani and H. Oya, *Geophys. Mongr.*, Vol. 161, (1989).

[11] V. S. Sonwalkar, and J. Harikumar, An explanation of ground observations of auroral hiss: Role of density depletions and meter scale irregularities, *J. Geophys. Res.,* **105**, 18,867 (2000).

[12] T. Miyake, Y. Omura, H. Matsumoto, Electrostatic particle simulations of solitary waves in the auroral region, *J. Geophys. Res.*, **105**(A10), 23239 (2000).

[13] C. F. Kennel, and H. E. Petschek, Limit on the stably trapped particle fluxes, *J. Geophys. Res.*, **71**, 1, (1966).

[14] U. S. Inan, F. A. Knifsend, and J. Oh, Subionospheric VLF "Imaging" of lightning-induced electron precipitation from the magnetosphere, *J. Geophys. Res.*, **95**, (A9), 17217 (1990).

[15] V. S. Sonwalkar, and U. S. Inan, Lightning as an embryonic source of VLF hiss, *J. Geophys. Res.,* **94**, 6986 (1989).

[16] N. P. Meredith, R. B. Horne, M. A. Clilverd, D. Horsfall, R. M. Thorne, R. R. Anderson, Origins of plasmaspheric hiss, *J. Geophys. Res.,* **111**, 705 (2006).

[17] J. L. Green, S. Boardsen, L. Garcia, S. F. Fung, B. W. Reinisch, Reply to "Comment on "On the origin of whistler mode radiation in the plasmasphere" by Green et al.," *J. Geophys. Res.,* **111**,(2006).

ELECTRON FLUX MAPS OF SOLAR FLARES: A REGULARIZATION APPROACH TO RHESSI IMAGING SPECTROSCOPY

ANNA MARIA MASSONE*

CNR - INFM LAMIA, via Dodecaneso 33, I-16145 Genova, Italy
**E-mail: massone@ge.infm.it; http://www.ge.infm.it/~ massone*

MICHELE PIANA

Dipartimento di Informatica, Università di Verona, I-37134 Verona, Italy

MARCO PRATO

Dip. di Matematica, Università di Modena e Reggio Emilia, I-41100 Modena, Italy

A. GORDON EMSLIE

Department of Physics, Oklahoma State University, Stillwater, OK 74078, USA

GORDON J. HURFORD

Space Sciences Laboratory, University of California at Berkeley, CA 94720, USA

EDUARD P. KONTAR

Department of Physics & Astronomy, The University, Glasgow G12 8QQ, UK

RICHARD A. SCHWARTZ

SSAI, Laboratory for Astronomy and Solar Physics
NASA Goddard Space Flight Center, Greenbelt, MD 20771, USA

Reuven Ramaty High Energy Solar Spectroscopic Imager (RHESSI) is a nine-collimators satellite detecting X-rays and γ-rays emitted by the Sun during flares. We describe a novel method for the construction of electron flux maps at different electron energies from sets of count visibilities measured by RHESSI. The method requires the application of regularized inversion for the synthesis of electron visibility spectra and of imaging techniques for the reconstruction of two-dimensional electron flux maps. From a physical viewpoint this approach allows the determination of spatially resolved electron spectra whose information content is fundamental for the comprehension of the acceleration mechanisms during the flaring events.

Keywords: Solar flares; RHESSI; visibilities; regularization methods

1. Introduction

Solar flares are rapid releases of huge amounts of energy originally stored in the solar atmosphere [1]. In correspondence of a flare, the solar plasma is heated up to 20 million degrees and electromagnetic radiation is emitted over the whole electromagnetic spectrum. According to most theoretical models, much of the released energy is used to accelerate electrons, which primarily emit hard X–rays, and ions, which primarily emit gamma rays, to very high energies. Therefore hard X–ray images of the flare emission at many photon energies are an essential tool for the comprehension of the acceleration mechanisms occuring during a solar flare. The realization of this imaging–spectroscopy approach to solar flares is at the basis of the mission of the NASA satellite Reuven Ramaty High Energy Solar Spectroscopic Imager (RHESSI) [2]. This imaging spectrometer utilizes nine rotating collimators with grids of different pitch in order to modulate the solar radiation and nine germanium detectors to measure the energy of each incident photon very precisely. Thanks to this hardware RHESSI is able to perform, in particular, hard X–ray imaging at an angular resolution in the range $2-7$ arcseconds, a temporal resolution of tens of milliseconds, in the energy range from 3 keV to 400 keV; and hard X–ray spectroscopy with a spectral resolution from 0.5 keV to 2 keV, in the same energy range. However it is well-established that, from a physical viewpoint, information of central interest in the study of solar flares are concerned with the phase space distribution of electrons in the plasma, of which the hard X–ray emission is the Bremsstrahlung radiation signature. Information retrieval on the electron distribution on the source from RHESSI data can be realized by applying regularization techniques for the reduction of noise amplification and the solution of remote sensing problems [3]. In particular, reliable reconstructions of the mean source electron distribution from RHESSI hard X–ray spectra have been obtained by means of classical zero order and first order Tikhonov approaches and by an ad hoc formulated technique realizing a triangular matrix row elimination with energy binning [4]. The next and more intriguing step to perform imaging spectroscopy at an electron distribution level, i.e. to produce spatial maps of the electron flux spectrum, is described in the present paper. This method utilizes, as input data, a set of count visibilities, i.e. of calibrated measurements of specific spatial Fourier components of the source distribution; then a 'count–to–electron' inversion algorithm is applied in the spatial frequency domain to obtain a set of electron flux visibility spectra which are processed with Fourier–based imaging methods to produce electron flux images at different electron energies. This combination of visibility data and regularization methodology allows the derivation of robust information on the spatial structure of the electron flux spectrum image, which is the quantity of key interest for the physics of solar flares.

The plan of the paper is as follows. In Sec. 2 visibility measurements are introduced. Section 3 describes the novel imaging spectroscopy algorithm in more details.

Section 4 applies the method to a set of visibilites measured during the solar flare of April 15, 2002 and our final conclusions are offered in Sec. 5.

2. Visibilities

RHESSI data are light curves, i.e. photon-induced counts recorded while the collimators rotate. For an extended source, the resulting modulated count rates depend on the source intensity, location and size. It follows that the essential of the data analysis task for RHESSI X–ray imaging is the inverse problem of determining the location, size and geometry of a source given a set of observed modulation profiles from the different collimators. Using the visibility technique the solution of such inverse problem can be addressed by Fourier methods which may provide accurate reconstruction of the X–ray source within a reduced computational time. A simplified description of this imaging approach is as follows [5]. For each time point of a modulation profile a roll bin can be defined and binned into several (twelve in the case of RHESSI) aspect phases measuring the position of the source with respect to some reference point in the collimator grid. Owing to the angular drift of the satellite axis, for different rotations, different phases correspond to the same roll angle. In order to increase the signal-to-noise ratio, roll angles and aspect phases are accumulated into roll and phase bins and the corresponding counts are stacked into nine histograms, one for each collimator. The last step for the visibility construction is to fit by a Fourier series the count profile in the histogram as a function of the phase bin. The visibility is therefore a complex number whose real and imaginary parts are proportional to the first two coefficients of the fitting Fourier series. It follows that, roughly speaking, a visibility at a specific count energy is a complex number representing the Fourier transform of the count flux at a particular point in the frequency plane. Because of the rotating modulation collimator design of RHESSI, these visibilities are sampled on cylindrical profiles with radii equal to the inverse spatial period of the imaging grids.

The use of visibilities as empirical data for astronomical image processing is advantageous for different reasons. Visibilities are fully calibrated data since they do not suffer any remaining instrument dependence other than spatial frequencies specifically related to the instrument itself. Visibilites are linear combinations of measured counts and therefore have well determined statistical errors. Background is automatically removed during visibility generation and indications of systematic errors may be provided by redundancy (for example, amplitudes for visibility azimuths differing by 180 deg should be the same). Finally, which is more important, the availability of visibilities at different energies and for all nine detectors allows the application of effective Fourier–based image processing methods to visualize the emitting X–ray sources.

3. Electron Flux Spectrum Images

The starting point for our approach is the following formal definition of visibility based on the discussion of the previous section. For each one of the nine RHESSI detectors, the visibility at spatial frequency (u, v) and count energy q is given by

$$V(u, v; q) = \int_x \int_y \sum D(q, \epsilon) I(x, y; \epsilon) e^{i(ux+vy)} \delta\epsilon dx dy \quad , \tag{1}$$

where $D(q, \epsilon)$ is the entry of the Detector Response Matrix (DRM) at count energy q and photon energy ϵ, $\delta\epsilon$ is the quadrature weight in the photon energy space and the sum in the integral is performed over all the photon energies. The DRM is a structured matrix accounting for many detector properties like the attenuator and blanket transmission, the response to modulated photons and the detector resolution. We now introduce the mean source spectral flux image $\overline{F}(x, y; E)$ as the spatial function

$$\overline{F}(x, y; E) = \frac{1}{N(x, y)} \int_{z=0}^{l} n(x, y, z) F(x, y, z; E) dz \quad , \tag{2}$$

where $n(x, y, z)$ is the local density of target particles within a source of line-of-sight depth l, $N(x, y) = \int_{z=0}^{l} n(x, y, z) dz$, and $F(x, y, z; E)$ is the differential electron flux spectrum. Our aim is to construct images at different electron energies E, whose pixel content is the value of $\overline{F}(x, y; E)$. To do this we first need to introduce a mathematical model for the physical process which relates hard X–ray emission to electron acceleration in the plasma. It is well-established [6] that such process is essentially collisional Bremsstrahlung which, in this imaging spectroscopy framework, is described by the Volterra integral equation of the first kind

$$I(x, y; \epsilon) = \frac{1}{4\pi R^2} \int_{\epsilon}^{\infty} N(x, y) \overline{F}(x, y; E) Q(\epsilon, E) dE \quad . \tag{3}$$

Here $I(x, y; \epsilon)$ is the photon spectrum image, $R = 1$ AU and $Q(\epsilon, E)$ is the Bremsstrahlung cross-section which we will assume according to formula 3BN in the Koch and Motz paper [7], i.e., isotropic, fully relativistic with Coulomb correction at small energies. Inspired by the count visibility spectrum defined by Eq. (1) we introduce the electron visibility spectrum

$$W(u, v; E) = \int_x \int_y N(x, y) \overline{F}(x, y; E) e^{i(ux+vy)} dx dy \quad . \tag{4}$$

A very simple computation exploiting the linearity of all the relations introduced so far, leads from Eqs. (1)–(4) to

$$V(u, v; q) = \int_q^{\infty} W(u, v; E) K(q, E) dE \quad , \tag{5}$$

where

$$K(q, E) = \frac{1}{4\pi R^2} \sum D(q, \epsilon) Q(\epsilon, E) \delta\epsilon \quad . \tag{6}$$

Equation (5) naturally inspires the following algorithm for the synthesis of electron flux spectrum images:

For each detector and each frequency pair (u, v) for which visibilities are available:

1 construct a count visibility spectrum $V(u, v; q)$, i.e. count visibilities versus the corresponding count energies;

2 apply a regularized inversion to obtain an electron visibility spectrum $W(u, v; E)$, i.e. electron visibilities versus the corresponding electron energies.

For each detector and each electron energy:

3 construct the set of corresponding electron visibilities.

For each electron energy:

4 apply some Fourier-based image processing method to construct the electron flux map.

In some more details, step 1 and step 3 of the algorithm are essentially re-ordering procedures: in step 1 we fix a point in the frequency space and take all visibilities with different count energies corresponding to that point; in step 3 we fix the energy (and the detector) and take all visibilities with different frequencies corresponding to that electron energy. Regularized inversion in step 2 is performed by Tikhonov zero-order regularization [8]. This approach allows us to address Eq. (5) as a rectangular problem and therefore to reconstruct $W(u, v; E)$ up to electron energies significantly bigger than the count energies in the data. The image synthesis in the final step 4 is realized by applying a Maximum Entropy Method (MEM). The computational cost of this algorithm consists in the cost of the inversion procedure (for each visibility: a singular value decomposition plus a root finding process for optimally determining the regularization parameter) plus the image synthesis through MEM (for each electron energy and each MEM iteration: two Fast Fourier Transforms). Therefore the total cost will be:

$$C \sim 2M[m^2 n + n^2 m + A] + mk4(L \times L)\log(L \times L) \quad , \tag{7}$$

where M is the number of visibilities, m is the number of electron energies, n is the number of count energies, A is the computational cost of the root finding algorithm (bisection method, which converges linearly), k is the iteration number for MEM and $L \times L$ is the image dimension.

4. Application to RHESSI Data

We apply the reconstruction algorithm described in the previous section to the analysis of RHESSI data recorded during the flare event of April 15 2002, in the time interval 00:01:20 – 00:02:20 UT. Detectors 3–9 provide the visibility data and the count energies involved in the analysis start from 10 keV with an energy binning of 2 keV. We apply the Maximum Entropy algorithm to the count visibilities and produce count flux images like the ones in Fig. 4 for a series of energy channels. These images show the presence of a loop already from the first energy channel; the loop shape deteriorates soon and at the channel 30–32 keV is no longer visible. Our algorithm is then applied to obtain the corresponding electron flux images, some of which are given in Fig. 4. Thanks to the application of regularized inversion on the Volterra Eq. (5) information in the electron domain are obtained at electron energies bigger than the count energies involved. In particular, we find that in the case of electron flux images the loop persists up to around 40 keV. The reliability of this approach and, in particular, of the regularization method applied, can be quantitatively assessed by comparing the signal-to-noise ratio (SNR) of the count visibility and electron visibility spectra. For example, Fig. 4 shows the SNRs associated to all the visibility spectra (real and imaginary part) of Detector 3. The regularization power reflects into a substantial stability of these SNRs, while a notable deterioration of the SNR of the electron visibility spectra would occur if naive inversions of the severely ill-posed problems were applied.

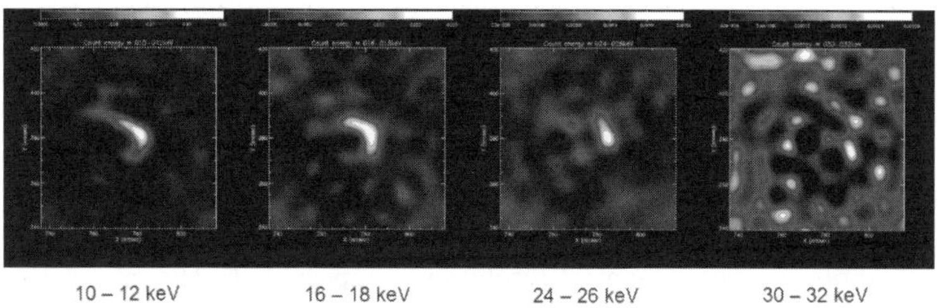

Fig. 1. Count flux maps obtained by means of Maximum Entropy applied to count visibility in the case of the event of April 15, 2002.

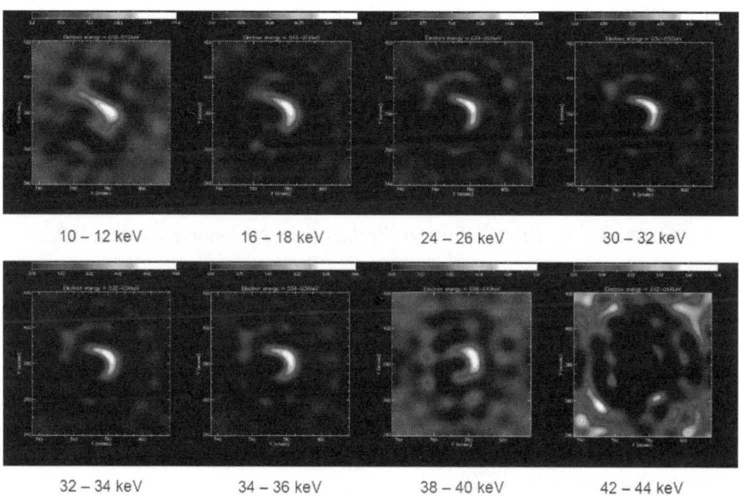

Fig. 2. Electron flux maps obtained by applying regularized inversion to the count visibility spectra and Maximum Entropy to the reconstructed electron visibility sets at different electron energies. The flare event is the same of Figure 1.

5. Conclusions

In this paper we presented a new image processing method able to produce electron flux images at different electron energies from the analysis of count visibilities

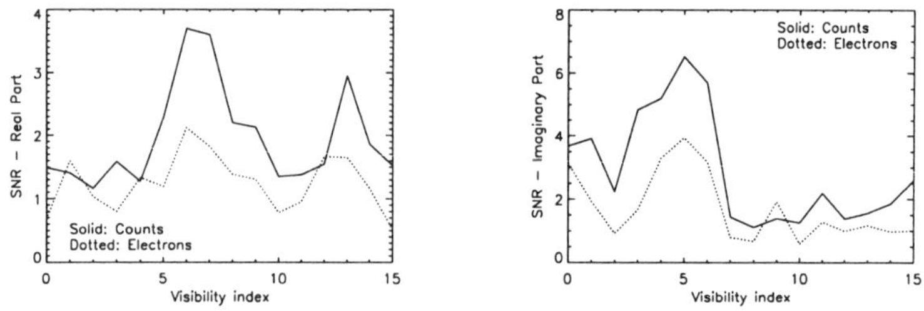

Fig. 3. Signal-to-noise ratios for all count (solid) and electron (dotted) visibility spectra of Detector 3: real part (left panel); imaginary part (right panel).

provided by RHESSI at different count energies. The method is based on the application of regularized inversion to construct electron visibility spectra from count visibility spectra and on a Fourier-based imaging method for the synthesis of an image from the sampling of its Fourier transform. The electron flux maps we obtain with this approach have a spatial resolution comparable to the one of the count flux maps synthesized from the count visibilities measured by RHESSI but, which is more important, they also vary smoothly with electron energy, thanks to the regularization procedure introduced in the method. Furthermore this approach can be applied to very easily derive electron flux spectra at different regions of the source and to infer spatially resolved information on the acceleration processes in the loop.

References

[1] P. A. Sweet, Mechanisms of solar flares, *Ann.Rev.Astron.Astrophys.* **7**, 149 (1969).

[2] R. P. Lin et al., The Reuven Ramaty High Energy Solar Spectroscopic Imager, *Solar Phys.* **210**, 3 (2002).

[3] I. J. D. Craig and J. C. Brown, *Inverse Problems in Astronomy*, (Adam Hilger, London, 1986).

[4] J. C. Brown, A. G. Emslie, G. D. Holman, C. M. Johns-Krull, E. P. Kontar, R. P. Lin, A. M. Massone and M. Piana, Evaluation of algorithms for reconstructing electron spectra from their bremsstrahlung hard X–ray spectra, *Astrophys. J.* **643**, 523 (2006).

[5] G. J. Hurford et al., The RHESSI imaging concept, *Solar Phys.* **210**, 61 (2002).

[6] J. C. Brown, A. G. Emslie and E. P. Kontar, The determination and use of mean electron flux spectra in solar flares, *Astrophys. J.* **595**, L115 (2003).

[7] H. W. Koch and J. W. Motz, Bremsstrahlung cross section formulas and related data, *Rev. Mod. Phys.* **31**, 920 (1959).

[8] M. Piana, A. M. Massone, E. P. Kontar, A. G. Emslie, J. C. Brown and R. A. Schwartz, Regularized electron flux spectra in the July 23, 2002 solar flare, *Astrophys. J.* **595**, L123 (2003).

PROBLEMS AND SOLUTIONS IN CLIMATE MODELING

ALFONSO SUTERA

Department of Physics, University of Rome La Sapienza,
P.le Aldo Moro 5, 00185 Rome, Italy E-mail: alfonso.sutera@roma1.infn.it

Although the dramatic increase in computational power, a complete description of the Earth's climate by means of solutions of the equations of motion is far from being achieved. Most of the problems arise because the solutions depend on the second order effects of the dynamic and thermodynamic instability processes that plague the system. These instabilities are such that the observed behavior strongly departs from the response of the atmosphere to the forcing associated with the energy input. In illustrating the nature of the problem, we shall discuss the Atmospheric General Circulation, i.e. the circulation obtained by averaging the atmospheric motion along the longitude. We will show, by considering both the observed motion and some theoretical models, that the solution requires the parameterization of the momentum and heat fluxes associated with the baroclinic instability of the full three dimensional field. We will discuss how the parameterizations strongly depend on the detailed nature of the external parameters. As a conclusion, some speculations on the nature of the closure needed for this problem will be offered.

Keywords: General Circulation, Baroclinic waves, Climate

1. Introduction

The etymological meaning of the term Climate stands for slope or inclination. It denotes a zone of the world between two lines of latitude with reference to the long-term atmospheric conditions such as wind, temperature, rainfall etc. Thus, if the atmosphere were in solid body rotation with the Earth, we would expect that the climate would vary smoothly as a function of latitude, with increasing relative momentum as the latitude increases. By the 17th century it was known that, around $30°N$ of latitude, the climate was dry with weak winds. South of this latitude winds were regular north-easterly (the Trade Winds) and to the north they were irregularly blowing from a westerly direction. This pattern appeared to mirror itself south of the equator with steady south easterly trade winds. When scientists tried to account for the general circulation, their discussion mainly regarded the trade winds. Because their steadiness, in fact, they were assumed to be the easiest ones to explain. Galileo (and for this matter also Keplero) saw the trade winds as a consequence of the failure of the Earth's gaseous envelop to keep up with the speed of the Earth's rotation. At low latitudes, where the Earth is rotating faster, the air is lagging behind, so that an earthbound observer would experience easterly winds. The opposite mechanism would occur aloft with the Earth lagging the atmosphere. In this frame, the trade winds were merely a manifestation of the Earth's rotation. Of course, rotation alone cannot account for the general motion of the atmosphere since, as we know, the differential heating imposed by the Earth's curvature must set up an equator to pole pressure gradient and, thereby, some kind of motion. Therefore, in understanding climate, a slightly different question may be posed: given a rotating atmosphere driven by a differential heating, what kind of trade winds will be generated? The success of any theoretical answer will depend on the degree of similarity of the achieved circulation with the observed one. The present paper is intended to present a set of results which illustrate the nature of the problem, the degree of accuracy of our theoretical knowledge and how far we are from a full understanding of the observed circulation. In concluding this introduction, the author feels the compulsion to apologize with the many, not cited, keen minds who have been challenged by the problem above and have offered intriguing solutions to it. However, reviews on the subject are not lacking. The interested reader may refer to Lorenz [5] for an historical account and to Lindzen [4] for the contemporary approach to the problem.

2. The Equations of Motion

The atmosphere is a thin layer of a gas mixture that includes chemical species interacting with radiation. Therefore, it is convenient to group the laws of motion into two classes: the basic hydrodynamic and thermodynamic laws that apply to a great variety of fluid systems and laws expressing the forces and the heating in terms of the current state of the atmosphere and its environment. The latter laws include radiation processes and the turbulent transfer of heat and matter. For any practical purpose, laws in both classes must be approximated, since they describe also phenomena that are of a secondary importance for the behavior of the large scale motion. These simplifications lead to the so called *primitive equations* (Holton [3]); they are:

$$\frac{du}{dt} = \frac{\tan\phi}{a} uv + 2\Omega \sin\phi v - \frac{1}{a\cos\phi}\frac{\partial \Phi}{\partial \lambda} + F_\lambda \tag{1}$$

$$\frac{dv}{dt} = -\frac{\tan\phi}{a} u^2 - 2\Omega \sin\phi u - \frac{1}{a}\frac{\partial \Phi}{\partial \phi} + F_\phi \tag{2}$$

$$\frac{\partial \Phi}{\partial p} = -\frac{RT}{p} \tag{3}$$

where Ω is the magnitude of the angular velocity vector $\mathbf{\Omega}$, a is the mean Earth's radius, α is the specific volume and $F_{(.)}$ is any body force in the appropriate direction. The Eulerian total derivative for any arbitrary scalar X is defined as:

$$\frac{dX}{dt} = \frac{\partial X}{\partial t} + \vec{V}\cdot\nabla X + \omega\frac{\partial X}{\partial p} \tag{4}$$

where ∇ is curvilinear version of the gradient operator. On the other hand, mass conservation requires:

$$\nabla \cdot V + \frac{\partial \omega}{\partial p} = 0 \tag{5}$$

The thermodynamic equation for the specific entropy and the gas law in terms of absolute temperature T are:

$$\frac{dT}{dt} = \kappa \frac{T\omega}{p} + \frac{Q}{c_p} \tag{6}$$

where

$$\omega = \frac{dp}{dt}, \tag{7}$$

where Q is the rate of heating of the gas envelop. Here, λ is the longitude, ϕ the geographic latitude, and z the vertical coordinate. They point eastward, northward and against the gravity vector respectively. The velocity vector is $\vec{V} = (a \cos\phi \frac{d\lambda}{dt}, a\frac{d\phi}{dt}, \frac{dp}{dt})$. The exact modeling of r.h.s of Eq. (6) requires the study of the second class of laws that in this paper will be approximated by a simple Newtonian cooling about some reference temperature:

$$\frac{Q}{c_p} = \frac{T - T_r}{\tau}, \tag{8}$$

where τ is a suitable time constant and T_r is an externally imposed temperature field. In the troposphere T_r is decreasing poleward.

The main approximations are: the flow is hydrostatic, only reversible thermodynamics transformations are allowed and the actual distance of an air particle from the rotation axis is the Earth's radius. The latter implies that the angular momentum conservation is applied as if all the atmosphere particles were at the Earth's surface. Most of these assumptions are consistent with the small aspect ratio of the fluid and with almost dry nature of the gas. Notice that Eq. (3) implies that it exists a single value monotonic relationship between pressure (or, by Eq. (7), any other thermodynamical variable) and height. Thus, we may use p as a vertical coordinate by changing the partial derivative along constant z to constant p and defining the vertical velocity accordingly. The lower boundary conditions, however, must be specified. They are awful if the surface pressure is not a constant. We can overcome this situation by using as a vertical coordinate $\sigma = \frac{p}{p_s}$, with p_s the surface pressure. In this case, in fact, $\sigma = 1$ is always the lower boundary, while a prognostic equation for p_s can be derived from Eq. (5).

The assumption that the flow is Boussinesq (α constant everywhere except when it combines with gravity, Pedlosky [6]) allows a first glance to the nature of the steady solutions of these equations. Setting, in fact, the l.h.s. of Eq. (1), Eq. (2), Eq. (5) and Eq. (6) to zero and considering any field independent on longitude, there are at least two classes of solutions depending on whether F_λ and F_ϕ are zero or not. In the first case v is zero so that no northward transport is present. In the second case, v is not zero so that, by virtue of continuity, a closed circulation may occur. Raising motion at the equator, for preserving the flow absolute angular momentum, will acquire relative momentum which is exchanged with the Earth's surface, where the return flow is and friction is largest. Thus, at the lower level the wind vector will acquire a southerly component at the lower layer, i.e. trade winds exist. This single, equator to pole, cell circulation is the celebrated Hadley cell and it is a consequence of angular momentum conservation. The most important outcome of this solution is that the associated climate departs from the one obtained by considering the first class of solutions where, for the absence of any heat transport, the imposed temperature field T_r remains unchanged. It appears obvious, therefore, that the exact knowledge of the atmospheric transport is of a paramount relevance for any argument involving the sensitivity of the climate to changes in T_r. Notwithstanding, there is an obvious flaw in this single cell model, namely it does not fully account for the deflective nature of the Coriolis force. If we consider angular momentum conservation alone, in fact, it is required that a portion of fluid at relative rest at equator (i.e. with an absolute velocity of about 400m/s), in its poleward displacement, must acquire a relative velocity of 200m/s at $30°N$ which is unobserved velocity. A more convincing picture is portrayed if the effect of the Coriolis force on particle motion is considered. In this case, in fact, the wind can increase conserving angular momentum up to a latitude wherein the r.h.s of Eq. (1) is zero. Thereafter it will satisfy Eq. (2), i.e a geostrophic balance, rather than an angular momentum principle. Notice that the switch between the two circulations occurs at a latitude where the pressure gradient is zero. It is known that this latitude corresponds to the tropical calms. So we must seek for an Hadley cell solution extending up to the tropical calms where the angular momentum conserving wind reach its maximum. In summary, the effect of the Hadley cell is to export heat from the equator to the tropical calm and therefore cannot produce the observed equator to pole temperature gradient. Northward of this latitude , it is observed that the surface wind shifts to a southwesterly direction. To balance this wind against dissipation, Ferrel envisioned an indirect cell: the pressure maximum required at the tropics by the Hadley cell implies a cyclonic circulation in the meridional plane (latitude-pressure), and, therefore, it gives a southerly component to u as observations require. This Ferrel cell may be obtained by very small value of $F_{(.)}$ (see Lindzen [4]). An easily perceived limitation of our reasoning, however, is that the symmetric solution outlined in so far requires a choice of a particular set of initial conditions (independent on longitude) and a specification of $F_{(.)}$. As a matter of fact, both choices appear quite arbitrarily set, since, regarding $F_{(.)}$, we have no theory, while the other lacks generality. Furthermore, the removal of the first choice (i.e. longitudinal independent perturbations), because the non-linear nature of Eq. (4), generates Reynolds stresses. They may balance otherwise the zonally averaged (i.e. along λ) momentum and induce other solutions. As we shall see, the implications of removing these arbitrary conditions have profound effects on the climate response to T_r.

3. Parameter Settings and Numerical Solutions

The equations of motion can be numerically solved if $F_{(.)}$ and T_r are specified. For the present paper we consider a diffusive approximation as:

$$F_\lambda = \nu \frac{\partial^2 u}{\partial p^2} \tag{9}$$

$$F_\phi = \nu \frac{\partial^2 v}{\partial p^2} \tag{10}$$

where $\nu = 1 m^2 s^{-1}$, a value for which the axisymmetric equations have no Ferrel cell (Lindzen [4]).

Our numerical solutions will be obtained by using the simplified GCM PUMA (Portable University Model of the Atmosphere, Fraedrich et al. [2]). PUMA is a global spectral model that solves the primitive equations on sigma levels. The non adiabatic and dissipative processes are represented by the Newtonian cooling and the Rayleigh friction respectively. The model is run at T21 resolution with 20 equally spaced sigma levels. T_r is:

$$T_r = T_0(\sigma) + \frac{\Delta T_r(\sigma)}{2}(\cos(2\phi) - \frac{1}{3}) \tag{11}$$

where T_0 and ΔT_r are listed in Table 1 for each σ level. It should be mentioned that T_r describes a stably stratified atmosphere subjected to a poleward decreasing temperature as the second Legendre polynomial P_2^0. It is a reasonable approximation for the actual radiative Earth's temperature for mean equinoctial conditions. The relaxation time scale τ has been set to a fairly accepted value of 15 days while a Raleygh friction (TFRC) is applied at the lowermost level simulating the momentum exchange occurring at the ground.

The sensitivity of the model to ΔT_r (with a step of $10K$) will be studied for three cases labeled as C1, C2, C3. C1 is the model response if axisymmetric initial conditions are considered. It is meant to quantify the previous qualitative statement about the Hadley circulation. C2 is as C1 only that the initial conditions allow longitudinal variation of the model variables. C3 is as the others except for ΔT_r at the upper model levels where it changes sign. This temperature reversal is typical for stratospheric conditions in the summer Hemisphere where poleward ozone transport induces a polar temperature higher than at the equator. By considering Eq. (2), it is easy to understand that it implies a decreasing u with height at these levels. For the case C1 the model achieves a steady state after 2 model-years, the other cases show a low frequency variability which become statistically steady after 4 model-years. Consider a numerical solution for one of the cases described above. We can define its time average (over the last of the 4 model-years). Moreover, let $\bar{A}$ be any zonally averaged fields at a steady (statistical) state. Eq. (2) written in p coordinate and zonally averaged, implies that we can define the Stokes stream function Ψ as:

$$\Psi = 2\pi a \int_0^{p_s} \bar{v} \frac{dp}{g}, \tag{12}$$

where the integral is extended from 0 to p. It is a measure of the circulation intensity in the meridional plane (i.e. ϕ, p plane). Where Ψ has positive value we have an anticyclonic circulation, i.e an Hadley cell, while for negative values we have a Ferrel cell. By Eq. (11) T_r is a symmetric with respect to the equator, i.e. in equinoctial conditions. To give an idea about the observed circulation, in Fig. 3 we report Ψ as obtained by considering the observed multiyear averaged $\bar{v}$ for April. The Hadley and Ferrel cells are clearly shown, while the small direct polar cell is indicative of some polar easterlies. Their relative extremes are respectively of 120, -40 and 0.5 $10^9 kg/s^{-1}$.

Table 1. Experimental setup for PUMA: The global mean restoration temperature $T_0(\sigma)$, the equator-pole temperature difference $\Delta T_r(\sigma)$, and the Rayleigh friction damping time scale TFRC.

Case	$T_0(\sigma)(K)$ top to bottom	$\Delta T_r(\sigma)(K)$ top to bottom	$TFRC(\sigma)(day)$ top to bottom
C1	265.14, 254.91, 246.36, 240.95, 237.56, 234.65, 235.24, 237.73, 243.54, 248.87, 253.41, 257.83, 261.72, 265.60, 268.96, 272.33, 275.36, 278.34, 281.06, 283.74	0.00, -3.05, -20.32, -26.19,-28.13, 28.26, 34.91, 40.37, 45.10, 49.19, 52.05, 54.69, 56.20, 57.70, 58.31, 58.91, 59.26, 59.56, 59.76, 59.94	0.0, 0.0, 0.0, 0.0, 0.0, 0.0, 0.0, 0.0, 0.0, 0.0, 0.0, 0.0, 0.0, 0.0, 0.0, 0.0, 0.0, 0.0, 0.0, 0.0, 0.1
C2	As for C1	0.00, 0.00, 0.00, 0.00, 0.00, 28.26, 34.91, 40.37, 45.10, 49.19, 52.05, 54.69, 56.20, 57.70, 58.31, 58.91, 59.26, 59.56, 59.76, 59.94	As for C1
C3	As for C1	As for C1	As for C1

Figure 2 summarizes our results in terms of maximum intensities of Hadley, Ferrel and polar cells as a function of the forcing by the equator-pole temperature difference at the surface ΔT. The remarkable differences between the axisymmetric solution and the others are:

1) the rather different slopes;

2) the emergence of Ferrel and polar cells, in accordance with the observations, for the cases C2 and C3;

3) Figure 2 b,c are unchanged if $\nu \neq 0$.

Thus, we may conclude that the axisymmetric solution, while is a probable state, it is unlikely to occur, since for any arbitrary longitudinal dependent perturbation the climate will be driven far away from the axisymmetric state. This is a manifestation of an unstable behaviour that equilibrates far away from the axisymmetric solution.

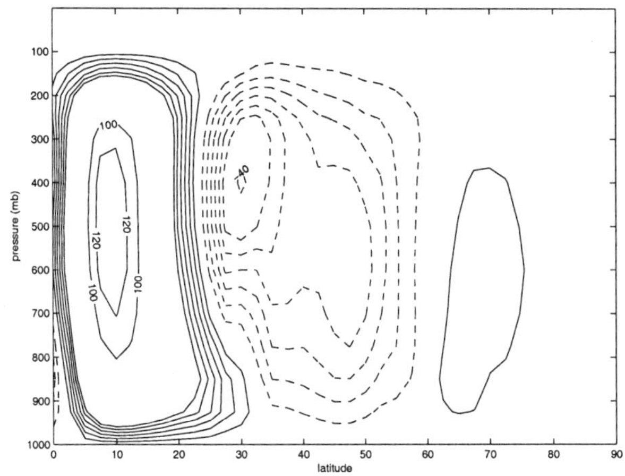

Fig. 1. Zonal mean cross section of the mass stream function for April climatological conditions. Units are $10^9 kgs^{-1}$. Contours are every $5 \cdot 10^9 kgs^{-1}$ within the range (-30, 30) and zero line is excluded; values greater than $100 \cdot 10^9 kgs^{-1}$ and less than $40 \cdot 10^9 kgs^{-1}$ are shown with labels. Solid lines denote positive values, dashed lines negative ones.

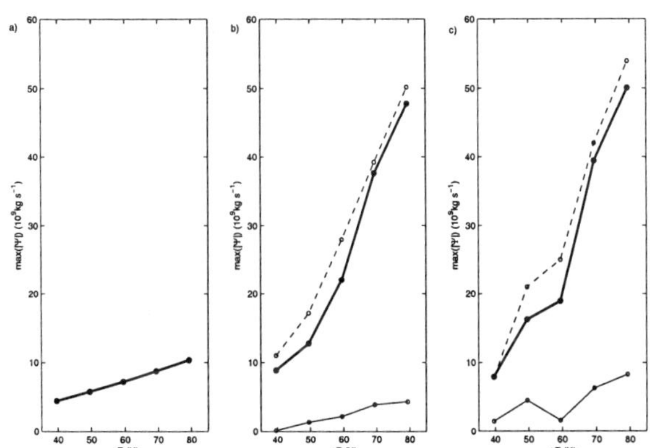

Fig. 2. Relative maximum values of the absolute mass streamfunction as a function of the forcing pole-equator temperature difference at the surface for a) C1, b) C2 and c) C3 cases. Thick lines refer to Hadley cell, dashed lines to Ferrel cell and thin solid lines to the polar cell. Units are $10^9 kgs^{-1}$.

Moreover, we notice that for $\Delta T < 70K$ C2 and C3 have a rather different behavior. While C2 shows an approximately linear dependence on ΔT. C3 shows a peculiar slope change in Ψ. It appears that, at variance with C2, C3 has a regime change for $50K < \Delta T < 70K$. This is confirmed by Fig. 3 where the zonal mean zonal wind u is shown in the meridional plane. Case C3 shows, in fact, that the maximum upper level westerly wind shifts from subtropical to mid-latitude position, while the secondary maximum goes the other way around. This has a dramatic impact on climate, since the momentum and heat transports implied by Fig. 2 would show such a similar dependence on ΔT.

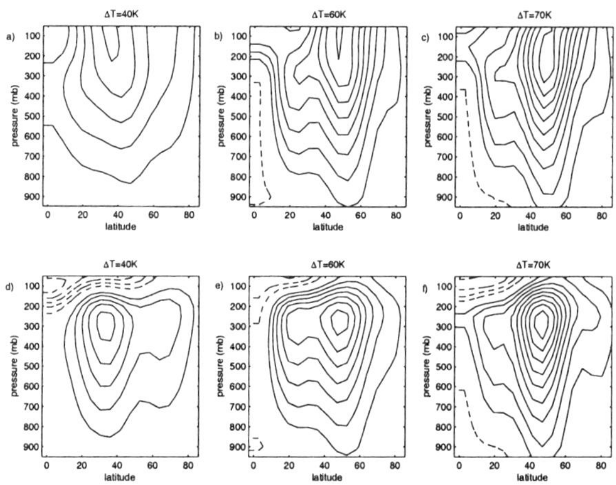

Fig. 3. Latitude-height (pressure) cross-section of the zonal Mean zonal wind averaged over the 4 model-year in ms^{-1}. Dashed lines are negative values and contour interval is $4ms^{-1}$ with zero line excluded. Plots are for cases C2 (a-c) and C3 (d-f) and different values of the equator-pole temperature difference in the troposphere. ΔT values reported in the figures refer to the forcing temperature difference at the surface.

4. An Heuristic Model

To shed some light on the model behaviour we may formulate an heuristic model as follows. The atmosphere is a baroclinic flow, i.e. the surfaces with constant entropy and constant pressure are not parallel. Thus, a constant entropy surface grazing the ground at the equator may be at lower pressure at the pole. Thus, a particle displaced adiabatically nortward from the equator will be rising to keep its entropy constant, despite the hydrostatic nature of the flow. By continuity a return flow is required. Again, because the deflective nature of the Coriolis force, a disturbance moving along a latitudinal circle will be forced back to its position but in doing so will deplete the potential energy of the zonally symmetric flow that can be converted into kinetic energy of the disturbance. Thus, the perturbation can grow until the total kinetic energy equals the rate at which it is dissipated. The constant entropy surfaces are the key element of this instability mechanism. More they are sloping, more energy can be transferred to the perturbation. A slope, however, is just a geometric constraint, i.e.the aspect ratio (height scale with respect the horizontal scale) of the gas envelop. Of course, the efficiency of this mechanism would depend also on the vertical gradient of the entropy surface. Since in our forcing specification the latter has remained unchanged, this aspect will not be discussed any further. Reversing the temperature gradient in the stratosphere, instead, amounts to reverse the northward slope of the isentropic surfaces so that the effective geometric scale is actually smaller when compared with the isothermal stratosphere. Next, consider a Boussinesq baroclinic, wavelike in the longitude direction, disturbance at a single level and let suppose that a single zonal wave is unstable with respect to the axisymmetric circulation set by T_r. Let u', v', w', T' its velocity components (along longitude, latitude and height) and temperature. It is easy to show that this disturbance transports meridionally heat and momentum if the correlations $\overline{u'T'}$, $\overline{u'v'}$ are not zero, i.e. the fields must have a phase difference along the longitude. For an easy understanding of the origin of the double-jet structure, let consider the zonally averaged equations of section 2 in a cartesian coordinate system for a Boussinesq flow at the steady state. Writing the equations Eq. (1), Eq. (2), Eq. (3) in the cartesian plane and neglecting the $F_{(.)}$ we obtain:

$$f\bar{v} = -\frac{\partial}{\partial y}\overline{u'v'} \tag{13}$$

$$f\bar{u} = -\alpha\frac{\partial}{\partial y}\bar{p} \tag{14}$$

$$\frac{\partial}{\partial z}\bar{p} = -\frac{g}{\alpha} \tag{15}$$

$$\frac{\partial}{\partial y}\overline{v'T'} = \frac{\bar{T} - T_r}{\tau} \tag{16}$$

$$\bar{p}\bar{\alpha} = R\bar{T} \tag{17}$$

where any vertical advection terms have been neglected and $f = 2\Omega sin(\phi_0)$ with ϕ_0 a given latitude. If the fluid is Boussinesq, for continuity, we also have:

$$\frac{\partial}{\partial x}u' + \frac{\partial}{\partial y}v' = 0 \tag{18}$$

Consider

$$u' = Re[\tilde{u}_{k,l}e^{ikx}cos(ly)] \tag{19}$$

$$T' = Re[\tilde{w}_{k,l}e^{ikx}sin(ly)], \tag{20}$$

then Eq. (18) implies:

$$v' \propto Re[\tilde{v}_{k,l}e^{ikx}sin(ly)]. \tag{21}$$

Now,

$$\overline{u'v'} \propto sin(2ly) \quad (22)$$
$$\overline{v'T'} \propto sin^2(ly). \quad (23)$$

It follows that

$$\frac{\bar{T} - T_r}{\tau} \propto sin(2ly) \quad (24)$$
$$\bar{v} \propto cos(2ly) \quad (25)$$

and by Eq. (15)

$$\bar{u} \propto cos(2ly) \quad (26)$$

Thus, if we choose $cos(ly)$ to be half the wave length the zonally averaged flow will have a full wave length and therefore $\bar{u}$ will have two extremes. A more detailed and rigorous treatment can be found in Bordi et al. [1]

5. Conclusions

A mechanistic view on the general circulation of the atmosphere has been presented. In particular, the prominent role that the longitudinal dependence of disturbances has in producing a reasonable circulation in the meridional plane for a dry atmosphere in hydrostatic equilibrium has been highlighted. Moreover, the sensitivity of this circulation to modification on the externally imposed equator to pole temperature gradient has been shown. For weak gradients, the state of the atmosphere circulation may undergo to a regime shift depending on the nature of radiative equilibrium operating in the stratosphere. Despite the simplified description of the interactions between atmospheric composition and radiation, the model indicates that strong non linearities may occur so that the climate system may be strongly sensitive to the external forcing. Given these unexpected results, it appears very urgent to study the nature of the dynamical transports responsible for this behavior. We notice that the origin of the trade wind has challenged scientists from the beginning of modern Science. In the author's opinion, beyond countless qualitative explanations offered in so far, any quantitative theory is still out of reach, even if the problem is cast in a simple framework as the one here considered. The lack of a definitive answer for this fundamental problem, let wonder if the solution of some recent problems, such as the causal nature of the observed, modest increase of the earth's surface temperature, may go beyond the simplistic approach based on simulations and the so called Consensus Science. The struggle of a few against a multitude signed the transition from the ptolemaic conception of the world to the copernican view. This should make one to think that simulations and Consensus Science may be not the main avenue to give definitive answers, even when scientists feel the pressure posed by societal demands.

Acknowledgments

I wish to thank my collaborators I. Bordi and M.Petitta for helping me during the completion of this work; K. Fraedrich for numerous conversations on the subject and for providing PUMA. I had the privilege to know Livio Scarsi and I dedicate this note to his memory.

References

[1] I. Bordi, Fraedrich K., Lunkeit F., Sutera A., *On non-linear baroclinic adjustment with the stratosphere* (Nuovo Cimento C , 29,497-518, 2006)

[2] K. Fraedrich, Kirk, E. and Lunkeit, F., *Portable University Model of the Atmosphere Deutsches Klimarechenzentrum, Report No. 16.* (http://www.mi.uni-hamburg.de/puma, 1998).

[3] J. R. Holton, *An Introduction to Dynamic Meteorology* (Academic Press, New York, 1992).

[4] R. S. Lindzen, *Dynamics in Atmospheric Physics* (Cambridge University Press, Cambridge, 1990).

[5] E. N. Lorenz, *The nature and theory of the general circulation of the atmosphere.* (World Meteorological Organization Monograph No. 218, TP 115, 1967).

[6] J. Pedlosky, *Geophysical Fluid Dynamics* (Springer-Verlag, New York, 1979).

NUMERICAL SIMULATIONS AND DIAGNOSTICS IN ASTROPHYSICS: A FEW MAGNETOHYDRODYNAMICS EXAMPLES

GIOVANNI PERES[1]*, ROSARIA BONITO[1], SALVATORE ORLANDO[2], and FABIO REALE[1]

1 - Dip. Scienze Fisiche ed Astronomiche, Sez. Astronomia, University of Palermo,
2 - Osservatorio Astronomico di Palermo "G.S. Vaiana", INAF,
Piazza del Parlamento 1, 90134 Palermo, Italy
** E-mail: peres@astropa.unipa.it*

We discuss some issues related to numerical simulations in Astrophysics and, in particular, to their use both as a theoretical tool and as a diagnostic tool, to gain insight into the physical phenomena at work.

We make our point presenting some examples of Magneto-hydro-dynamic (MHD) simulations of astrophysical plasmas and illustrating their use. In particular we show the need for appropriate tools to interpret, visualize and present results in an adequate form, and the importance of spectral synthesis for a direct comparison with observations.

Keywords: Magneto-Hydro-Dynamics; modeling; diagnostics; Astrophysics.

1. Introduction

The analysis of astronomical observations always rests on an interpretation of the phenomenon, sometimes even only a rather simple one as the assumption that the emitted spectrum originates from an isothermal plasma. On the other hand, given the impossibility to perform experiments under controlled conditions but, rather, to observe astrophysical phenomena and objects, plus the fact that most astrophysical objects are characterized by complex phenomena involving several physical effects at play simultaneously and interacting in complex ways, one is forced either to the risk of ignoring important effects by considering oversimplified models or to the need of building up realistic but often complex models.

The need for models which involve several physical effects of comparable importance (e.g. thermal conduction, radiative losses, some heating agent, magnetic fields etc.) often cannot be treated analytically and typically leads to a numerical treatment of the model itself; sometimes even requiring high performance computing (HPC).

Such models are often used as a theoretical tool to understand the basic physics involved in phenomena, sometimes through parameters-space exploration or, more directly, through an appropriate switching on/off of the various terms in the equations and comparing the different solutions. Models can also be used in a less obvious way as tools to develop and tune-up new diagnostics, by identifying some important observable signature of a specific phenomenon or effect, or even as a diagnostic tool through a detailed comparison with observations.

However standard fitting procedures, e.g. through minimization of χ^2, are not viable with these models for several reasons: for instance each "point" in the parameter space may require a large amount of computing resources; on the other hand parameter derivation through inversion techniques in many cases has been proved mathematically unstable. The choice, thus, in many cases is forward modelling, i.e. compute (i.e. model) a set of cases guessing the mathematical domain of the right parameters of the model, obviously putting the guess for the best value at the center of your domain. Typically the comparison of the various (often computer demanding) calculations will bring a comprehension of the phenomenon and the related physics and probably show which case is closer to, or matches, observations.

However we have found that a comparison with models requires great care, given the complex nature and appearance of the phenomena. First of all in order to understand the results of a model one requires powerful tools to organize, visualize and handle the model results, often in the form of large table of numerical results on, as is the case of MHD models, density, temperature and velocity plus magnetic fields at each point and time. The need for powerful "visualization tools" is common to all the area of HPC, and there is a general effort to develop and to provide them; some teams which develop powerful codes take care to provide a basic visualization tool along with their code, as it happens for the FLASH code (see next sections) which we use and have helped to develop.

Matching models to observations is an even more demanding feat: one has to match appearance, evolution (when possible), observed features including spectra as collected by instruments. The need to match the emission, and how it is detected by instruments, calls for powerful spectral synthesis codes, capable of producing the emission and spectrum from each part and from the globality of the modeled object or phenomenon. Over the years, along with the numerical models, our group has had to develop some visualization tools and complex tools for comparison with observations (e.g. in X-rays, in radio etc.).

In the following we will show two different examples pertaining to two fields of Astrophysics: the morphology and physical characteristics of one class of Supernovae Remnants and the X-ray emission recently detected in protostellar jets. This method, however, has been widely applied to many more cases, such as the study of solar magnetically confined plasma and the modeling of stellar flares.

2. Supernovae Remnants

The characteristics and the chemical abundances of the interstellar medium (ISM) are strongly influenced by supernovae (SN) and by supernovae remnants (SNRs). However, many features of the interaction between SNR shock fronts and ISM are influenced by several factors, among which the multi-phase characteristics of the medium, its density and temperature, the intensity and direction of the ambient magnetic field. These factors cannot be easily constrained, thus somehow limit our understanding of the ISM, of SNR dynamics and of their interaction.

In particular the bilateral supernova remnants (BSNR, also called "barrel-shaped," or "bipolar"), see [1], are benchmarks for the study of large scale SNR-ISM interactions, since small scale effects appear to be irrelevant. The BSNRs are characterized by two opposed radio-bright limbs separated by a region of low surface brightness; in general, these remnants appear asymmetric, distorted and elongated with respect to the shape and surface brightness of the two opposed limbs.

In spite of the interest around BSNRs, a satisfactory and complete model which explains the observed morphology and the origin of the asymmetries does not exist. The predictions of ad-hoc models have consisted so far of a qualitative estimate of the BSNRs morphology, with no real estimate of the ISM density interacting with the shock.

Orlando and collaborators [2], in order to explain BSNR at an adequate level, have used detailed physical modeling, high-level numerical implementations and extensive simulations. They investigated whether the morphology of BSNR observed in the radio band could be mainly determined by the propagation of the shock through a non-uniform ISM or, rather, across a non-uniform ambient magnetic field. To this end, they modeled the propagation of a shock generated by an SN explosion in the magnetized non-uniform ISM with detailed numerical MHD simulations, considering two complementary cases of shock propagation: 1) through a gradient of ambient density with a uniform ambient magnetic field; 2) through a homogeneous isothermal medium with a gradient of ambient magnetic field strength.

A proper use of modeling, synthetic "observations" derived from models and comparison of observations with models and derived for critical values of the parameters strongly constrained the models of the observed phenomenon.

The authors used FLASH, a very accurate and advanced multi-dimensional MHD code for astrophysical plasmas [3] designed to make efficient use of massively parallel computers with the message-passing interface (MPI) for interprocessor communications. The core of FLASH is based on a directionally split Piecewise Parabolic Method (PPM) solver to handle compressible flows with shocks and the code solves the MHD equations on a block-structured adaptive mesh. To this end, FLASH uses the PARAMESH package [4] for the parallelization and the adaptive mesh refinement portion of the code. FLASH has been developed mainly at the Center for Astrophysical Thermonuclear Flashes (the FLASH center) at the University of Chicago. Our group collaborates with the FLASH center to upgrade the code with new numerical modules and to test it. We have already adapted FLASH for applications to coronal plasmas, to supernova remnants, and to proto-stellar jets. In the simulations presented here and in the following section we have used FLASH with customized numerical modules that treat optically thin radiative losses and thermal conduction.

From the simulations, Orlando et al. [2] synthesized the synchrotron radio emission, making different assumptions about the details of acceleration and injection of relativistic electrons. The scope was to perform simulations for different plasma configurations and different magnetic fields orientations, to synthesize radio emission from the simulations postulating various electrons injection directions with respect to the magnetic field and the different orientations of the system with respect to the line of sight; and to compare the various "observations" synthesized from models with good observations to obtain constraints on the models and, anyhow, insight.

The authors found that asymmetric BSNRs are produced if the line-of-sight is not aligned with the gradient of ambient plasma density or with the gradient of ambient magnetic field strength; they also derived useful parameters to quantify the degree of asymmetry of the remnants that may provide a powerful diagnostic of the microphysics of strong shock waves through the comparison between models and observations. BSNRs with two radio limbs of different brightness can be explained if a gradient of ambient density or, most likely, of ambient magnetic field strength is perpendicular to the radio limbs. BSNRs with converging similar radio arcs can be explained if the gradient runs between the two arcs. Figure 6 shows good examples of simulations which yield two different apparent morphologies, along with the corresponding two examples of observed morphology well matched by simulations.

3. Protostellar Jets

The early stages of the star birth manifest various mass ejection phenomena, including collimated jets which can travel through the interstellar medium at supersonic speed (at several hundreds of km/s or even at $\approx$ 1000 km/s), forming shock fronts at the interaction front between the jet and the unperturbed ambient medium. In the last 50 years these features have been studied in detail in the radio, infrared, optical and UV bands, and are known as Herbig-Haro (hereafter HH) objects.

Recently, see for instance [7], X-ray emission from HH objects has been detected with both the XMM-*Newton* and *Chandra* satellites in a few young stellar objects (YSO); the X-ray emission, given the relatively low temperature ($\approx 10^6$ K) derived from spectra and, in some cases, the evident misplacement with respect to the originating stars appears to be unrelated to the typical X-ray emission phenomena of stars, typically tied to coronae. Protostellar jets and the circum-proto-stars are usually low temperature (and low-energy) environments and X-ray emission, therefore, has been a surprise.

Thus understanding the X-ray emission from protostellar jets is important to understand which physical mechanism of star and planet formation leads to X-rays thermal emission. Furthermore, X-rays (and ionizing radiation in general) may affect the environment of young stellar objects and, in particular, the physics and chemistry of the accretion disk and its planet-forming environment. The ionization state of the accretion disk around young stellar objects will determine its coupling to the ambient and protostellar magnetic field, and thus, among other things, influence its turbulent transport. This, in turn, will affect the accretion rate and the formation

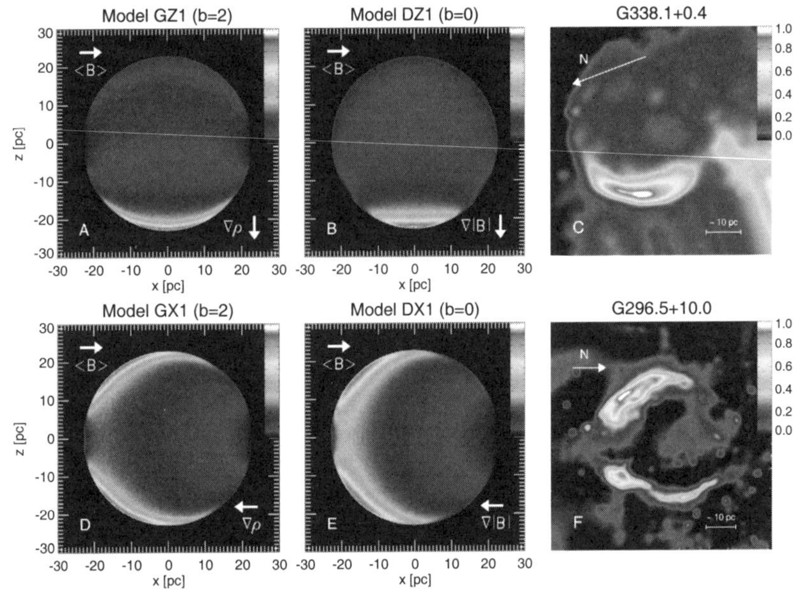

Fig. 1. Synchrotron radio emission (normalized to the maximum of each panel), at $t = 18$ kyrs since the SN explosion, synthesized from models assuming a gradient of ambient plasma density (panels A and D) or of ambient magnetic field strength (panels B and E) when the LoS is aligned with the y axis. All the models assume quasi-perpendicular particle injection. The directions of the average unperturbed ambient magnetic field, and of the plasma density or magnetic field strength gradient, are shown in the upper left and lower right corners of each panel, respectively. The right panels show two examples of radio maps (data adapted from [5] and [1]); the arrows point in the north direction) collected for the SNRs G338.1+0.4 (panel C) and G296.5+10.0 (panel F). The color scale is linear and is given by the bar on the right. Adapted from [2]

of structures in the disk and, therefore, the formation of planets. X-rays also can act as catalysts of chemical reactions in the disk's ice and dust grains, thereby significantly affecting its chemistry and mineralogy. The ability of the forming star to ionize its environment will therefore significantly affect the outcome of the formation process.

Several models have been proposed to explain the X-ray emission from protostellar jets, but the emission mechanism is still unclear. In this perspective, we developed a detailed hydrodynamic model of the interaction between a supersonic protostellar jet and the ambient medium; our aim is to explain the detailed physics that may lead to the observed X-ray emission. Our model takes into account optically thin radiative losses and thermal conduction effects. We use the FLASH code also in this case.

A first paper [6], presented an exploratory study with a set of results concerning a jet less dense than the ambient medium, with a density contrast $\nu = n_a/n_j = 10$ (where n_a is the ambient density and n_j is the density of the jet); this model showed

X-ray emission in good agreement with the observed X-ray emission from HH 154 [7]. Bonito and collaborators [6] have shown the validity of the physical principle on which our model is based: a supersonic jet traveling through the ambient medium produces a shock at the jet/ambient medium interaction front, leading to X-ray emission in good agreement with observations. A second paper studied the effects of varying the control parameters characterizing the jet dynamics, more specifically, the ambient-to-jet density ratio, $\nu = n_a/n_j$, and the Mach number, $M = v_j/c_a$ (c_a is the ambient sound speed). Through the exploration of a wide range of the control parameter space, [8] tried to determine the range of these parameters that can give rise to X-ray emission consistent with observations. The scope included:

- to constrain the jet/ambient medium interaction regimes leading to the X-ray emission observed in Herbig-Haro objects in terms of the emission by a shock forming at the interaction front between a continuous supersonic jet and the surrounding medium;
- to derive detailed predictions to be compared with optical and X-ray observations of protostellar jets;
- to get insight into the protostellar jet's physical conditions.

A wide set of two-dimensional hydrodynamic numerical simulations, in cylindrical coordinates, modeling supersonic jets ramming into a uniform ambient medium were performed. The model explains the observed X-ray emission from protostellar jets in a natural way and, in particular, shows that a protostellar jet less dense than the ambient medium well reproduces the observations of the nearest Herbig-Haro object, HH 154, and allows to make detailed predictions of a possible X-ray source proper motion ($v_{sh} \approx 500$ km s^{-1}) detectable with *Chandra.* The simulations of jets as dense as, or more dense than, the ambient cannot reproduce observations. Also we found that a careful account of the interstellar and circum-jet absorption of X-rays leaves the X-ray emission of the tip of the jet, in excellent agreement with observations since it hides the very soft emission from the rest of the jet and surrounding region. Furthermore, our results suggest that the simulated protostellar jets which best reproduce the X-rays observations cannot drive molecular outflows.

4. Conclusions

Models are fundamental to understand the physics of astrophysical objects because these objects are complex and controlled experiments on them are not possible. Models are not only important to provide physical insight but can give fundamental diagnostics of complex phenomena. It is very important, however, to supplement these models with appropriate tools to handle, present and interpret model results as well as tools for proper comparison with observations.

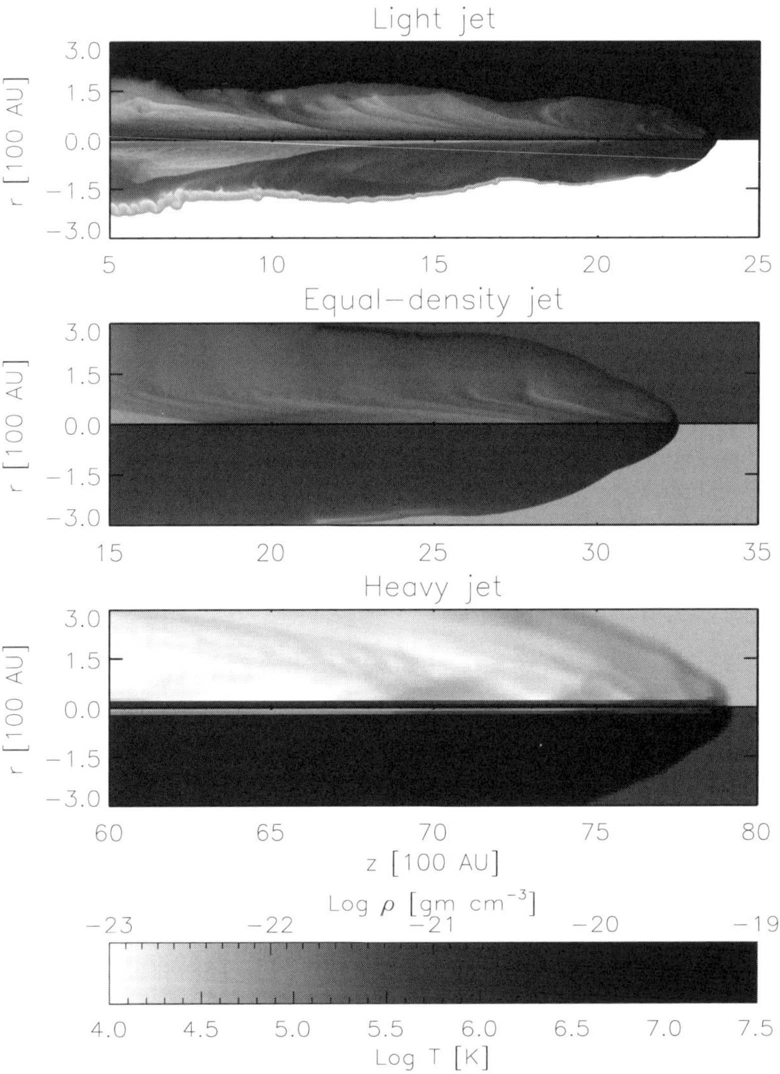

Fig. 2. Two-dimensional mass density (upper half-panels) and temperature (lower half-panels) cuts in the $r-z$ plane after 20 years since the beginning of the jet/ambient medium interaction for the best cases of light (upper panels), equal-density (middle panels) and heavy jets (lower panels). Adapted from [8].

5. Acknowledgements

We acknowledge partial support by the Agenzia Spaziale Italiana. The software used in this work was in part developed by the DOE-supported ASC/Alliances Center for Astrophysical Thermonuclear Flashes at the University of Chicago, using modules

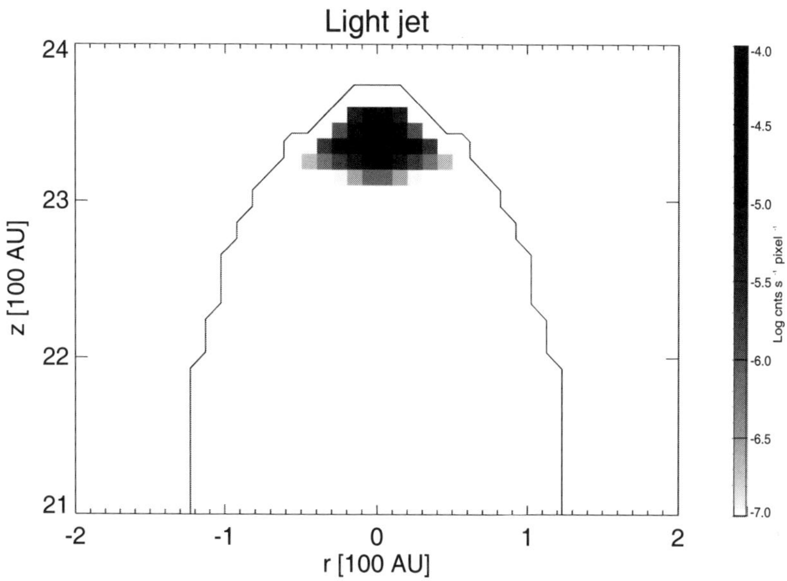

Fig. 3. Synthesized X-ray emission, in logarithmic scale, as predicted to be observed with ACIS-I, for the light jet simulation, 20 years since the beginning of the jet/ambient medium interaction. At a distance $D \approx 150$ pc, 100 AU corresponds to about 0.7 arcsec. Adapted from [8].

for thermal conduction and optically thin radiation constructed at the Osservatorio Astronomico di Palermo. The calculations were performed on the cluster at the SCAN (Sistema di Calcolo per l'Astrofisica Numerica) facility of the INAF Osservatorio Astronomico di Palermo and at CINECA (Bologna, Italy). This work was partially supported by grants from CORI 2005, by Ministero Istruzione Universit e Ricerca and by INAF.

References

[1] Gaensler, B. M. *ApJ*, **493**, 781, (1998)
[2] Orlando S., F. Bocchino, F. Reale, G. Peres, O. Petruk, On the Origin of Asymmetries in Bilateral Supernova Remnants, *A&A* in press (2007)
[3] Fryxell et al. *ApJS* **131**, 273, (2000)
[4] MacNeice et al. *Comp. Phys. Commun.*, **126**, 330, (2000)
[5] Whiteoak, J. B. Z., Green, A. J., *A&AS* **118**, 329, (1996)
[6] Bonito, R.; Orlando, S.; Peres, G.; Favata, F.; Rosner, R., *A&A* , 424, L1, (2004)
[7] Favata, F., Fridlund, C. V. M., Micela, G., Sciortino, S., and Kaas, A. A. *A&A*, 386, 204, (2002)
[8] Bonito, R.; Orlando, S.; Peres, G.; Favata, F.; Rosner, R., *A&A* **462**, 645, (2007)
[9] AMS-LaTeX Version 2 User's Guide (American Mathematical Society, Providence, 2004).
[10] B. W. Bestbury, *J. Phys. A* **36**, 1947 (2003).

NUMERICAL SIMULATIONS OF MULTI-SCALE ASTROPHYSICAL PROBLEMS: THE EXAMPLE OF TYPE Ia SUPERNOVAE

Friedrick K. RÖPKE

Max-Planck-Institut für Astrophysik,
Karl-Schwarzschild-Str. 1, D-85741 Garching, Germany
** E-mail: fritz@mpa-garching.mpg.de*
www.mpa-garching.mpg.de/~fritz

Vastly different time and length scales are a common problem in numerical simulations of astrophysical phenomena. Here, we present an approach to numerical modeling of such objects on the example of Type Ia supernova simulations. The evolution towards the explosion proceeds on much longer time scales than the explosion process itself. The physical length scales relevant in the explosion process cover 11 orders of magnitude and turbulent effects dominate the physical mechanism. Despite these challenges, three-dimensional simulations of Type Ia supernova explosions have recently become possible and pave the way to a better understanding of these important astrophysical objects.

Keywords: Hydrodynamical Simulation; Ia Supernovae; Computational Grid

1. Introduction

Astrophysics naturally features problems on large scales, which often can be addressed with the methods of hydrodynamics. The number of particles is huge and the interactions are in many cases (with the important exception of gravity) short-ranged. This allows the description of the systems in terms of thermodynamical variables. From the formation of planets to the evolution of large-scale structure in the Universe, hydrodynamical methods have been successfully applied to astrophysical problems on various spatial scales.

Astrophysical problems usually challenge numerical techniques and computational resources due to their pronounced multi-scale character. Physical processes take place on vastly different time scales. Moreover, the range of spatial scales involved is typically far beyond the capabilities of today's supercomputers. Therefore, approximations and numerical modeling are inevitable.

Here, we will discuss the a typical astrophysical scenario – the thermonuclear explosion of a white dwarf (WD) star which is believed to give rise to a Type Ia supernova (SN Ia) explosion. A comprehensive treatment of this scenario would involve the modeling of the formation of the progenitor system, its stellar evolution, its approach to the explosive state, the ignition of the explosion, the explosion stage itself, and the evolution of the remnant. But we are far from dealing with problems of such complexity. The different stages of the evolution of the system are characterized by very different timescales and distinct physical mechanisms. For instance, the stellar evolution of the progenitor system may take more than a billion years while the actual explosion takes place on a timescale of seconds. Therefore, different methods are applied to address these stages. Stellar evolution is usually treated in hydrostatic approaches, the evolution towards the ignition of the thermonuclear explosion takes about a century and needs special hydrodynamical approximations, while the explosion process is modelled via a combination of hydrodynamics, turbulence modelling and treatment of nuclear reactions.

2. Astrophysical Model

The favored astrophysical model of SNe Ia is the thermonuclear explosion of a WD star composed of carbon and oxygen [1, 2]. This end stage of stellar evolution for intermediate and low-mass stars is stabilized by the pressure of a degenerate electron gas, because after nuclear burning of hydrogen and helium it fails to trigger carbon and oxygen burning. A WD is a compact object which would be eternally stable, cool off, and disappear from observations. However, since many stars live in binary systems, it is possible that it accrets material from its companion. There exists a limiting mass for stability of a degenerate object like a carbon/oxygen WD (the Chandrasekhar mass $\sim 1.38 M_{\odot}$) beyond which it becomes unstable to gravitational collapse. Approaching the Chandrasekhar mass, the density in the core of the WD reaches values that eventually trigger carbon fusion reactions.

This leads to about a century of convective burning. Finally, however, a thermonuclear runaway occurs and gives rise to the formation of a burning front, usually called a thermonuclear flame. This flame propagates outward burning most of the material of the star and leading to an explosion – a process that occurs on timescales of seconds.

Hydrodynamics allows for two distinct modes of flame propagation [3]. While in a subsonic deflagration the flame is mediated by the thermal conduction of the degenerate electrons, a supersonic detonation is driven by shock waves. Observational constraints rule out a prompt detonation for SNe Ia [4] and the flame must therefore start out in the slow deflagration mode [5]. The flame propagation has to compete with the expansion of the star due to the nuclear energy release. Once the dilution due to expansion has lowered the fuel density below a certain threshold, no further burning is possible. The energy release up to this point needs to be sufficient to gravitationally unbind the WD star and to lead to a powerful explosion. This is only possible if the propagation velocity of the deflagration flame is accelerated far beyond the speed of a simple planar flame (the so-called laminar burning speed).

It turns out that this can be achieved by the interaction of the flame with turbulence. This turbulence is generic to the scenario. Burning from the center of the star outward, the flame leaves light and hot nuclear ashes below dense and cold fuel. This inverse density stratification in the gravitational field of the WD is buoyancy unstable. Consequently, burning bubbles form and float towards the surface. Shear flows at the interfaces of these bubbles lead to the generation of turbulent eddies. By wrinkling the flame these increase its surface and the net burning rate is enhanced. Thus, the flame accelerates. Whether this acceleration suffices to yield the strongest SNe Ia observed, is currently debated [6, 7]. It has been hypothesized that a transition of the flame propagation from subsonic deflagration to supersonic detonation in later stages of the explosion may occur and provide an the ultimate speed-up of the flame [8].

3. Challenges

The astrophysical scenario of SNe Ia described in the previous section obviously poses great challenges numerical modeling. Many of the problems found here are typical for a broad range of astrophysical phenomena. The contrast between the time scales of the actual explosion to that of the ignition process (let alone the stellar evolution of the progenitor) is only part of the scale-problem. The spatial scale ranges in the explosion as well as in the pre-ignition phase are huge. Both processes are dominated by turbulence effects with integral scales not much below the radius of the star ($\sim$2000 km). A typical Reynolds number is as high as 10^{14} and consequently the Kolmogorov scale is less than a millimeter. Turbulence effects with Reynolds numbers far beyond anything occurring on Earth are common in astrophysics. The scales of the objects and typical velocities are huge but at the same time the viscosities of astrophysical fluids are not extraordinarily high. This indicates that neither a full temporal nor spatial resolution in a single numerical approach is possible. Therefore the problems are usually broken down into sub-problems which can be treated with specific approximations and numerical techniques. Moreover, astrophysical equations of state are often more complex than those found under terrestrial conditions and are in some cases not even well-known.

4. Governing Equations

Hydrodynamical problems in astrophysics can often be treated with the Euler equations with gravity as external force which have to be augmented by a description of nuclear reactions and an appropriate astrophysical equation of state. This set of equations is obtained when reaction and diffusive transport phenomena are neglected:

$$\frac{\partial \rho}{\partial t} = -\boldsymbol{\nabla} \cdot (\rho \boldsymbol{v}), \tag{1}$$

$$\frac{\partial \boldsymbol{v}}{\partial t} = -(\boldsymbol{v}\boldsymbol{\nabla}) \cdot \boldsymbol{v} - \frac{\boldsymbol{\nabla} p}{\rho} + \boldsymbol{\nabla}\Phi, \tag{2}$$

$$\frac{\partial \rho e_{\text{tot}}}{\partial t} = -\boldsymbol{\nabla} \cdot (\rho e_{\text{tot}} \boldsymbol{v}) - \boldsymbol{\nabla}(p\boldsymbol{v}) + \rho \boldsymbol{v} \cdot \boldsymbol{\nabla}\Phi + \rho S, \tag{3}$$

$$\frac{\partial \rho X_i}{\partial t} = -\boldsymbol{\nabla} \cdot (\rho X_i \boldsymbol{v}) + \boldsymbol{r}_{X_i} \tag{4}$$

$$\boldsymbol{r}_{X_i} = f(\rho, T, X_i), \tag{5}$$

$$p = f_{\text{EOS}}(\rho, e_{\text{int}}, X_i), \tag{6}$$

$$T = f_{\text{EOS}}(\rho, e_{\text{int}}, X_i), \tag{7}$$

$$S = f(\boldsymbol{r}), \tag{8}$$

$$\Delta\Phi = 4\pi G \rho. \tag{9}$$

Mass density, velocity, pressure, total energy, internal energy, mass fraction of species i, temperature, reaction rates, chemical source term, and gravitational potential are denoted by ρ, $\boldsymbol{v}$, p, e_{tot}, e_{int}, X_i, T, $\boldsymbol{r}$, S, and Φ, respectively. Index i runs over $1 \ldots N$, where N is the number of species contained in the reacting mixture.

The equation of state is indicated by f_{EOS}. For astrophysical objects matter may occur under extreme conditions and the equation of state may differ significantly form that of terrestrial matter. This is the case for our example as the equation of state of the WD star is dominated by an arbitrarily relativistic an degenerate electron gas. The nuclei form an ideal gas of nuclei, and radiation and electron-positron pair creation/annihilation contribute to the equation of state as well.

In many cases effects like heat conduction and diffusion play a major role. In principle this holds for the example considered here, too. On the smallest scales, the flame is mediated by the heat conduction of the degenerate electrons. However, these scales cannot be resolved in full-star simulations and the treatment of flame propagation we will discuss below parametrizes it in a way such that the above set of equations is sufficient to describe the astrophysical scenario.

5. Modeling Approaches

Different methods for modeling the hydrodynamics make different approximations and are suitable to certain sub-problems or simplifications of the problems. First and pioneering approaches to simulate SNe Ia, for instance, assumed spherical symmetry [9]. These neglect instabilities and turbulence effects, which have to be parametrized in such simulations, but they allow for efficient Lagrangian discretization schemes. Consistency and independence of artificial parameters can, however, only be reached in multi-dimensional simulations. Eulerian discretizations are preferred here.

Depending on the time scales of the physical phenomena under considerations, certain simplifications can be made to the hydrodynamical equations. While in the explosion simulations usually the set of equations spelled out above is applied, the pre-ignition convection and ignition processes, as well as the propagation of deflagration flames on small scales, are strongly subsonic phenomena. The magnitude of steps allowed in the numerical simulations is set by the fastest motions that contribute to the mechanism. Therefore, for subsonic phenomena it would be an overkill (and in some cases it would also be numerically unstable) to follow sound waves. Therefore these are filtered out in anelastic and low-Mach number approaches applied specifically to the pre-ignition and ignition phase of thermonuclear supernovae. Allowing for much larger time steps than the sound crossing time over a computational grid cell, such approaches facilitate the numerical study of phenomena taking place on time scales of minutes and hours. These approaches are described in detail elsewhere [10, 11] and will focus on the traditional implementation of hydrodynamical processes below.

The problem of spatial scales is approached in two strategies. While off-line small-scale simulations of otherwise unresolved phenomena test the assumptions of large-scale models [12–16], these in turn rely on models for unresolved effects. The outstanding challenge in the latter is the description of turbulence effects and a promising strategy for addressing these in astrophysical simulations is the application of subgrid-scale turbulence models. Since in SNe Ia the propagation of the deflagration flame is dominated by turbulence effects, such models are applied here and an example will be discussed below.

The nucleosynthesis in astrophysical events is a rich phenomenon which can involve hundreds of isotopes and reactions between them. While it is possible to run extended nuclear reaction networks concurrently with one-dimensional astrophysical simulations, they are prohibitively expensive in three-dimensional approaches. Therefore, such simulations usually apply greatly simplified treatments of nuclear reactions in order to approximate the energy release. In this way the dynamical effects of nuclear burning can be treated without large errors. However, in order to compare the results of astrophysical simulations with observables such as spectra and light curves[a] (the only way to validate astrophysical models), details of the chemical structure of the object are to be known. One approach to this issue is to advect a number of tracer particles with the hydrodynamical simulation which record the evolution of the thermodynamical conditions, and to feed this information into extended nuclear reaction networks in a postprocessing step [17, 18].

6. Numerical Methods

There exists a large number of standard techniques for solving the Euler equations in hydrodynamical simulations. In astrophysics, a widely used finite-volume approach that discretizes the integral form of the equations, is the piecewise parabolic method [19, 20] – based on a higher-order Godunov scheme.

[a]the temporal evolution of the luminosity of the event, usually restricted to a range of wavelengths set by an observational filter

The selection of the geometry of the computational grid needs special consideration. Although spherical coordinates seem best suited for many astrophysical objects featuring an average spherical symmetry, these are afflicted with coordinate singularities. Therefore, currently there seems to be a trend towards Cartesian set-ups.

The challenge of incorporating phenomena that occur on scales unresolved in simulations has to be addressed by modeling. For the example of large-scale SN Ia simulations this applies to the propagation of the thermonuclear flame and to turbulence. As described above, both are connected. For modeling turbulence on unresolved scales, a Large Eddy Simulation (LES) ansatz is chosen. Flow properties on resolved scales are used to determine closure relations for a balance equation of the turbulent velocity q at the grid scale [21].

The structure profiles of a thermonuclear flame at high and intermediate fuel densities extent typically over less than a centimeter. These scales cannot be resolved in simulations capturing the evolution of the entire WD star (radius 2000 km and expanding). Therefore, in these simulations, the flame is treated as a mathematical discontinuity separating the nuclear fuel from the ashes. A numerical technique to represent the propagation of this discontinuity is the level set method [22, 23]. It associates the flame surface $\Gamma(t)$ with the zero level set of function G:

$$\Gamma(t) := \{\boldsymbol{r} | G(\boldsymbol{r}, t) = 0\}.$$

For numerical convenience, we require G to be a signed distance function to the flame front, $|\nabla G| \equiv 1$ with $G < 0$ in the fuel and $G > 0$ in the ashes. The equation of motion is then given by

$$\frac{\partial G}{\partial t} = (\boldsymbol{v}_\mathrm{u}\boldsymbol{n} + s_\mathrm{u})|\nabla G|.$$

Here, $\boldsymbol{v}_\mathrm{u}$ is the fluid velocity ahead of the flame, s_u is the effective flame propagation velocity with respect to the fuel, and $\boldsymbol{n} = -\nabla G / |\nabla G|$ is the normal to the flame front. This equation ensures that the zero level set (i.e. the flame) moves in normal direction to the flame surface due to burning and additionally the flame is advected with the fluid flow. The burning speed s_u has to be provided externally in this approach since the burning microphysics is not resolved. While the flame propagation proceeds with the well-known laminar flame speed in the very first stages of the explosion, it quickly gets accelerated by interaction with turbulence. By virtue of the implemented subgrid-scale model the turbulent burning speed of the flame can easily be determined. In the turbulent combustion regime that holds in most parts of the supernova explosion, it is directly proportional to the turbulent velocity fluctuations. This is the way in which the multidimensional LES approach to flame propagation in thermonuclear supernovae avoids tunable parameters in the description of flame propagation.

In order to take the expansion of the WD into account in the simulation, one has to adapt the computational grid accordingly. One option is adaptive mesh refinement, which, however, suffers from the usually volume-filling turbulent flame structure. Therefore, a refinement all over the domain would be necessary and the gain in efficiency from this method is marginal. An alternative is to use a computational grid with variable cell sizes. This grid can be constructed to track the expansion of the star [24] or the propagation of the flame inside it (or both [25]).

7. Three-dimensional Type Ia Supernova Simulations

Applying the techniques discussed above, three-dimensional simulations of deflagration thermonuclear burning can be performed [6, 21, 26] (the incorporation of a delayed detonation stage is also possible with slight modifications of the methods [27]). The goals of such simulations is to determine whether an explosion of the WD can be achieved in the model and whether the characteristics of such an explosion meet observational constraints. As a direct link from the simulation of the pre-ignition convection studies and the flame ignition simulations to the explosion models is still lacking, the flame ignition is introduced by hand in configurations motivated by off-line studies [10, 28, 29].

To illustrate the typical flame evolution in deflagration SN Ia models, the full-star model presented by [26] shall be described here. The flame was ignited in a number of randomly distributing spherical flame kernels around the center of the WD. This resulted in a foamy structure slightly misaligned with the center of the WD (shown in Fig. 1). Such multi-spot ignition models are motivated by the strongly turbulent convective carbon burning phase preceding the ignition (but alternatives such as asymmetric off-center ignitions have also been considered). Starting from this initial flame configuration, the evolution of the flame front in the explosion process is illustrated by snapshots of the $G = 0$ isosurface at $t = 0.3\,\mathrm{s}$ and $t = 0.6\,\mathrm{s}$ in Fig. 1. The development of the flame shape from ignition to $t = 0.3\,\mathrm{s}$ is characterized by the formation of the well-known "mushroom-like" structures resulting from buoyancy. This is especially well visible for the bubbles that were detached from the bulk of the initial flame. But also the perturbed parts of the flame closer to the center develop nonlinear Rayleigh-Taylor features. During the following flame evolution, inner structures of smaller scales catch up with the outer "mushrooms" and the initially separated structures merge forming a more closed configuration (see snapshot at $t = 0.6\,\mathrm{s}$ of Fig. 1). This is a result of the large-scale flame advection in the turbulent flow, burning, and the expansion of the ashes. After about $2\,\mathrm{s}$ self-propagation of the flame due to burning has terminated in the model. The subsequent evolution is characterized by the approach to homologous (self-similar) expansion. The resulting density structure at the end of the simulation is shown in the $t = 10\,\mathrm{s}$ snapshot of Fig. 1.

The goal of such simulations is to construct a valid model for SNe Ia which meets the constraints from nearby well-observed objects. Such models can then be used to test and refine the methods that are used to calibrate cosmological distance measurements based on SN Ia observations [30], which pioneered the new

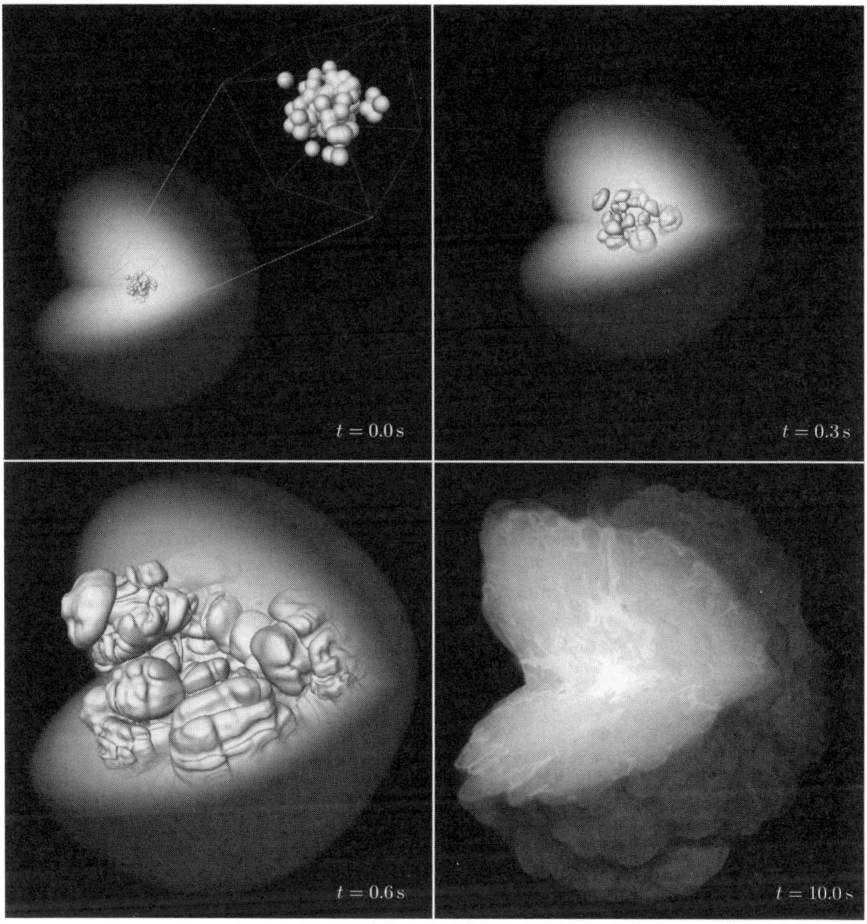

Fig. 1. Snapshots from a full-star SN Ia simulation starting from a multi-spot ignition scenario. The density is volume rendered indicating the extend of the WD star and the isosurface corresponds to the thermonuclear flame. The last snapshot corresponds to the end of the simulation and is not on scale with the earlier snapshots.

cosmological standard model with an accelerated expansion of the Universe [31, 32] pointing to a dominant new "dark" energy form.

References

[1] F. Hoyle and W. A. Fowler, *ApJ* **132**, 565 (1960).
[2] W. Hillebrandt and J. C. Niemeyer, *ARA&A* **38**, 191 (2000).
[3] L. D. Landau and E. M. Lifshitz, *Fluid Mechanics*, Course of Theoretical Physics, Vol. 6 (Pergamon Press, Oxford, 1959).
[4] W. D. Arnett, J. W. Truran and S. E. Woosley, *ApJ* **165**, 87 (1971).

[5] K. Nomoto, D. Sugimoto and S. Neo, *Ap&SS* **39**, L37 (1976).
[6] M. Reinecke, W. Hillebrandt and J. C. Niemeyer, *A&A* **391**, 1167 (2002).
[7] V. N. Gamezo, A. M. Khokhlov and E. S. Oran, *Phys. Rev. Lett.* **92**, 211102 (2004).
[8] A. M. Khokhlov, *A&A* **245**, 114 (1991).
[9] K. Nomoto, F.-K. Thielemann and K. Yokoi, *ApJ* **286**, 644 (1984).
[10] M. Kuhlen, S. E. Woosley and G. A. Glatzmaier, *ApJ* **640**, 407 (2006).
[11] J. B. Bell, M. S. Day, C. A. Rendleman, S. E. Woosley and M. Zingale, *J. Comp. Phys.* **195**, 677 (2004).
[12] F. K. Röpke, J. C. Niemeyer and W. Hillebrandt, *ApJ* **588**, 952 (2003).
[13] F. K. Röpke, W. Hillebrandt and J. C. Niemeyer, *A&A* **420**, 411 (2004).
[14] F. K. Röpke, W. Hillebrandt and J. C. Niemeyer, *A&A* **421**, 783 (2004).
[15] W. Schmidt, W. Hillebrandt and J. C. Niemeyer, *Combust. Theory Modelling* **9**, 693 (2005).
[16] M. Zingale, S. E. Woosley, C. A. Rendleman, M. S. Day and J. B. Bell, *ApJ* **632**, 1021 (2005).
[17] C. Travaglio, W. Hillebrandt, M. Reinecke and F.-K. Thielemann, *A&A* **425**, 1029 (2004).
[18] F. K. Röpke, M. Gieseler, M. Reinecke, C. Travaglio and W. Hillebrandt, *A&A* **453**, 203 (2006).
[19] P. Colella and P. R. Woodward, *J. Comp. Phys.* **54**, 174 (1984).
[20] B. A. Fryxell, E. Müller and W. D. Arnett, *Hydrodynamics and nuclear burning*, MPA Green Report 449, Max-Planck-Institut für Astrophysik (Garching, 1989).
[21] W. Schmidt, J. C. Niemeyer, W. Hillebrandt and F. K. Röpke, *A&A* **450**, 283 (2006).
[22] S. Osher and J. A. Sethian, *J. Comp. Phys.* **79**, 12 (1988).
[23] M. Reinecke, W. Hillebrandt, J. C. Niemeyer, R. Klein and A. Gröbl, *A&A* **347**, 724 (1999).
[24] F. K. Röpke, *A&A* **432**, 969 (2005).
[25] F. K. Röpke, W. Hillebrandt, J. C. Niemeyer and S. E. Woosley, *A&A* **448**, 1 (2006).
[26] F. K. Röpke and W. Hillebrandt, *A&A* **431**, 635 (2005).
[27] F. K. Röpke and J. C. Niemeyer, *A&A* **464**, 683 (2007).
[28] D. Garcia-Senz and S. E. Woosley, *ApJ* **454**, 895 (1995).
[29] L. Iapichino, M. Brüggen, W. Hillebrandt and J. C. Niemeyer, *A&A* **450**, 655 (2006).
[30] M. M. Phillips, *ApJ* **413**, L105 (1993).
[31] A. G. Riess, A. V. Filippenko, P. Challis, et al., *AJ* **116**, 1009 (1998).
[32] S. Perlmutter, G. Aldering, G. Goldhaber, et al., *ApJ* **517**, 565 (1999).

NUMERICAL SIMULATIONS IN ASTROPHYSICS: FROM THE STELLAR JETS TO THE WHITE DWARFS

FRANCESCO RUBINI* and LUCA DELZANNA

Dipartimento di Astronomia e Scienza dello Spazio, Universitá di Firenze, Largo E. Fermi 2, Firenze, 50125, Italy
** E-mail: rubini@arcetri.astro.it*

JOSEPH A. BIELLO

Department of Mathematics, University of California, Davis, California, 95616, USA

JAMES W. TRURAN

Department of Astronomy, University of Chicago Chicago, IL, 60637, USA

Astrophysical phenomena ranging from supersonic outflows from young stars to pre-runaway burning convection in white dwarfs can be described using fluid hydrodynamic or magnetohydrodynamic models. Despite their basic simplicity, the numerical solution of these models is still a challenge. A reliable description of these phenomena requires a careful representation of the full range of processes present in the astrophysical environment such as sound waves, magnetic viscosity and thermal conductivity which, in turn, yield an enormously wide range of temporal and spatial scales. Resolving such scales is still well beyond modern computational capability.

In this work we show that the effects of taking unrealistic physical parameters and correspondingly unrealistic numerical resolution to describe space or time, is very much dependant on the problem being considered. There are cases when reliable results can be achieved, even when important approximations are used to model the environment; the numerical simulation of the emissivity properties of stellar jets is one of them. In other problems, such as burning convection in pre-runaway white dwarfs, poor numerical resolution yields completely different scenarios.

Keywords: Stellar Jets, Pre-runaway White Dwarfs

1. The Origin of the Optical Knots in Stellar Jets

1.1. *Observations, physical model and numerical simulations*

Stellar jets are supersonic flows from young stars that have been widely imaged by the Hubble Space Telescope (HST) [1, 2]. Bright knots observed in emission lines from a wide range of λ, that move at velocity of 0.7 to 1.0 times the flow speed are spectacular tracers for stellar jets. The current interpretation of the origin of such knots involves the presence of pulsation in the ejection mechanism [3–6], though the underlying physical reason for such pulsation has not been identified. We have used shock-capturing numerical schemes to solve the Navier-Stokes equations that model the interaction between the jet gas and the interstellar medium (ISM). Transport equations for both molecular hydrogen N_{H_2} and ions N_{HII}, including the corresponding source terms, have been taken into account. Such models enable us to simulate the jet emission properties in H_α and $[SII]$ lines and to make a comparison between simulated and observed knots. Below left is an image from HST for the jet HH-111. On the right are written the equations that are solved in the numerical simulation.

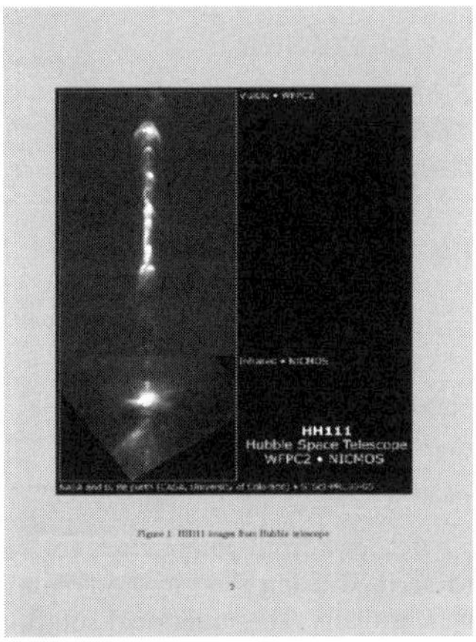

$$\frac{\partial \rho}{\partial t} + \nabla \cdot (\rho \mathbf{u}) = 0$$

$$\frac{\partial}{\partial t}(\rho \mathbf{u}) + \nabla \cdot (\rho \mathbf{u}\mathbf{u} + p\mathbf{I}) = \mathbf{0}$$

$$\frac{\partial}{\partial t}(\rho E) + \nabla \cdot [\rho H \mathbf{u}] = -S_{rad} - S_{ion} - S_{H_2}$$

$$\frac{\partial N_{HII}}{\partial t} + \nabla \cdot (N_{HII}\mathbf{u}) = \mathcal{N}_{ion} - \mathcal{N}_{rec}$$

$$\frac{\partial N_{H_2}}{\partial t} + \nabla \cdot (N_{H_2}\mathbf{u}) = -\mathcal{N}_{diss}$$

1.2. *Results and comments*

We show that over a wide range of parameters the internal oblique shocks (IOS) that arise from the pressure gradient between the jet and the ISM are responsible for the formation of the emitting knots. We also show that such knots are not static, but show a degree of proper motion even when non-pulsating inflow conditions are imposed [7]. The following figure shows snapshots of a supersonic jet arising from a nozzle at the left of the numerical domain. The jet bores its way into the ISM with an inflow velocity of order 200 km/s and an internal jet pressure of 20 times the ISM pressure. Optical knots that closely resemble the observed knots in the picture by the Hubble telescope of the jet HH-111, form and are visible in the $[SII]$ lines.

Our conclusion is that IOS with both stationary initial conditions and pulsating inflow conditions are plausible and cooperate in different settings. We also show that the results are not affected by the approximations used to describe the properties of the environment. In this problem numerical simulations are actually able to provide the gross features of the flow, and detailed information about the underlying physical mechanisms, even when a small numerical Reynolds number (and resolution) is taken.

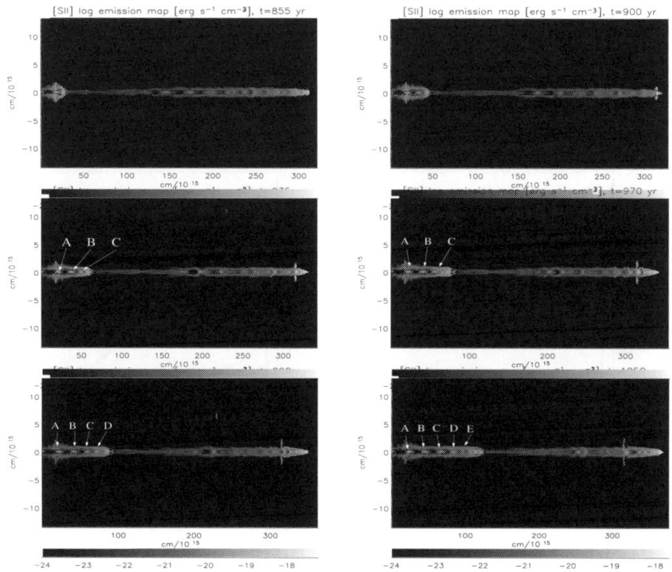

Fig. 1. [SII] optical knots of a jet travelling in the wake of an older jet. Snapshots are taken from t=760 yr to t=1050 yr

2. Numerical Simulations of pre-runaway CO White Dwarfs

Another challenging problem is illustrated in the next figure.

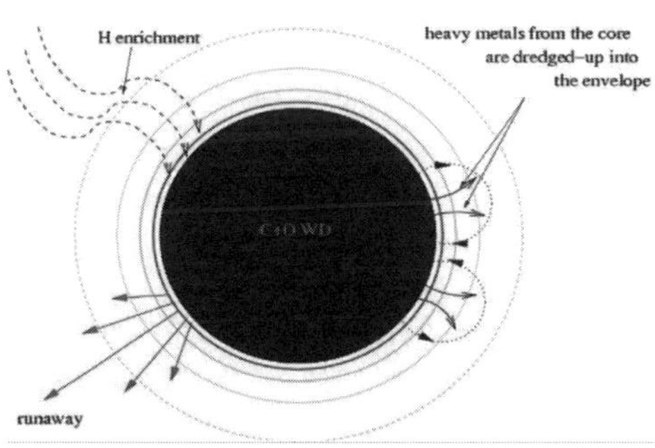

Fig. 2. The nova evolution, from the enrichment phase to the runaway

Hydrogen from the companion star piles-up in the CO white dwarf atmosphere. Density, pressure and temperature grow in the immediate vicinity the star surface, where heavy metals from the core are dredged-up and mixed with the atmospheric gas. When T approaches 10^8 K thermonuclear reactions make the temperature grow to 2×10^8 K, and the runaway takes place. Numerical simulations begin with some distribution of density, pressure, temperature and species abundances involved in the CNO hot cycle (from 19 to 32, depending on the network accuracy) in hydrostatic equilibrium with the gravitational field [8].

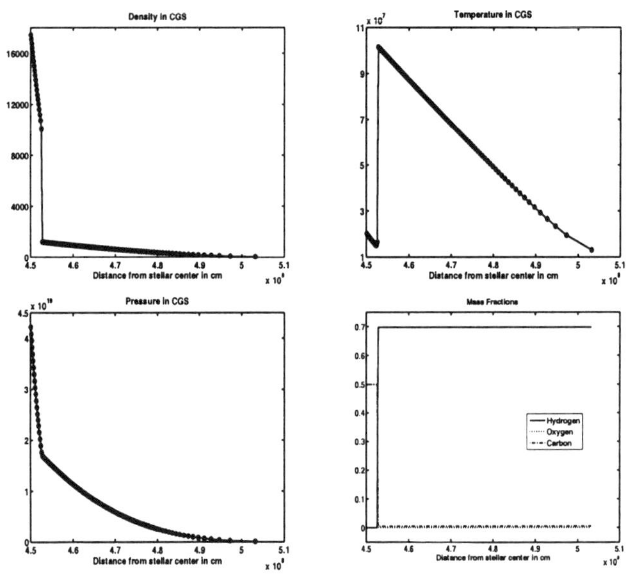

Fig. 3. From top, left, clock-wise: density, temperature, abundances, pressure as functions of the height. These profiles are computed via 1-D equilibrium codes.

2.1. *Physical models, numerical tools and FAQ*

- The temporal evolution of pressure, density, velocity and abundances in the stellar atmosphere are calculated by solving the Navier-Stokes equations, coupled with the 19 species network responsible for nuclear energy generation. Since convection is a major mechanism in the energy transport from below, thermal conductivity and kinetic viscosity are also taken into consideration.
- Spectral (Fourier transforms) and quasi-spectral (6-th order compact differences) approximations are used to calculate derivatives in the horizontal, periodic direction and in the vertical direction, respectively. Such a choice guarantees that solutions as precise as possible, as long as the velocity remains sub-sonic.
- When the run-away takes place, one-third of the atmosphere consists of CO originating from the core of the star.
  Question: Are convective motions arising from the initial temperature gradient responsible for CO dredge-up ?

2.2. *Results*

Most results available in the literature have been obtained by solving the ideal Navier-Stokes equations via second-order accurate, high viscosity numerical methods. Despite the lack of physical viscosity and thermal conductivity such simulations actually use very low numerical Rayleigh numbers, $R_a = \frac{g\alpha L^3 \Delta T}{\nu\sigma}$, because of the high numerical kinematic viscosity ν and thermal conductivity σ actually used.

In the next figure we sketch the temporal evolution of the horizontally averaged temperature profile calculated by Glasner et al. 1997 [8]. The temperature steadily grows from the initial conditions, and the run-away is achieved after $\approx$ 200 s.

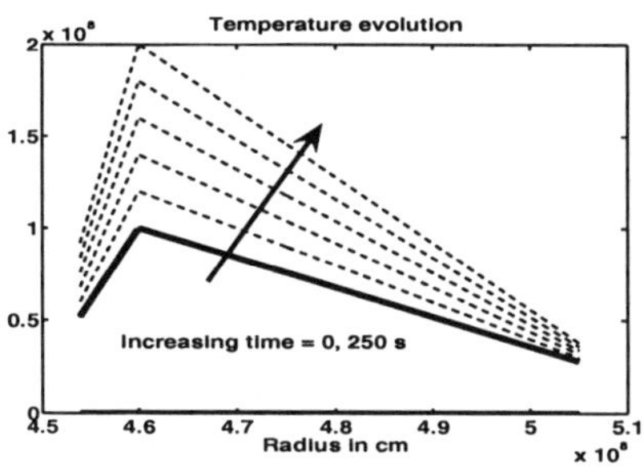

Fig. 4. From 8: T as a function of time and radius

According to these calculations convection is too weak to drive the CO mixing.

Preliminary results from our calculations, obtained by using a numerical method with very small numerical viscosity, yield different results which can be grouped in two classes: small and large Rayleigh number simulations.

2.3. *The small Rayleigh number case*

The next figure shows snapshots of the temperature profile in a non-ideal, viscous simulation where physical viscosity and thermal conductivity are such that $R_a \approx$ 10000, a relatively small number compared with the value in white dwarfs, though large enough to allow for convective instabilities to develop. Convective cells form, on scales of order 30 km and have velocities of order 10^7 m/s, large enough to provide some important CO mixing. In the first 500 seconds the temperature profile remains smooth. In fact, during this period the turbulent timescale is shorter than the average nuclear reaction timescale, and turbulence is able to mix the temperature gradients, smoothing away any local excess of energy. At t=570 s nuclear reactions occur faster than turbulent mixing causing local hotspots of energy to form and degenerate in the run-away.

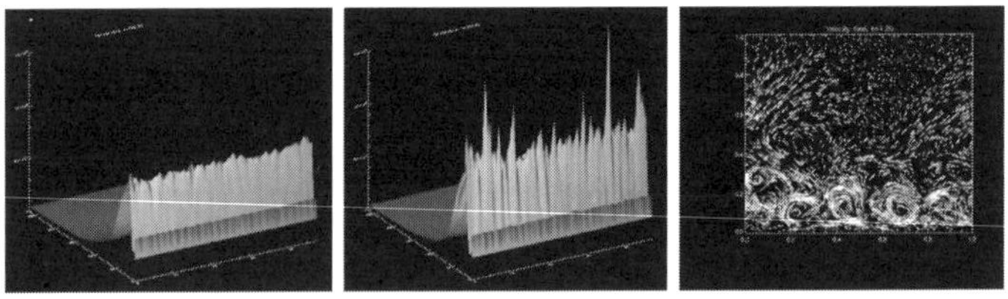

Fig. 5. $R_a = 10000$ Temperature field at t=390 (left) and t=570 (middle) s, with convective cells in the velocity pattern (right)

2.4. *The large Rayleigh number case*

A completely different scenario takes place in the large Rayleigh number simulations, $R_a \approx 100000$. In this case convective instabilities are strong enough to create a very well mixed interior: a nearly constant entropy equilibrium is achieved as the initial temperature profile is flattened by convective energy transport from base to the stellar atmosphere. In the new equilibrium the average temperature drops from the initial 100 MK to $\approx$ 40 MK. This value is quite far from the nuclear ignition values, and the runaway does not take place.

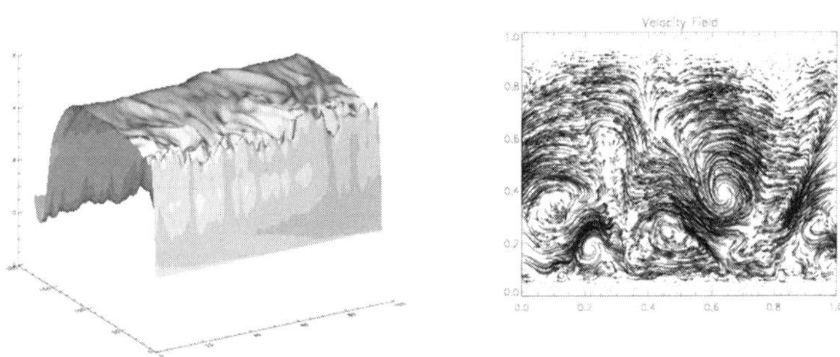

Fig. 6. $R_a = 100000$. Temperature field (left) and velocity pattern (right) at t=200 s

In this problem the numerical simulations are still far from the physical parameter regime and themselves have not begun to converge to a consistent pattern of mixing. Using increasingly realistic values for viscosity and thermal conductivity still yields quite different scenarios. It also apparent that further investigations should be done to provide better initial conditions for these simulations. In this setting the question "how much of CO is dredged-up by convection" remains poorly posed, since the answer heavily depends on the strength of the convection itself, or on the Rayleigh number used in the calculation.

3. Conclusions

We have performed numerical simulations of two different scenarios, the investigation of the optical knots in stellar jets and white dwarf evolution to the runaway. We have shown that the reliability of the solutions to these problems depends both on the problem and the numerical tool.

Dynamical problems, such as the propagation of supersonic flows from a nozzle, are less sensitive to the choice of the initial parameters and numerical resolution.

Instability problems from initial equilibria, instead, are more 'numerical parameter dependent', and require more accurate exploration of the numerical parameter space in order to yield solutions which are stable in the sense of having converged to a behavior consistent with the physical setting.

Acknowledgments

This project has been supported by the ASCI Flash Group of the University of Chicago.

The author also wishes to thank Maria Luisa Carfora for useful talks and suggestions, and for her long-lasting inspiring personal support.

References

[1] Ray, T. P., Mundt, R., Dyson, J. E., Falle, S. A. E., & Raga, A. C. 1996, *ApJ* **468**, L103;
[2] Reipurth, B., Balley, J., 2002, *ApJ*, **580**, 336;
[3] Raga, A.C., Kofman, L., 1992, *ApJ* **386**, 222;
[4] Stone J.M., Norman M., 1993, *ApJ* **413**, 210;
[5] Suttner G., Smith D.M., Yorke H.,W., Zinnecker, H., 1997, *A&A*, **318**, 595; Castets A., Grenoble;
[6] Falle, S. A. E. G., Raga, A. C., 1995, *Mon. Not. R. Astronom. Soc.* **272**, 785;
[7] Rubini, F., Giovanardi, C., Lorusso, S., Leeuwin, F., Bacciotti, F., 2004, *Astrophysics and Space Science*, **293**, 181 ;
[8] Glasner, S. A., Livne, E., 1997, *ApJ*, **475**, 754;

STATISTICAL ANALYSIS OF QUASAR DATA AND VALIDITY OF THE HUBBLE LAW

SISIR ROY[1,2@], JOYDIP GHOSH[1], MALABIKA ROY[1] and
MENAS KAFATOS[1]

1- Center for Earth Observing and Space Research, College of Science, George Mason University, Fairfax, VA, 22030-4444 USA
2- Physics and Applied Mathematics Unit, Indian Statistical Institute, Calcutta-700035, India
[@]Email: sisir@isical.ac.in

The aim of this paper is to study the associations, if any, among distributions of physical parameters from multivariate data with general and arbitrary truncations based on different and varied observational selection criteria. We have taken samples from Véron-Cetty(V-C) quasar catalogue(2006) and the Sloan Digital Sky Survey(SDSS, DR3) which indicate the linearity (in log$z$) of the Hubble law up to a small redshift $z(\leq 0.3)$ limit but non-linearity for higher redshift for both data sets. This may raise new queries not only for the observations of extragalactic sources, like, quasars but also in the cosmological debate about the role of the environment surrounding quasars and that of host galaxies as in the Dynamic Multiple Scattering(DMS) due to induced correlation phenomena, especially when higher redshift$(z \geq 0.3)$ are concerned.

Keywords: Quasar data, test of independence, Hubble law, correlation induced phenomena.

1. Introduction

Among many different ways of testing models of cosmological sources, especially quasars, one is through the investigations of the distributions, ranges and more importantly the correlations among the relevant physical characteristics, such as luminosity, spectra, redshifts or cosmological distances. The linearity of the Hubble relationship [1] that exists between apparent magnitude(m) and redshift(z) has been studied in detail for several years, and found to be valid for galaxies at low redshifts [2, 3]. But in the case of the extragalatic sources, for example quasars, clear deviation is observed towards high redshift$(z \geq 0.3)$. In this work, we have followed the same method of non-parametric tests, developed by Efron et. al. [4–6] to study quasar redshift data. This study points to the probable solutions and sheds new light in understanding the role of various cosmological models, especially for high redshifts. Our study contains, first, the data, taken from the entire set of Véron-Cetty(V-C) catalogue [7] for quasar redshifts and, next, prepared with the consideration of various selection criteria, based on, say, radio properties, color, optical characteristics etc. We follow the above mentioned statistical tools of Efron et al. and performed the regression analysis in each case and obtain the validity of the Hubble relation [8] for small redshifts($z \leq 0.295$), but large deviation at higher redshifts($z \geq 0.295$), i.e., not compatible with present status of the standard cosmological models.

The primary objects with their corresponding i-PSF magnitudes of quasars are considered for statistical analysis, from Sloan Digital sky survey(SDSS) (Fig.1(ii)), according to the SDSS group [9, 10]. In our work, the plot of (m) and (z)(Fig.1(ii)) shows similar truncation lines as found in V-C(Fig.1(i)). This motivated us to follow the same procedures as in the V-C quasar data and the results of the analysis reach to the same conclusions as in the case of V-C quasar data [8].

Presently, researchers find difficulty in understanding the large spread observed

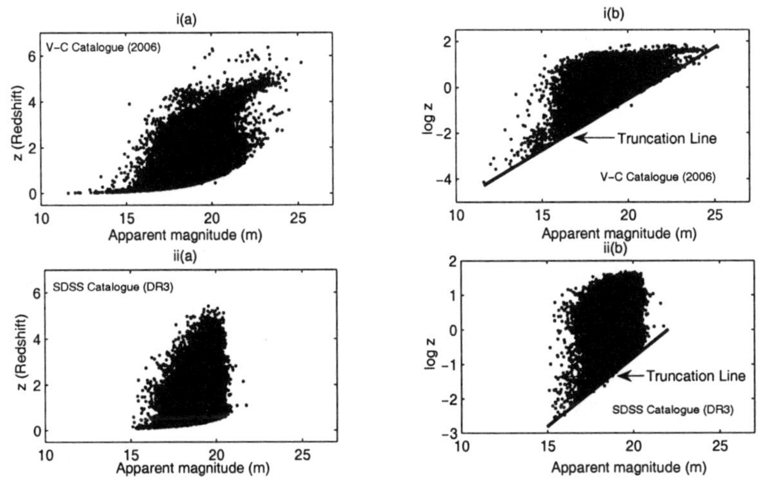

Fig. 1. Apparent magnitude (m) versus redshift(z) in linear scale (a) and in log-scale(b) with corresponding truncation line for (i) V-C(2006) [7] and (ii) SDSS(2005) data release 3. [9].

in the redshifts($z \geq 0.3$) from the point of view of standard cosmological models. Ferland and Botoff [12] considered the importance of the surrounding medium, though from another point of context. Baldwin [13] discussed the importance of the optimum condition needed in the medium in the formation of the spectrum and the corresponding different observed line widths. Here, we propose that the local environments around the sources as well as that of the host galaxies i.e., the surrounding medium play the key role in understanding the non-linearity at high redshifts. The present authors [14, 15] have proposed and developed the Dynamical Multiple Scattering(DMS) theory to explain the shift and broadening of the spectral lines both at laboratory and cosmological scales. The main idea, here, is that the frequency of electromagnetic waves, originating from of the central source, i.e., the continuum, will be shifted during its propagation through the turbulent and

inhomogeneous medium, occurring due to the presence of diverse physical parameters, like ionization potential, electron density, refractive index, dielectric constant. These are intrinsic to the fluctuating medium, where micro turbulence and other types of statistical fluctuations occur.
In section II, we describe statistical tests for various data sets taken from the V-C(2006) and SDSS catalogues. Section III contains discussion on the possible explanations and implications of results obtained. Our results provide the basis of application of the DMS framework on the quasar problem the comparison with the statistical regression and relation to the standard cosmological models are described. Finally, a brief discussion regarding the overall problem is provided in section IV.

2. Statistical Analysis of Data from the V-C(2006) and SDSS Quasar Catalogues

The impossibility of direct distance measurements to quasi stellar objects prevents one to validate any direct relationship between distance and redshift where, the measurable quantities are the apparent magnitude (m), redshift (z) which, as related to the luminosity function or, even the probability distribution of absolute magnitude (M), as predicted by a given cosmology and the angular diameter of the object. However, one can estimate the probability distribution of M, independent of the location of the source, which, however, is accepted widely for small redshifts. But for high redshifts where the problem of nonlinearity exists, the evolutionary effects are considered in standard cosmological models(details in references [5, 6]) as a probable solution. Koranyi [2] studied the validity of luminosity and the Hubble law in detail, however, criticizing the quadratic redshift-distance relation, in the chronometric cosmology, proposed by Segal et al. [16].
Our analysis indicates that the Hubble like linear distance-redshift relation is valid for low $z, i.e., for(z \leq 0.295)$ but a large deviation from linearity for $z \geq 0.295$ is present. This fact is hard to be understood within the context of standard cosmological models, with evolutionary effects as well as as in the chronometric cosmology [16]. It means the parameters exhibit dispersions and lack of a priori knowledge about specific cosmological models which demands a more critical studies, besides others, from the statistical point of view before these results can be taken as validation of cosmological models.
Conventionally, astronomers work with absolute magnitude M, connected to a particular cosmological model and apparent magnitude m, where, the distance modulus($d \simeq m - M$), is related to the luminosity distance. This, again, is related to redshift z of the source, first established by Hubble[1], through the relation:

$$d \simeq (cz)/(\mathbf{H_0}) \simeq 3000h^{-1}\text{Mpc} \quad \text{for} \quad 10^{-2} \geq z \leq 10^{-1} \tag{1}$$

$\mathbf{H_0}$ being the Hubble constant which can also be derived from the cosmological theory, if the universe is assumed to be homogeneous and isotropic. Various authors [2, 3] discussed limits of validity of the Hubble relation which has already been tested for galaxies and supernovae [21, 22]. Crudely speaking, one can use the above relation for $d << 300h^{-1}$Mpc, or for $z << 10^{-1}$, which, however, may be considered as a first order approximation to the formula for the luminosity distance, as a function of redshift z in the Friedman model [17].

Different types of statistical arguments and tests have been applied on the data gathered by astronomers, to extract important statistical characteristics considering the truncation of different categories and the luminosity function, based on different cosmological models and the linearity observed in the Hubble's law in the case of galaxies. Efron et al. [4–6] applied their newly developed nonparametric statistical method on the analysis of various samples for galaxies and quasars though with much smaller samples, available that time. They proposed the evolutionary phenomena as the possible causes for the observed correlations and the nonlinearity found in the Hubble law, in the higher redshifts of quasars.

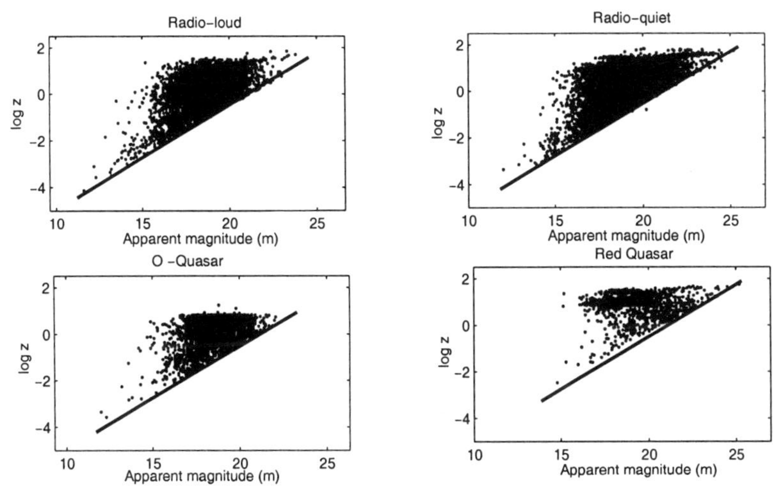

Fig. 2. Apparent magnitude (m) versus redshift(z) in log-scale with corresponding truncation line for (a) Radio loud (b) Radio quiet(c) Red and (d)Optical Véron-Cetty quasar data.

2.1. *Truncation relationship*

Both V-C(V-C(2006) and SDSS(DR3,2005), like other redshift surveys, provide a pair of measurements (z_i, m_i) besides other astronomical observational measurements with various type of observational biases. Among them, often ignored, one of the most common biases introduced, is by limiting m_i in the surveys which can be written as (z_i, m_i) for $i = 1, 2, 3.....\mathcal{N}$ with $m_i \leq m_0$, termed as truncation of data. This implies that it is impossible to get the information regarding the existence of (y_i, z_i) if it falls outside the region R_i where, due to experimental con-

straints, the distribution of each y_i is truncated to a known interval R_i depending on z_i.
To answer the statistical question i.e.,*Is the sample of observed points* (z_i, m_i) *in the truncated data set of the survey concerned consistent with the hypothesis* H_0*[i.e., Null-hypotheses] which demands that* z_i *and* m_i *are statistically independent?*- Efron and Petrosian [4] proposed a nonparametric method to investigate this issue in detail using a small sub-sample of quasar data. Here, the absolute magnitude M_i can be estimated if we assume a particular cosmological model. The data set (z_i, M_i) can then be re-expressed satisfying a truncated relationship of the form:

$$M_i \leq m_0 - 5\log d + C \quad \text{,where,} \quad C = \text{constant} \tag{2}$$

i.e. in general mathematical form $\log y \geq ax + b$ for some constants a and b.
In case of V-C data, we have taken samples using different intrinsic characteristics like, radio, color or optical properties etc. with corresponding selection criteria, including the total amount of data present in the catalogue. Consequently, the different truncation lines, obtained for different selection based samples are obtained, as shown in Fig2.
For each set, the same type of linear truncation but with different values of constants are obtained. To do this, a few points were discarded from each set, negligible compared to the size of the data considered.

Table 1: Results of non-parametric association tests.

Parameters	Data Size(N)	τ_w	p-value(p_1)	τ_{w1}	p-(value)(p_2)
V-C(2006) Radio quiet	75950	53.861	0	60.339	0
V-C(2006) Radio loud	8639	25.29	1.96x10^{-141}	25.26	4.50x10^{-141}
V-C(2006) Red quasar	1537	0.307	0.7584	0.4578	0.647
V-C(2006) Total data	84590	61.22	0	66.527	0
SDSS(2005) (Data Release 3)	46226	31.83	1.285x10^{-222}	36.489	8.286x10^{-292}

Note: N is the number of data points in each sample. The correlation τ and the probability value p for rejection of independence between apparent magnitude(m) and redshift(z) are given, following the one-sided truncation method employed by Efron & Petrosian [4].

2.2. *Permutation tests of independence*

In Table 1 we summarize all of our results, performed following the methodology of Efron et al [4]., in detail. This has been pointed out in another paper, to be published in the same proceedings [18]. The corresponding p-values for all these data samples (except in the case of Red quasars) clearly indicate the existence of associations between m and z values.
The equation of the truncation line(T) and correspondingly, those of regression

analysis((R_1 for $z \leq 0.3$ and R_2 for total data have been calculated for each set, from which, as an example, only two are mentioned here(i.e., total V-C(2006) [7] and SDSS(DR3) [9]quasar data):
Taking $x = m; y = \log z$,

- $T = y = -9.48 + 0.45x$;
  $R_1 = m = 19.79 + 2.00\log z$ for $(z \leq 0.3)$;
  $R_2 = m = 19.27 + 0.675\log z - 0.19(\log z)^2 + 0.14(\log z)^3$ for $V-C(2006)$ quasars(total).
- $T = y = -8.81 + 0.4x$;
  $R_1 = m = 21.14 + 2.3\log z$ for $(z \leq 0.3)$
  $R_2 = m = 18.92 + 0.04\log z - 0.04(\log z)^2 + 0.28(\log z)^3$ for SDSS(DR3, i-PSF quasars)(2005).

The results clearly show that for low redshifts ($z \leq 0.295$), the Hubble law ,i.e. linearity in $\log z$ is maintained, but for high redshifts($z \geq 0.295$), a non-linearity in z arises.

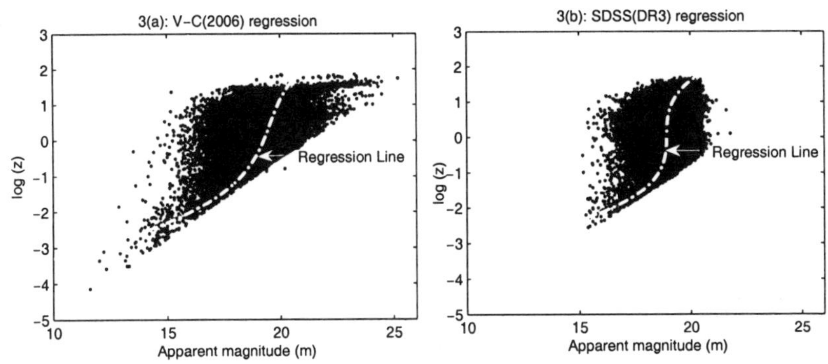

Fig. 3. Regression line(-.-.) for apparent magnitude(m) versus redshift(z) in V-C(2006) and SDSS(data release 3(2005)) total quasar data.

3. Possible Implications and Discussion

Conventionally, the Hubble relation is attributed as due to the Doppler mechanism for the shift of spectral lines. The above statistical analysis for both the the V-C(2006) and for the SDSS(DR3, i-PSF) data clearly rejects the hypothesis of independence between the apparent magnitude (m) and redshift (z), especially at higher redshift($z \geq 0.295$). With respect to the cosmological aspects, this deviation, observed in our analysis, might be due to following reasons:
(a) The presence of rapid luminosity evolution as emphasized by Efron and Petrosian [5, 6], may gradually be more prevalent for the intermediate range of redshifts i.e., $z\epsilon(0.295; 2.995)$.
(b) The effect of environment should be considered as a contributing factor in

addition to the Doppler effect. This would have important astronomical and, consequently, cosmological implications.

According to the currently established standard models of quasars, spectral lines are emitted from the broad line regions(BLR) as well as from the narrow line regions(NLR), i.e., from two types of clouds, named after the characteristics of lines and with respect to the position of those clouds relative to the torus surrounding the central black hole (central engine). Due to the gravitational attraction, matter from the dust torus is heated, causing it to radiate with a broad, non-thermal power spectrum. This radiation excites gaseous clouds which emit line radiation. However, not all of the spectral ranges show the theoretically expected slope in the linear part of the Hubble diagram. This obviously points towards the developmental effects arise in corresponding spectral ranges. It is worth mentioning that Wold et al. [19] have made a survey of quasar environments at $0.5 \leq z \leq 0.8$ and concluded that the quasars are located in a variety of environments. Hutchings [20], following his recent observations, pointed out that the environment that triggers QSO activity almost certainly changes with cosmic time and with the nature and the characteristics of the emission line gas around at higher redshift QSOs.

More and more results have been published regarding the possible impacts of the nature of environments associated with, especially high redshift quasars. The nonlinearity or deviation in the Hubble diagram (i.e., the existence of a *bulge*) has already been mentioned by several authors in connection to Supernovae [21, 22]. It clearly points to the possibility of environmental effects on the observational aspects of quasar-like astronomical objects. Their role should be considered in explaining the observed bulge and associated other criteria, i.e., the Hubble relation, especially at higher redshift in addition to the Doppler effect or gravitation. For example, molecular gases [12] as well as the presence of micro turbulence are observed in these environments which could have a substantial effects on the spectrum we observe.

At this point, we propose the effects of local environments around quasars playing important roles in understanding this large deviation [11]. The present authors [14, 15] developed the Dynamical Multiple Scattering (DMS) theory based on the induced correlation phenomena, first introduced by Wolf [23], to explain the shift and broadening of the spectral lines for distant extra galactic astronomical objects like quasars. This mechanism mimics the Doppler effect even in the absence of any relative motions between source and the observer. Here, the shift and broadening depends on the intrinsic characteristic parameters of the medium through which the light is being propagated. Moreover, the physical association of galaxy and quasars as observed by Arp et. al. [24] has natural explanation with DMS theory [25]. The laboratory experiments help us to estimate these parameters related to medium for the shift and broadening of the spectral lines. Based on DMS, we have fitted the lines which can accommodate the above deviation from linearity(Fig.4). The possible ranges of these parameters [11] were estimated which gives certain idea about the amount of the contribution from this phenomenon based on the intrinsic characteristics of the medium. This can be verified from the future observations in quasar

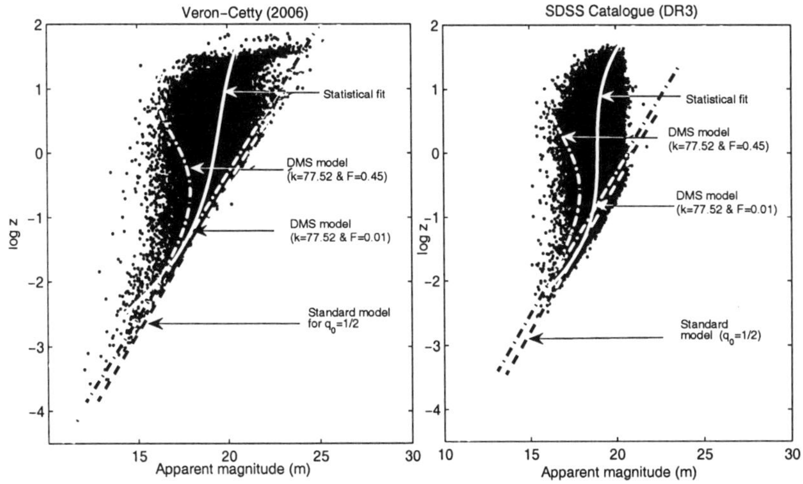

Fig. 4. The Hubble curve obtained from V-C(2006) [7] and SDSS(2005) [9] total quasar data :(a) Standard cosmological model(q_0 = 1/2)(white & black –), (b)Statistical fitting(solid white -), (c) Contribution due to DMS mechanism with maximum and minimum values of intrinsic parameters(F & k [11])(-.-.) of the medium surrounding the quasars.

astronomy. It then, can immediately raise very important issues like the validity of cosmological models, at the present state or in totality, especially, at high redshifts. Consequently, it may lead to the estimation of age of the universe. In other words, the induced correlation mechanism in the medium, in the DMS theory, may throw a possible explanation towards this puzzle. In essence, our statistical analysis may open up a new vista in modern cosmological debate.

Acknowledgements Two of the authors (S.Roy and M.Roy) greatly acknowledge the kind hospitality and financial support from Center for Earth Observing and Space Research, College of Science, George Mason University, USA, during this work. Authors are grateful to Jack Sulentic, University of Alabama and Jogesh Babu, Penn State University, USA for their constant encouragement and valuable suggestions. We are also indebted to D. P. Schneider, Penn State University, USA for valuable suggestions in selecting the data of SDSS for statistical analysis.

References

[1] E. Hubble, *ApJ*, **84**, 270(1936).
[2] D.M. Koranyi,*Astronom.J.*, **477**, 36-46(1997).
[3] P. Coles, F. Lucchin, *Cosmology : The Origin and Evolution of Cosmi Structure*, 2nd Edition, (John wiley & Sons, Ltd.), p77 (2002).
[4] B. Efron & V. Petrosian, *Astrophys J.*, **399**, 345-352(1992).
[5] B. Efron & V. Petrosian, *J. Am. Stat. Assoc.*,**94**, N447, 824-834(1999).
[6] A. Maloney & V. Petrosian, *Astrophys.J.*, **518**, 32-43(1999).
[7] VERONCAT-Véron Quasars and AGNs(V2006),*HEASARC Archive*(2006).

[8] S. Roy et al.,*Statistical analysis of Quasar data and Hubble law, astro-ph/0605356.*
[9] D.P. Schneider et al., *The Sloan Digital sky survey Quasar Catalogue III. Third Data Release, arXiv:astro-ph/0503679v 30 March.*
[10] C. Stoughton et al., *AJ*, **123**, 485(2002).
[11] S. Roy et al., *Dynamic Multiple Scattering, frequency shifting and Possible effects on Quasar astronomy, astro-ph/0701071.*
[12] M. Botoff and G. Ferland, *Astrophys. J.*, **568**, 581-591(2002).
[13] J. Baldwin et al., *Astrophys.J.*, **455**, L119-L122(1995).
[14] S. Datta, S. Roy, M. Roy and M. Moles, *Phys.Rev.A*, **58**, 720(1998).
[15] S. Roy, M. Kafatos, S. Datta,*Phys.Rev.A*, **60**, 273(1998).
[16] I.E. Segal I.E. & J.F. Nicoll, *Procd. Nat. Acad. Sci.(USA)*, **89**, 11669-11672(1992).
[17] A.A. Friedman, *Zeittschrift für Physik*, **10**,377-386(1922).
[18] S. Roy et al., Non-parametric tests for Quasar Data and Hubble Diagram, to be published in the proceedings of *Data Analysis in Astronomy: modelling and simulation in Science, Erice, Sicily,Italy*(2007).
[19] M. Wold, M. Lacy, P.B. Lilje, S. Serjeant, in *"QSO Hosts and their Environments", QSO environments at Intermediate Redshifts and Companions at Higher Redshifts*, eds. Márquez eta al., *Kluwer Academic/Plenum Publishers*(2001).
[20] J.B. Hutchings, in *"QSO Hosts and their Environments", QSO environments at Intermediate Redshifts and Companions at Higher Redshifts*, ed. by Márquez eta al., Kluwer Academic/Plenum Publishers(2001).
[21] J. Huchra, M. Davis, D. Latham and J.Tonry, *Astrophys.S*, **52**, 89(1983).
[22] S. Perlmutter , P. Schmidt Brian., *astro-ph/0303428 and references there in.*
[23] E. Wolf, D.V.F. James,*Rep.Prog.Phys. and references there in*, **59**,771(1996).
[24] H.C. Arp, *Quasars, Redshifts and Controversies*,(Berkley, CA: Intersteller Media)(1987).
[25] S. Roy, M. Kafatos and S. Datta, *Astron.Astrophys.*, **353**, 1134-1138(2000).

NON-PARAMETRIC TESTS FOR QUASAR DATA AND HUBBLE DIAGRAM

SISIR ROY[1,2@], DHRUBAJIT DATTA [2], JOYDIP GHOSH[1], MALABIKA ROY[1] and MENAS KAFATOS[1]

1- Center for Earth Observing and Space Research, College of Science, George Mason University, Fairfax, VA, 22030-4444 USA
2- Physics and Applied Mathematics Unit, Indian Statistical Institute, Calcutta-700035, India
[@]*Email: sisir@isical.ac.in*

The scatter plot for apparent magnitude(m) and redshift (z) in the case of quasars as compiled in Véron-Cetty(V-C) Catalogues(2003 and 2006) clearly indicate that the data are truncated. We use non-parametric tests as developed by Efron and Petrosian to analyse these truncated data. Our analysis rejects the null hypothesis in case of both catalogues of quasar data. Further analysis shows linearity (in $\log z$) of the Hubble law up to very small $z(\leq 0.3)$ values but non-linearity for higher redshifts. The results raise new possibilities not only in the observational astronomy but for the entire cosmological debate.

Keywords: Non-parametric tests; truncation line; quasar data; apparent magnitude and redshift; Hubble law.

1. Introduction

Efron and others[1−3] considered different types of statistical arguments and tests on the truncated data gathered by astronomers to extract important statistical characteristics. For example, it is often necessary to establish a permutation test of independence to give a statistical description of the bivariate distribution of two physically important parameters. These workers considered different types of selection criteria, especially the truncation of different categories, using rank statistics, and applied to magnitude limited galaxy and quasar surveys with the main purpose to describe the evolution of QSO luminosity function, and the physical evolution of individual objects, based on different cosmological models. Here, the truncation implies that it is not possible to get the information regarding the existence of (y_i, z_i) if it falls outside the region R_i where, due to experimental constraints, the distribution of each y_i is truncated to a known interval R_i depending on z_i. Many experimental situations and constraints lead to truncated data, for example, determination of the distances to astronomical objects which remains one of the great challenges of observational astronomy. For cosmological distances, the problem is particularly difficult, because, one of the critical tasks of redshift survey data is to determine the distribution of the luminosity of the distant extragalactic sources, especially quasars and their corresponding distribution with cosmological epoch.

However, the parameters exhibit dispersions and lack of a priori knowledge about specific cosmological models. This demands a critical study from the statistical point of view before these results can be taken as validation of specific cosmological models. In section II, we carry out the non-parametric tests, following the techniques of Efron et al.[1], but with much greater amount of data. Section III deals with the regression techniques and β distribution, employed for studying the linearity of Hubble's law, especially in the case of high quasar redshifts ($z \geq 0.3$). Some hints about the nonlinearity of Hubble's law, specially at high redshift values is given which follows the Efron and Petrosian's views. The cosmological implications are discussed in section IV.

2. Statistical Analysis of Quasar Data from Véron-Cetty Catalogues

We have performed association tests the with Véron-Cetty(V-C) (2003)[4] catalogue for quasar redhsifts and then the same was done with the more recently published version of the catalogue(2006)[5]. Both the catalogues clearly show the same one sided truncation between m and z.

2.1. *Truncation Relationship*

The V-C Catalogue[4,5] like other redshift surveys, provides a pair of measurements (z_i, m_i) as well as other astronomical observational measurements. Various type of observational biases are ignored. One of the most common biases introduced in this survey, is by limiting m_i. We can write the data set as (z_i, m_i) for $i = 1, 2, 3.....\mathcal{N}$ with $m_i \leq m_0$. This is known as truncation of data, from the statistical point of view. This refers to minium/or maximum values on a probability distribution whose main purpose is to constrain the sample space to a set of "plausible" values. The process of determination of the physical characteristics from truncated data can be traced back starting with first observations of stars in the disc of our galaxy. A more detailed discussion is available in a review article[6]. In general, most of the methods can be put into two categories: parametric and non parametric. In the former, there are certain shortcomings, sometimes because of erroneous assumptions about the forms of the distributions[7]. Among many variations of nonparametric methods employed, here, we have adopted the method, described in detail by Efron and Petrosian[2] and independently by Tsai[8] using a small sample of data on redshift surveys of galaxies and quasars. The complete description of the analysis is as follows:

(i) The first step is to determine whether two variables are correlated or they are independent , and (ii) if correlated , to find a way to account for the evaluation of that correlation. So, one of the main issues in analyzing this set of astronomical data is to answer the following statistical question:

Is the sample of observed points (z_i, m_i) in the truncated data set consistent with the hypothesis H_0, i.e., (Null-hypotheses) which demands that z_i and m_i are statistically independent ? From the V-C data(2003)[4](Fig.1(a) with z and m) and in Fig.1(b)

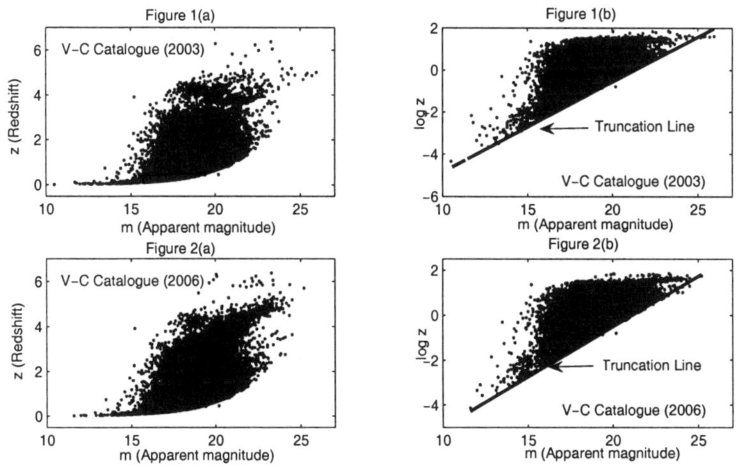

Fig. 1. Apparent magnitude (m) versus redshift(z) in linear scale (a) and in log-scale(b) with corresponding truncation line, from Véron-Cetty(V-C) (2003)(1) and (2006)(2) catalogues.

(with logz and m), and, for the same from V-C data(2006)[5] in Fig.2(a) and Fig.2(b), it appears that in the scatter plot of redshift(z) against apparent magnitude(m), there exists at least one sided truncation. The idea of truncation, here, is used in the sense that the observations (z_i, m_i) are observable if some condition or mathematical relation is satisfied, say,

$$\log z \geq am + b \quad \text{for appropriate values of } a \text{ and } b \tag{1}$$

From equation(1), we find that the linear truncation(lower envelope) curve is given by $a = 3/7$ and $b = -64/7$ and there are only 18 data points among the 48683 points(V-C data[4]) for which

$$\log z \geq 3/7m - 64/7 \tag{2}$$

and the same for the V-C(2006) catalogue is

$$\log z \leq (0.4482)m - 9.48 \tag{3}$$

where, 91 points discarded out of 84592 data points which is negligible compared to the data size.

2.2. *Permutation Tests of Independence*

Next, to perform the test of independence for truncated data, suppose the data consist of a random sample of n pairs from the joint distributions of

$$\text{data} = (x_i, y_i), \quad i = 1, 2, ...\mathcal{N} \tag{4}$$

where this is to be used to test the null hypothesis of independence i.e., $H_0 : x$ and y are independent. Following Efron and Petrosian[1], let us take x and y values such that they can be considered as observed pairings of $x_{(i)}$ and (y_j) such that there are $n!$ ways to choose the pairing vector $j = (j_1, j_2, ..., j_n)$ and all $n!$ ways are equally likely under the null hypothesis, without any loss of information. These n ordered x-values are denoted by $\mathbf{x} = (x_{(n)}, x_{(n-1)}, ...x_{(1)})$ where, $x_{(n)} < x_{(n-1)} < ... < x_{(1)}$ and, likewise, $\mathbf{y} = (y_{(n)}, y_{(n-1)}, ..., y_{(1)})$, with $y_{(n)} < y_{(n-1)} < ... < y_{(1)}$. The data of eq.(4) can again be described by the observed pairings between $(x_{(i)})$ and $(y_{(j)})$ as

$$\text{data} = \{(x_{(i)}, y_{(j)}), \quad i = 1, 2, ..., n\} \tag{5}$$

and for the truncated data, we assume that pairs (x_i, y_j) to be observable only if they satisfy the truncation relationship

$$y \leq u(x) \quad \text{where } u(x) \text{ is monotonic function of } x \tag{6}$$

For example, in the case of V-C(2003) truncation line, i.e. eq.(2), $x = m$, $y = -\log z$, we get $u(x) = (-3/7)x + 64/7$.

Now, with the help of the permutation test of independence hypothesis, H_0, a comparison is made between the observed values of the statistic t with the permutation distribution of t i.e., with N permutation values (data*)(details in ref.(1)). The general convention is to discard H_0, if t is found extremely large or too small compared to permutation distribution of t. In the present case, the test for the accept of independence, when calculated has been calculated as $|t_i(\text{data})| \leq 1.96$, where, the rejection probability of the permutation test is taken to be approximately 0.05. This means the level of 0.05 permutation test rejects H_0 i.e., when $\leq 5\%$ of the time H_0 is true.

To find out test statistic(data), the rank-based statistics have been applied which is a powerful and robust method, easy approximation can be provided for the relevant computation of permutation distribution. The detail of this method of procedure can be found in ref.(1). In short, R and N are defined as

$$R_i = \text{rank of } y_{ji} \text{ in } Y_i \quad \text{and} \quad N = \prod_{i=1}^{n} N_i \tag{7}$$

so that R_i can have values (i) 1 for the smallest member of Y_i (ii) N_i for largest value of $y_{(j_i)}$ (for details, see ref.(1)). The normalized rank statistic is then given by

$$T_i = \frac{(R_i - E_i)^{\frac{1}{2}}}{V_i} \quad \text{where} \quad E_i = \frac{(N_i + 1)}{2}; \quad V_i = \frac{(N_i - 1)}{12} \tag{8}$$

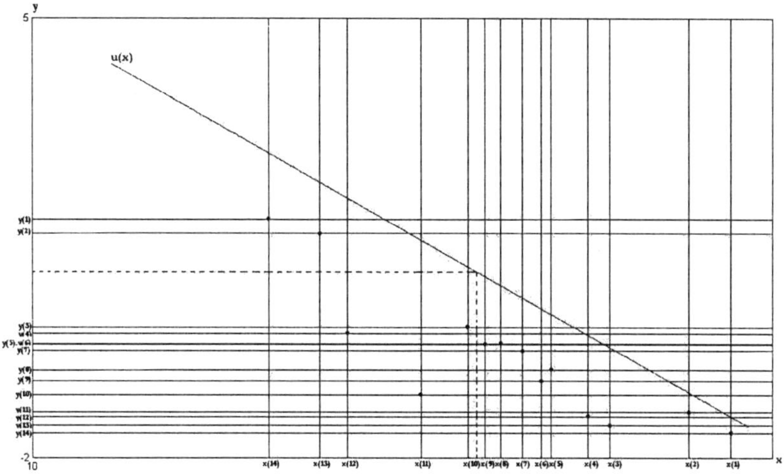

Fig. 2. Hypothetical truncated sample of size 14; the points indicate the observed quasars and the solid lines denote the abscissa and ordinates of the points; $u(x)$ indicates the truncation boundary, above which points are not observed. Boundary of the eligible set for $J_{10} = 3, 4, 10$ is shown by the dashed line. It shows clearly that for $i = 1, 2, ..., 14$ the number of points in the eligible sets are $N_i = 3, 4, 6, 7, 8, 7, 6, 5, 4, 3, 2, 3, 2, 1$ respectively. For defination of J_{10} and N_i, see ref.(1).

It has expectation 0 and variance 1 under H_0.
Usually, for convenience, the test statistic for independence is based on the linear combination of the normalized ranks T_i. Then, by taking $w = (w_1, w_2, ...w_n)$ as the vectors of weights, the test statistic can be expressed as,

$$t_w(\text{data}) = \frac{\sum_{(i=1)}^{n} w_i T_i}{\sqrt{\sum_{(i=1)}^{n} w_i^2}} \tag{9}$$

with mean 0 and variance 1 under H_0. The value of t_w for the whole set of data has been calculated from the above equation(9) as $|t_w(\text{data})| = 30.3195$.
However, in order to clarify the test more powerfully, we applied another test, considering a big probability of rejection under a alternative hypothesis H_1 to H_0 where $w_i = \frac{x_i - x_{\min}}{u_i - u_{\min}}$ and is denoted as locally the most powerful test(details in ref. 1). In this case the value is found to be $|t_{w_1}(\text{data})| = 30.08154$.
In both cases, the extremely large value of $t_w(\text{data})$ clearly rejects the hypothesis of independence. The $p -$ value is then calculated to be ~ 0 where $p -$ value is the maximum level of significance under which the null hypothesis (here, independence of two variables) is accepted. Here, as the p-value is almost 0 , we arrived at the conclusion that redshift z and apparent magnitude m **are not independent**. This, in another way suggests essentially a dispersion from Hubble relation, especially, from $z \geq 0.3$ onwards indicating a strong correlation between redshift and luminosity under conventional assumption about the nature of their redshifts. The

same analysis, done for V-C quasar data[5], gives rise

$$|t_w(\text{data})| = 61.22 \quad \text{and} \quad |\text{t}_{\text{w}_1}(\text{data})| = 66.5273$$

for more powerful test.

3. Regression analysis of Data and Hubble Law

To get the best fitting of these data we have employed the regression analysis technique. Subsequently, we investigate the conditions under which we can obtain the Hubble relation. The scatter plot of z vs. apparent magnitude m in V-C(2003) is considered first, as illustrated in Figure 1(a,b). Our regression analysis shows that we can use the following relation between m and z:

$$\log(m - 12) = -4.528 + 16.542z^{1/4} - 13.891z^{1/2} + 3.884z^{3/4} \tag{10}$$

for $z \epsilon (0, 7)$ i.e., for the whole range of z.

The following observation motivated us to analyze the data in a different manner. Here, we observed that in the region [0.2950; 2.995], the truncation distribution of the variable $\frac{(m-12)}{(f(z)-12)}$ can be well approximated by Beta distribution with parameters α and β being some functions of z. So, given the truncation, the above conditional distribution can well be approximated by Beta$(\alpha(z), \beta(z))$ where $f(z) = (7/3)\log(z) + 64/3$. In fact, we performed the above analysis for $z = 0.2950, 0.3050, 0.3150, \ldots, 2.9950$. This information has been used to calculate the expected value of m, given z for the last said values of z. Corresponding regression analysis gives rise to

$$E(m|z) = 19.484 + 0.886ln(z) - 0.783[\log(z)]^2 \tag{11}$$

for the region (0.2950 . . . 2.9950) and obtain 95% tolerance interval with coverage probability 0.95 in the similar fashion. Tolerance interval $A\%$ with coverage probability γ means that $A\%$ of the future observations will fall in the said interval with probability γ. In the specified region, we get a minimum value for apparent magnitude $(m_l(z)$ and a maximum magnitude$(m_u(z))$, for a given z where

$$m_u = 16.8 + 7.6263z - 4.162z^2 + 0.80z^3; \quad m_l = 12.51 + 5.576z - 1.686z^2 \tag{12}$$

These are shown in Fig.3. For the region (0. . . 0.2950), we use our general regression techniques to find out the prediction equation as

$$m = 20.060 + 2.139 \log z \tag{13}$$

and the prediction interval as

$$20.060 + 2.139 \log z \pm 1.9631\sqrt{0.4573}(1.0122 - 0.1132 \log z + 0.2937(\log z)^2 \tag{14}$$

respectively. Here, the prediction interval means that given z, the value of m will fall in that interval with probability 0.95. The similar conclusions have been reached also for the V-C(2006).

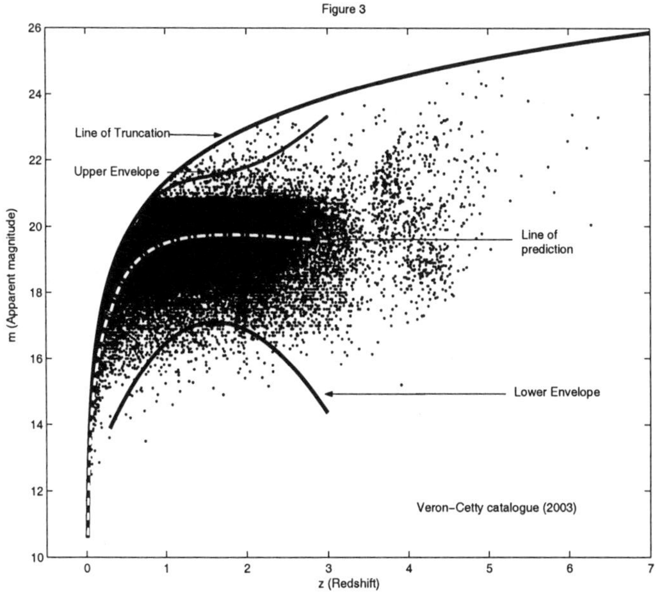

Fig. 3. Regression curves for m vs z. with Véron-Cetty data(2003)

4. Possible Cosmological Implications

Our statistical analysis clearly rejects the hypothesis of independence between the apparent magnitude(m) and redshift(z) for the entire set of both the V-C quasar catalogue(2003)(excepting around 18 data points) and of V-C(2006)(91 points rejected) which lie below the truncation line. From the regression analysis(taking V-C(2003)) we find a relation between m and $logz$ similar to the Hubble relation for small $z < 0.295$ and a clear deviation from the Hubble relation for large z $z > 0.295$. We performed the similar analysis for new V-C catalogue(2006), applying more constraints i.e., considering selection criteria, as suggested by an anonymous referee which will be described in another paper of these proceedings. Moreover, similar analysis is also under process to analyze the SDSS data[10], which will be reported in other paper. We emphasize that the line of prediction calculated here, together with prediction limit differs from the standard $m - z$ relation for cosmological models[11]. This may well be due to the local quasar environments which will be discussed in the second paper in this proceeding, and would have serious implications for any attempts to deduce the age of the universe using the Hubble law.

Acknowledgements

Two of the authors (S.Roy and M.Roy) greatly acknowledge the kind hospitality and financial support from Center for Earth Observing and Space Research , College of Science, George Mason University, USA, during part of this work.

References

[1] B. Efron and V. Petrosian, "A simple test of Independence for Truncated data with Applications to Redshift Surveys", *Astrophys J.* **399**:345-352(1992).
[2] B. Efron and V. Petrosian V., "Nonparametric Methods for Doubly Truncated Data", *J. Am. Stat. Assoc.***94**, N447, 824-834(1999).
[3] A. Maloney and V. Petrosian, "The Evolution and Luminosity function of Quasars from complete Optical surveys *Astrophys.J.*, **518**, 32-43(1999).
[4] VERONCAT-Véron Quasars and AGNs(V2003),*HEASARC Archive*(2003).
[5] VERONCAT-Véron Quasars and AGNs(V2006),*HEASARC Archive*(2006).
[6] V. Petrosian, in *Statistical Challenges in Modern Astronomy*, eds. E.D.Feigelson & G.J.Babu (Springer-Verlag, New York, 1992), p 173.
[7] V. Petrosian, "New Statistical Methods for Analysis of Large surveys: Distributions and Corrections", *arXiv:astro-phys/0112467* (1999).
[8] W. Tsai, "Testing the independence of truncation time and failure time" *Biometrika*, **77**, 169 (1990).
[9] P.K. Bhattacharya, H. Chernoff and S.S. Yang,"Nonparametric estimation of the slope of a truncated regression", *Ann.Statist.*, **11**, 505-514 (1983).
[10] "The Sloan Digital Sky Survey Quasar Catalogue III; Third Data Release", *astro-ph/0503679(DR3 data acess)* (2005).
[11] J.V. Narlikar, *Introduction to Cosmology*, (Jones & Bertlett Publishers, Inc., Boston, 1983).

DOPING: A NEW NON-PARAMETRIC DEPROJECTION SCHEME

DALIA CHAKRABARTY* and LAURA FERRARESE

School of Physics & Astronomy, University of Nottingham,
Nottingham, NG7 2RD, U.K.
**E-mail: dalia.chakrabarty@nottingham.ac.uk*
www.nottingham.ac.uk/physics/

We present a new non-parametric deprojection algorithm DOPING (Deprojection of Observed Photometry using and INverse Gambit), that is designed to extract the three dimensional luminosity density distribution ρ, from the observed surface brightness profile of an astrophysical system such as a galaxy or a galaxy cluster, in a generalised geometry, while taking into account changes in the intrinsic shape of the system. The observable is the 2-D surface brightness distribution of the system. While the deprojection schemes presented hitherto have always worked within the limits of an assumed intrinsic geometry, in DOPING, geometry and inclination can be provided as inputs. The ρ that is most likely to project to the observed brightness data is sought; the maximisation of the likelihood is performed with the Metropolis algorithm. Unless the likelihood function is maximised, ρ is tweaked in shape and amplitude, while maintaining monotonicity and positivity, but otherwise the luminosity distribution is allowed to be completely free-form. Tests and applications of the algorithm are discussed.

Keywords: Galaxies: photometry, luminosities, radii, ...

1. Introduction

The preliminary step involved in the dynamical modelling of galaxies, concerns the deprojection of the observed surface brightness distribution into the intrinsic luminosity density, as has been practised by [1–3], among others. Deprojection though, is a non-unique problem, unless performed under very specific configurations of geometry and inclination, as discussed by [4–7] and others.

Over the years, several deprojection schemes have been advanced and implemented within the purview of astronomy; these include parametric formalisms designed by [8], [9] and [10], as well as non-parametric methods, such as the Richardson-Lucy Inversion scheme developed by [11] and [12] and a method suggested by [13]. While the parametric schemes are essentially unsatisfactory owing to the dependence of the answer on the form of the parametrisation involved, the non-parametric schemes advanced till now have suffered from the lack of transparency and in the case of the Richardson-Lucy scheme, lack of an objective convergence criterion.

Here, we present a new, robust non-parametric algorithm: Deprojection of Observed Photometry using an INverse Gambit (DOPING). DOPING does not need to assume axisymmetry but can work in a triaxial geometry with assumed axial ratios, and is able to incorporate radial variations in eccentricity. Although the code can account for changes in position angle, this facet has not been included in the version of the algorithm discussed here. In a future contribution, (Chakrabarty & Ferrarese, *in preparation*), DOPING will be applied to recover the intrinsic luminosity density of about 100 early type galaxies observed as part of the ACS Virgo Cluster Survey, as reported in [14].

The paper has been arranged as follows. The basic framework of DOPING is introduced in Section 2. This is followed by a short discourse on a test of the algorithm. An application to the observed data of the galaxy vcc1422 is touched upon in Section 4. Another application of DOPING is discussed in Section 5. The paper is rounded up with a summary of the results.

2. Method

The outline of the methodology of DOPING is presented below.

(1) The plane of the sky ($x - y$ plane) projection of the galaxy is considered to be built of the observed isophotes that we consider to be concentric and elliptical, such that the ith isophote has a semi-axis extent of a_i along $\hat{\mathbf{x}}$. The inputs to DOPING are the brightness I_i and the projected eccentricity e_p^i that define the ith isophote, where $i \in N$, $i \leq N_{data}$.
(2) We set up $\rho = \rho[\xi(x, y, z)]$, where ξ is the ellipsoidal radius for the geometry and inclination of choice.
(3) We identify pairs of (x_i, y_i) that sit inside the elliptical annulus between the ith and the $i + 1^{th}$ isophotes.
(4) At the beginning of every iterative step, the density distribution $\rho(x_i, y_i, z)$, over the line-of-sight coordinate z, is updated in size and amplitude, $\forall i$, subject to the only constraints of positivity and monotonicity. The scales over which this updating is performed are referred to as scl_1 and scl_2.
(5) This updating is continued till the maxima in the likelihood is identified by the inbuilt Metropolis algorithm; the likelihood is maximised when the observed brightness distribution is closest to the projection of the current choice of the density. Regularisation is provided in the form of a penalty function that is set as the product of the smoothing parameter α and a function of the Laplacian of ρ.
(6) The spread in the models in the neighbourhood of the maximal region of the likelihood function, is used to formulate the (± 1-σ) errors on the estimated density.

3. Tests

Prior to the implementation of the algorithm, it is extensively tested using analytical models. For one of these test, the results of which are presented in Figure 1, the surface brightness distribution is extracted by projecting the analytically chosen luminosity density distribution of a toy oblate galaxy. The projection is performed along the LOS coordinate z, under the assumption of an intrinsic minor to major axis ratio that goes as $\sqrt{1/(1+r^2/r_c^2)}$, (where r is the spherical radius and the scale length r_c is $0''.5$). Thus, by construction, this toy galaxy is rounder inside the inner $0''.5$ and outside this radius, it quickly (by $3''$) flattens to a disky system with eccentricity of about 0.99.

This toy galaxy is viewed edge-on (at 90°) for this test. The brightness distribution is then ported to DOPING and the deprojection is carried out under a chosen geometry+inclination configuration; the recovered luminosity density is compared to the true density of this model.

In Figure 1, the true density of this toy oblate galaxy is shown in open circles, along the photometric major (left panel) and minor (right) axes. The open triangles are used to depict the density recovered by DOPING, under the assumptions of face-on viewing angle (i) and triaxiality, with the LOS extent set to double the photometric major axis. Similarly, when the galaxy is viewed at i=20°, with the LOS extent set to half the photometric major axis, the recovered density distribution is plotted along the azimuths of 0° and 90° in crosses. In both cases, the two photometric axes are set as related in the way suggested by the projected eccentricity (e_p) data. In the former case, the value of α that is used in the penalty function, is 10 times higher than in the latter case.

As expected, when the LOS extent is set longer than in the test galaxy, it leads to a smaller density than the true density while a shorter LOS extent is betrayed in higher recovered densities. Also, when the test galaxy, modelled as triaxial, is not viewed along one of the principle axes, then as expected, isophotal twist is recovered. This results in a steeper drop in the projection of the recovered density in the case of i=20° (not shown here) than in the brightness data.

Like all other deprojection algorithms, DOPING also requires a seed or a trial density distribution to begin with. The parameters in the Metropolis algorithm, namely the temperature and the scale lengths scl_1 and scl_2 (see Section 2), are chosen to ensure robustness of the algorithm.

4. Applications

We demonstrate the applicability of DOPING in real galaxies by deprojecting the surface brightness profile of vcc1422 (IC3468), which is a Virgo Cluster dwarf elliptical, covered by the ACS Virgo Cluster Survey [15]. We choose this galaxy since being about 8 magnitudes fainter at the centre than the test galaxy considered above, it illustrates the efficacy of DOPING over a wide range of brightness. Photometrically, it is evident that this galaxy has a small central component, (a nucleus extending to about $0''.3$), that sits on top of a more extended component. The projected ellipticity of this outer component meanders its way up from about 0.12 at about $0''.4$ to $0''.3$ at about $1''.6$, to jiggle down to about 0.22 at about $120''$.

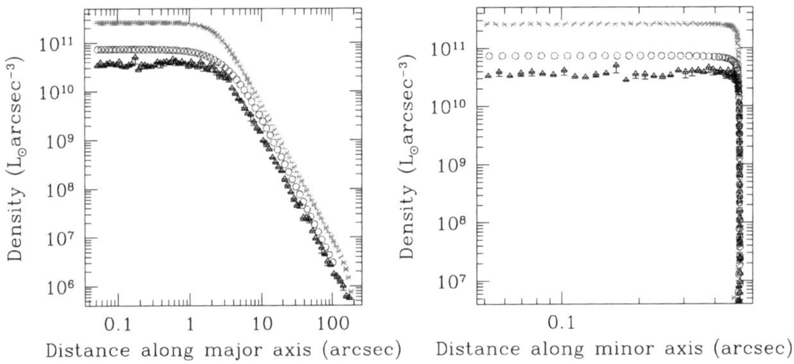

Fig. 1. Deprojection of a model brightness profile under assumptions of (i) ratio of extent along LOS to major axis coordinate=2; i=0° (ii) ratio of LOS extent to extent along $\hat{\mathbf{x}}$ = 0.5; i=20°. The recovered density distributions are shown at azimuths of 0° (right panel) and 90° (left panel), in crosses for case (ii) and open triangles for case (i). The true density of this toy galaxy is shown in open circles. The degeneracy in the deprojection exercise, as a function of inclination and geometry is brought out in this figure.

An experiment was conducted to bring out the importance of including this information about the variation in e_p. To ease such an exercise, the contribution of the nucleus to the surface brightness measurement was subtracted and the resulting brightness profile was then deprojected under different conditions. When all this variability in the projected shape of the galaxy is incorporated into the deprojection technique, the density recovered along the photometric major axis $\hat{\mathbf{x}}$ is depicted in open circles in Figure 2. This is compared to the density obtained under the assumption that the whole galaxy admits a single e_p of 0.25. In both cases, oblateness and edge-on viewing are assumed. Since the two profiles significantly differ, the comparison brings out the importance of including the details of the variation in the eccentricity, for this mildly eccentric system. With a more radically varying e_p profile, this difference would only increase.

DOPING has the capacity for deprojecting a multi-component system such as vcc1422; in these cases, the seed for the sought density is chosen such that it reflects the existence of all the components. Thus, in the case of vcc1422, inside $0''.3$, the seed should bear signatures of both the components, while outside it, only the contribution from the more extended of the two components is required. The result of this deprojection is shown in Figure 3. The deprojection was performed under the assumptions of oblateness and an edge-on inclination.

5. 3-D Morphology of Galaxy Clusters

A project is underway (Chakrabarty, Russell, de Philippis, *in preparation*) to decipher the degree of prolateness of a galaxy cluster, by deprojecting its X-ray bright-

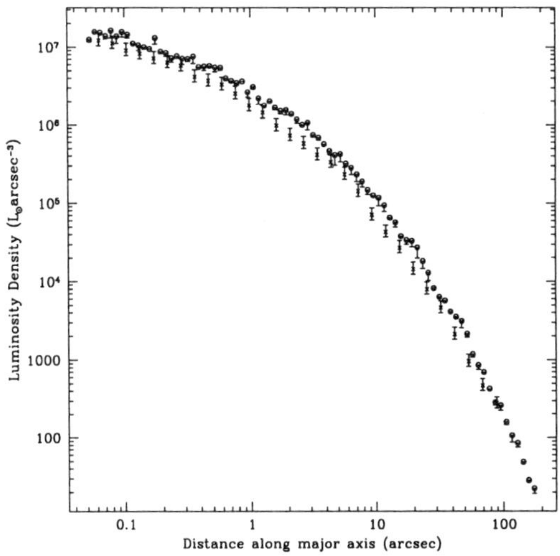

Fig. 2. The recovered density distribution of the nucleus subtracted brightness distribution of vcc1422, plotted along the $\hat{\mathbf{x}}$, obtained under the assumptions of i=90° and oblateness. The profile obtained from including the information about the changes in e_p with x is marked with open circles, while the deprojection carried out under the assumption of a constant e_p of 0.25 is represented in crosses.

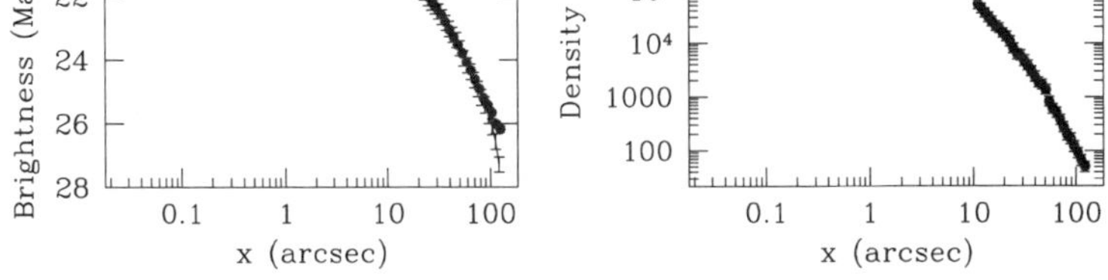

Fig. 3. Deprojection of the surface brightness data of the nucleated galaxy vcc1422. The recovered density is presented in the right while its projection has been overlaid on the observed brightness profile (in grey). Again, an oblate geometry and edge-on viewing were adopted.

ness distribution. The deprojection is carried out using DOPING, under the extreme geometry+inclination configurations that are allowed for the observed (uniform) e_p of the cluster. A comparison of the density profiles along $\hat{\mathbf{x}}$, recovered from the deprojection runs carried out at different inclinations, for the same geometry, allows us to quantify the "prolateness parameter" that indicates how much more prolate one system is compared to another.

In Figure 4, the recovered density profiles for the Abell clusters A1835 and A1413 are depicted; our analysis indicates that A1835 is more prolate than A1413 since its prolateness parameter is estimated to be $0.75^{+0.06}_{-0.05}$ as compared to that of 0.63 ± 0.06 for A1413.

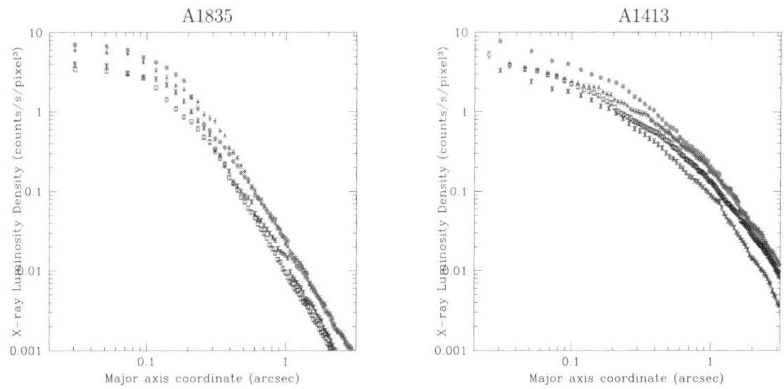

Fig. 4. Density profiles along $\hat{\mathbf{x}}$, recovered by deprojecting the *Chandra* X-ray brightness distribution of clusters Abell 1835 (left) and Abell 1413 (right), under the assumptions of (i) oblateness and i=90°, shown in crosses (ii) oblateness and i=i_{min}; in filled circles (iii) prolateness and i=90°; in filled triangles (iv) prolateness and i=i_{min}; in open circles. Here i_{min} is the smallest inclination allowed under oblateness, for a measured (uniform) e_p (=$\sin^{-1} e_p$).

6. Summary

In this paper, we have introduced a new non-parametric algorithm DOPING that is capable of inverting observed surface brightness distributions of galaxies and galaxy clusters, while taking into account variations in the intrinsic shapes of these systems. The potency of DOPING is discussed in the context of a test galaxy in which the eccentricity is made to change radically with radius. The code is also successfully applied to obtain the luminosity density distribution of the faint nucleated galaxy vcc1422. Lastly, a novel use is made of the capability of DOPING to deproject in general geometries, in determining the intrinsic shape of a galaxy cluster in terms of a prolateness parameter. It is envisaged that implementing a measure of the LOS extent of a cluster from Sunyaev Zeldovich measurements, will help tighten the estimate of prolateness of the cluster.

References

[1] Krajnovi, D., Cappellari, M., Emsellem, E., McDermid, R., de Zeeuw, P. T., 2004, *Monthly Notices of the Royal Astronomical Society*, 357, 1113.
[2] Kronawitter, A., Saglia, R. P., Gerhard, O., Bender, R., 2000, *Astronomy & Astrophysics*, 144,53.
[3] Magorrian, J., Tremaine, S., Richstone, D., Bender, R., Bower, G., Dressler, A., Faber, S. M., Gebhardt, K., Green, R., Grillmair, C., 1998, *Astronomical Journal*, 115, 2285.
[4] Gerhard, O. E., & Binney, J. J., 1996, *Monthly Notices of the Royal Astronomical Society*, 279, 993.
[5] Kochanek, C. S., & Rybicki, G. B., 1996, *Monthly Notices of the Royal Astronomical Society*, 280, 1257.
[6] Rybicki, G. B., 1987, *IAU Symposium*, 127, 397.
[7] van den Bosch, F. C., 1997, *Monthly Notices of the Royal Astronomical Society*, 287, 543.
[8] Bendinelli, O., 1991, *Astrophysical Journal*, 366, 599.
[9] Palmer, P. L., 1994, *Monthly Notices of the Royal Astronomical Society*, 266, 697.
[10] Cappellari, M., 2002, *Monthly Notices of the Royal Astronomical Society*, 333, 400.
[11] Richardson, W.H., 1972, *Jl. of Optical Society America*, 62, 55
[12] Lucy, L. B., 1974, *Astronomical Journal*, 79, 745.
[13] Romanowsky, A. J., & Kochanek, C. S., 1997, *Monthly Notices of the Royal Astronomical Society*, 287, 35, RK.
[14] Côté, P., Blakeslee, J. P., Ferrarese, L., Jordan, A., Mei, S., Merritt, D., Milosavljevi, M., Peng, E. W., Tonry, J. L., West, M. J., 2004, *Astrophysical Journal, Supplement*, 153, 223.
[15] Ferrarese, L., Côté, P., Blakeslee, J. P., Jordan, A., Mei, S., Merritt, D., Milosavljevi, M., Peng, E. W., Tonry, J. L., West, M. J., 2006, *Astrophysical Journal*, in press.

QUANTUM ASTRONOMY AND INFORMATION*

CESARE BARBIERI

University of Padova, Astronomy Department, vicolo Osservatorio 3, Padova, Italy
E-mail: cesare.barbieri@unipd.it

Future Extremely Large Telescopes, very fast photon detectors and extremely accurate and stable clocks will push the time resolution and time tagging capability of astronomical observations towards the limit imposed by Heisenberg uncertainty principle, thus paving the way to a novel Quantum (or Photonic) Astronomy. The capability to time tag the arrival time of each photon to the 10 picosecond level for hours of observations, over several narrow bandpasses, resolved in polarization state and direction of arrival on the focal plane of the telescope, will generate an impressive amount (up to 100 Terabyte a night) of multidimensional data. It is the purpose of the present paper to expound this new frontier of Information Technology applied to Astronomy.

Keywords: Quantum Astronomy; Information Technology

1. Introduction

Astronomy has always played a fundamental role in the advance of science, from chemistry to physics and even to biology. This primary role has been both in discovering natural phenomena not known in the terrestrial laboratory, and in promoting new technologies to overcome the limitations of the time. As a consequence, astronomy has also been an active promoter and a privileged user of advanced technologies, for example of optics, detectors, computers, information technology. This multifaceted role will certainly be played also in the future, specifically for the theoretical advancement and practical utilization of quantum optics, quantum information and quantum computing, leading to a novel astronomy to the quantum limits, which we might call quantum astronomy or photonic astronomy, paralleling the tendency to rename the whole science of optics to photonics to reflect the increasingly dominant role of photon physics. Our knowledge of the Universe almost entirely depends on the interpretation of properties of electromagnetic radiation ("light"). There are exceptions, such as neutrinos, cosmic rays, meteorites and other extraterrestrial materials, in-situ studies of planetary bodies, hopefully soon gravitational waves. However, the far greater part of our understanding of the Universe rests upon observing and interpreting properties of light from celestial sources. Quantum

*This work is partly supported by the Italian Ministry of University and Research through a PRIN 2006 grant.

mechanical properties of light useful in the astronomical applications can be found in [1], [2] and [3]. In order to explore this new realm of astrophysical information, the time resolution and time tagging capabilities of detectors must be pushed towards the picosecond region. Indeed, from Heisenberg's uncertainty principle, the time interval Δt between two distinct detections and the precision $\Delta\lambda/\lambda$ with which the wavelength of each photon is measured, must satisfy the following relationship:

$$\Delta t \approx \frac{\lambda}{2\pi c}\left(\frac{\lambda}{\Delta\lambda}\right)$$

For visible light, a 1A wide filter implies a time-tagging capability of the order of 10 picoseconds, to be maintained by a continuously running clock for the entire duration of the observation (may be hours for faint stars). Today clocks (in particular optical clocks) are on the verge of reaching the required performances. Detectors capable to go beyond the 100 picosecond limit are already available, and new ones are being developed.

Astronomers on their side are very actively studying Extremely Large Telescopes (ELTs) with diameters of 30 to 50m. We shall show how these ELTs offer an enormously increased sensitivity for studying astrophysical phenomena on sub-nanosecond timescales. In essence, this gain originates because celestial variable sources are normally not periodic, and their timescales are both unknown and unstable, so that in limiting applications one must study not simply light curves, but statistical functions associated with the photon stream, e.g. power spectra and autocorrelations. The amplitude of such second-order functions increases with the square of the collected light intensity: doubling the telescope diameter increases the area four times, and the second-order correlated signals by a factor of sixteen. Higher-order correlations increase even more steeply with telescope size. The consequences of these considerations are far-reaching: when discussing the scientific use of ELTs, one may tend to extrapolate current observing programs to qualitatively similar ones, merely with the quantitative difference of going to fainter sources. However, the much higher photon rate insured by the ELTs, and modern Quantum optical technology, might bring in fundamentally novel and qualitatively different fields which have so far not been accessible to astronomers. This new window of sub-nanosecond astrophysics might have the same impact of the opening of the wavelength window by space telescopes.

In other words, photons can be much more complex entities, and carry much more information, than so far appreciated in the astronomical community. Since our understanding of the Universe is based upon light, we should exploit every opportunity to extract additional information from it. The larger photon gathering capability of ELTs dictates a novel appraisal of the information content carried by photons of cosmic origin, and Quantum optics offers such an opportunity. To exploit these exciting possibilities, a huge volume of multidimensional data must be acquired, stored in well organized databases, distributed to a wide community, and finally analyzed with sophisticated algorithms. No specific solution will be indicated for these challenging problems connected with detectors, clocks, memories, etc., our main goal here being to make known the existence of this new Information Technology problematic associated to Astronomy.

2. Properties of Light in Astronomical Observations

Astronomical telescopes are equipped with many auxiliary instruments with different operating modes, producing different types of data, from photometric series to images to spectra. However, under a closer examination it appears that they all measure properties of light that can be deduced from the first-order correlation function of light, $G^{(1)}$, for two coordinates in space $\mathbf{r}$ and time t as in the following equation, where E is the electrical field, $< \quad >$ denotes time average, and * denotes complex conjugate:

$$G^{(1)}\left[\mathbf{r}_1, t_1; \mathbf{r}_2, t_2\right] = \left\langle E^*\left(\mathbf{r}_1, t_1\right) E\left(\mathbf{r}_2, t_2\right)\right\rangle$$

When $\mathbf{r}_1 = \mathbf{r}_2$, $t_1 = t_2$, $G^{(1)}$ yields the field intensity irrespective of the spectrum or geometry of the source. When $\mathbf{r}_1 \neq \mathbf{r}_2$, $t_1 = t_2$, $G^{(1)}$ gives the spatial autocorrelation function measured by phase (Michelson type) interferometers. When $\mathbf{r}_1 = \mathbf{r}_2$ but $t_1 \neq t_2$, $G^{(1)}$ becomes the autocorrelation function with respect to time, whose Fourier transform yields the spectrum.

These spatial and temporal first-order coherences can be traced back to single photon detection, which is the process of projecting (collapsing) the wave function on a one-photon state. Thus, one might say that existing astronomical instruments are limited to study such one-photon states. However, light can carry additional information (we make reference to the works of Glauber [4–6]). The detection of two photons by two spatially and/or temporally displaced detectors is the collapse of the wave function on the two-photon state, which is not simply the sum of two one-photon states. Spatial and temporal distribution of arrival times are entangled in two and many photon measurements. Consider for instance the temporal distribution of photon arrival times (see e.g. [7, 8]). These statistics can be random, as in chaotic, maximum-entropy thermal radiation, and then the Bose-Einstein statistics predicts bunching in time. The statistics might be quite different if the radiation is not in a maximum-entropy state, perhaps originating in stimulated laser-type emission (an idealized laser emits light whose photons are evenly distributed in time: in contrast to thermal emission there is no bunching). Studies of such quantum properties of light are not yet pursued in Astronomy, which usually merges all properties of radiation of a certain wavelength and polarization state into the quantity "intensity". However, with ELTs and modern detectors and clocks higher-order spatial and/or temporal correlation functions could become measurable. Second-, and higher-order degrees of coherence reflect correlated properties of two or more photons in the light from the source. For example, in stimulated emission there is a causal coupling between a first photon that is stimulating the emission of the next photon; stimulated emission thus is a property that cannot be ascribed to a single photon, there must always be at least two. The second-order signal is thus proportional to the conditional probability that a second photon is detected immediately after the first; therefore, this signal increases as the square of the observed light intensity, namely as the fourth power of telescope diameter. Although its strict definition involves quantum-mechanical operators, a simplified expression of the second order coherence function can be given in terms of intensities:

$$g^{(2)}\left[\mathbf{r}_1, t_1; \mathbf{r}_2, t_2\right] = \frac{\langle I(\mathbf{r}_1, t_1)\, I(\mathbf{r}_2, t_2)\rangle}{\langle I(\mathbf{r}_1, t_1)\rangle \langle I(\mathbf{r}_2, t_2)\rangle}$$

For $\mathbf{r}_1 \neq \mathbf{r}_2$ but $t_1 = t_2$, we obtain the Hanbury Brown - Twiss intensity interferometer (HBTII) which measures the angular sizes of stars, but in a different way than a phase interferometer. For $\mathbf{r}_1$=$\mathbf{r}_2$ but $t_1 \neq t_2$ we have an intensity-correlation spectrometer, which measures the width of spectral lines.

To our knowledge, HBTII [9, 10] has been the only astronomical instrument which could not be described by statistics of one-photon arrivals, requiring instead properties of two-photon wave functions. The interpretation of HBTII is still open to debate, see for instance Scarcelli et al. [11]. Limiting ourselves to astronomical applications, a fresh appraisal has been given by Ofir and Ribak [12, 13]. Not only micro-arcsec resolutions are possible, but also the phases can be recovered, leading to a full reconstruction of the image of the source, at the expenses of great computational complexity. Therefore, HBTII is quite valuable when discussing astronomical observations to the quantum limit, and we shall consider its properties in a later paragraph. We underline her that the need for accurate timekeeping at both sites $\mathbf{r}_1$ and $\mathbf{r}_2$ originates from the requirement $t_1 = t_2$.

Higher-order coherences increase even more rapidly with telescope size, as shown in Table 1. This very steep dependence makes the future ELTs enormously more sensitive for high-speed astrophysics and quantum optics than even the largest existing telescope.

Table 1. Sensitivity of telescopes to first, second and fourth order correlation functions (TNG: Telescopio Nazionale Galileo (Canaries), VLT: ESO Very Large Telescope (Chile), LBT: Large Binocular Telescope (Arizona), E-ELT: future ESO Extremely Large Telescope)

Telescope Diameter (m)	Intensity $< I >$	Second-order correlation $< I^2 >$	Fourth-order correlation $< I^4 >$
3.5 (TNG)	1	1	1
8.2 (VLT, LBT)	5.5	30	900
17.0 (MAGIC)	24	550	3×10^5
40 (E-ELT)	130	1.7×10^4	2.9×10^8

In the table, the intensity $<I>$ can be referred to physical units, by taking into account that in the visual band the flux of photons per unit bandwidth and unit time on the TNG from a star of apparent magnitude V is:

$$N_{ph}(V) \approx 10^3 \times 10^{-0.4V} \times Area(cm^2) \approx 10^8 \times 10^{-0.4V} \quad ph \cdot s^{-1} \cdot A^{-1}$$

(Vega, namely Alpha Lyrae, one of the brightest stars, has V=0). The second and fourth order correlation functions in the table are instead normalized to an arbitrary constant. The calibrated photon flux provides the opportunity for a quick calculation of the exposure time needed to detect quantum effects, simply as time needed to detect deviations from a Poisson distribution. The results, which are based on current detector and coating quantum efficiency and reflectivity, are given in Table 2. Optimal coatings, detectors and atmospheric quality, better statistical analysis, would certainly improve on those very preliminary figures, but the main message is that quantum astronomy is really photon starved! Therefore, no wonder that, while effects of non-linear optics are familiar in the laboratory, they are not yet familiar to astronomers.

Table 2. T(2) =Time needed to detect second-order quantum effects on a 40m telescope

V	T(2) s
10.00	1
12.50	10^2
15.00	10^4

Of course, observing in regular high-time resolution mode, where such narrow bandpasses are not required and integration is allowed for microseconds or longer would permit to reach much fainter stars.

3. The Intensity Interferometer

The original HBTII at Narrabri (Australia, see [10]) was composed of two 6.5m mirrors which could be separated from a minimum value d =10 m to a maximum value d = 188 m. The fluctuations in the anode currents were carried by high-frequency coaxial cables to the central station, where they were amplified by wide-band amplifiers (10-110 MHz) and combined in the linear multiplier (correlator). The main observing program was started in 1965, and led to the measurement of the diameters of 32 hot stars. Thus, in HBTII there is no interference of light. The interference is between the electric currents from the two photomultipliers, as measured by the correlator in a frequency range Δf from 10^7 to 10^8 Hz. Notice that the phase difference between the two correlated components is *not* the phase difference of light waves at the two detectors, but it is the difference of the relative phases of two Fourier components. Therefore, the correlation does not depend on the optical path difference between the two apertures. Intensity interferometry can thus be carried out very effectively also in the blue and visible.

The original HBTII was entirely analog. When photon counting (counting is a well defined process, while the arrival of a single photon may not be), the Signal to Noise ratio S/N (namely, the excess counts over the fluctuations of the random coincidences) can be written as:

$$\frac{\mathrm{S}}{\mathrm{N}} = K_{instr} \cdot q_e \cdot A \cdot n_0 \cdot \sqrt{T/\tau_c} \cdot g(\mathrm{sky})$$

S/N increases proportionally to quantum efficiency$q_e$of the detector, to the area A of each telescope, to n_0 (the photon flux per unit area per unit time per unit bandwidth), to the square root of the exposure time T divided by the electric bandwidth τ_c. The term $n_0/\sqrt{\tau_c}$ insures that S/N is independent of the optical bandwidth. The optical quality of the telescope enters through the amount of spurious light background $g(\mathrm{sky})$. Therefore, better detectors, wider electric bandwidths (τ_c as short as possible), larger telescopes of better optical quality, darker sites minimizing the influence of the sky background, will lead to sensitivities orders of magnitude higher than the original HBTII. Telescopes such as VLT and LBT, or the 17m Cerenkov light collector MAGIC can achieve in 3 hours the same limiting S/N the original Narrabri interferometer reached with week-long integrations.

We have already pointed out that HBTII is realized without light combination, but by correlating photon counts from two separate telescopes. This can be done only if an accurate common time reference is available to both detectors. Thus, one can envision the exciting possibility of performing Very Long Baseline Optical Intensity Interferometry (VLBOII), analogous to the radio (phase resolved) VLBI.

4. Intensity-Correlation Spectroscopy

Intensity-correlation spectroscopy (or photon-counting correlation spectroscopy) can be considered as the temporal equivalent of spatial intensity interferometry. The interferometry, i.e. the cross-correlation of the optical fluctuations, is then performed at the same spatial location, but with a variable temporal baseline (as opposed to variable spatial locations at the same instant of time in the spatial intensity interferometer). In any spectroscopic apparatus, the spectral resolution is ultimately limited by the Heisenberg uncertainty principle: $\Delta E\ \Delta t \geq \hbar/2$. Thus, to obtain a small value of the uncertainty in energy ΔE, the time to measure it, Δt, must be correspondingly long. Tilting the diffraction grating parallel to the direction of light propagation, or making light to travel many times back and forth inside the Fabry-Perot filter, are conventional methods of increasing the time light spends inside the instrument to produce higher spectral resolutions. In intensity-correlation spectroscopy, instead of mechanical devices one can use electronic timing of light along its direction of propagation. Consider light within some optical passband measured with very high time resolution. Intensity fluctuations in this light will arise from beating together of the various components of the electric field. In the general case of an arbitrary spectrum, the power content of spectral features inside the passband can be recovered, analogous to the recovery of the power content of spatial structures in intensity interferometry. This enables spectral resolutions corresponding to that of an instrument with physical size equal to the temporal delays, which can be orders of magnitude beyond those feasible with classical spectrometers.

5. Approaching the Heisenberg Limit: High Time Resolution Astrophysics

With high time-resolution detectors, astrophysical phenomena on successively shorter timescales have been discovered. Historically, perhaps the most notable discovery was that of pulsars, enabled by the development of high-time-resolution techniques in radio astronomy. More recent findings include the structure of gamma-ray burst afterglows and quasi-periodic oscillations on millisecond scales in compact accretion sources. Generally, the environments of compact objects are likely places for very rapid phenomena: the geometrical extent is small, the energy density high, the magnetic field strengths enormous, and a series of phenomena, ranging from magneto-hydrodynamic turbulence to stimulated synchrotron radiation, may take place. Some processes may extend over scales shorter than few kilometers, and there is no immediate hope for their spatial imaging. Insights can be gained through studies of their short time-scale instabilities, such as hydrodynamic oscillations or magneto-hydrodynamic flares.

Quantum phenomena will fully manifest in the optical band below the 10 picosecond barrier, but we hope that they can be discernible, albeit with reduced intensity, below 1 nanosecond. We have proposed [2, 3] a very high time resolution astronomical photometer (QuantEYE), able to make a decisive step toward the Heisenberg limit, having the power to detect and examine statistics of photon detection times. In our solution, originally conceived for the 100m OWL telescope of ESO, the telescope pupil was divided in 10×10 sub-pupils, each imaged on a Single Photon Avalanche Photodiode (SPAD) operating in Geiger mode, and able to time tag each incoming photon to better than 100 picoseconds. QuantEYE could thus cope with several GHz photon count-rates. A 50 Terabyte storage unit insured a full night of data acquisition. These instrumental capabilities would enable detailed studies of exotic phenomena such as variability close to black holes, surface convection on white dwarfs, non-radial oscillation spectra in neutron stars, fine structure on neutron-star surfaces, photon-gas bubbles in accretion flows, free-electron lasers in the magnetic fields around magnetars. Photon-counting correlation spectroscopy would reach spectral resolutions exceeding $R = 10^8$. Furthermore, given two distant ELTs, QuantEYE would permit a modern realization of the HBTII.

The QuantEYE concept can be adapted to more conventional high-speed astrophysical problems, even using small telescopes. We are currently realizing AquEYE (the Asiago Quantum Eye), a first scaled-down 2×2-channel prototype for the Asiago 182 cm telescope. As detectors, we have selected the 50 μm SPAD developed by Cova and collaborators [14] . Their q_e in the visible band is better than 45%, the dead time is around 70 nanoseconds, the time tagging capability is better than 40 picoseconds, and the afterpulsing probability is less than 1%. The electronics is based on boards developed for CERN. The core is a TDC board which acquires and processes the pulses coming from the four SPADs. The board tags each event with a precision of 35 ps per channel, and transfers all time tags to a mass storage device, saving the scientific data for further studies. Real-time correlations will be displayed to the observer. Regarding the crucial question of generating and maintaining a very accurate time for minutes and even hours, our present solution will tie the start/stop commands to UTC by means of the GPS (or the Galileo Navigation Satellite System) signal, so that the data can be referred to a common time scale adopted by all telescopes on the ground or in space (we recall the scientific interest of correlations with X-ray and Gamma ray timing). In addition, an internal high quality oscillator is needed, to overcome the jitter of ± 25 nanoseconds or so, affecting both GPS and GNSS.

6. Ultimate Data Rates

Our final considerations are relative to the analysis of the huge, multidimensional database generated by an ideal quantum instrument mounted on an ELT. Present day technology (detectors, clocks, memories, computers, algorithms) come close to have this ideal quantum instrument, so we feel justified to make a jump into the future. Our ideal quantum-optical spectrophotometer will be a spatially and spectrally resolving photon-counting instrument, which for each spatial and spectral element can determine the polarization state of each detected photon, and time tag its arrival time with picosecond accuracy, continuously for hours of uninterrupted observation.

The need for good spatial resolution is intrinsic to almost all astronomical applications. Using again the example of η Car, the quantum events are likely to take place in a specific region of this extended object, perhaps quickly migrating to nearby regions, so we need to distinguish where the interesting photons are coming from, at each particular moment. Moreover, an extended field of view allows to control the behaviour of nearby stars and of the sky. The spectral resolution is also necessary, e.g. to distinguish the different behavior of the continuum and of spectral lines. The ability to single out a specific polarization state is intrinsic to quantum measurements, and in addition polarized components could be present, ensuing from intrinsic mechanisms or scattering processes, and time variable. Another important characteristic of the ideal instrument is its ability to cope with very high and very low photon rates (e.g. as high as 1 GHz, and as low as 100 Hz from the sky). An instrument like QuantEYE would indeed have the largest dynamic range of all Astronomy, more than 10 orders of magnitude from the brightest to the faintest sources ($\Delta V \approx 25$), being limited only by photon counting statistics.

To be specific, and at the same time fairly realistic in technological demands, consider a 40m ELT (effective collecting area $\approx 1 \times 10^7$ cm^2), with an optical quality capable to concentrate most of the stellar light in a circle of diameter 0.3 arcsec on the sky. Our detector have 100x100 imaging elements covering 10×10 arcsec in the sky (0.1 arcsec per pixel, 3×3 pixels per star), at 1 spectral and 2 polarization channels. For instance, using Volume-Phase Holographic (VPH) gratings we could achieve at the same time a very narrow spectral bandwidth and Vertical/Horizontal polarization splitting with high efficiency. Each pixel has $q_e = 50\%$, and is capable of photon-counting with 10 ps time tagging precision, at a maximum photon rate of 10 MHz. An accurate clock is capable to maintain the time accuracy to 10 picoseconds for hours of observations.

On such system, the average count rate on the pixels illuminated by Vega will be $\approx 5\times10^8$ ph·s^{-1}·A^{-1}·px^{-1}, that is fifty times higher than the detector can handle. So, the linear regime starts at $V \approx 5$. After 1 second of observing such bright star, we would have stored in the memory the time tags of $\approx 10^7$ detected photons, for each of the 9 pixels, and for each polarization state. On the other extreme, a star like the pulsar in the Crab ($V \approx 15$) would have produced 10^3 ph·s^{-1}·A^{-1}·px^{-1}. In a dark site, the sky background ($V = 20$ sq·arcsec^{-1}) corresponds to 0.1 ph·s^{-1}·A^{-1}·px^{-1}. Of course, to detect interesting phenomena, the observing time will be much longer than 1 second, and the same requirement of much longer times is imposed by correlation spectroscopy. Due to the ensuing need to count photons for hours (3600 seconds being 3.6×10^{14} ticks of our clock), time stamping requires a depth of 32 bits (with one roll-over which is easy to achieve; moreover, only relative times need to be stored). The upper dimension of the data base associated with the bright star in linear regime would thus be 2(pol)$\times$9(pix)$\times 10^7 \times$32(bits) $\approx$ 1 GByte per second per Angstrom of bandpass.

Moreover, in normal Very High Time Resolution Astrophysics (see e.g. 15) extremely narrow bandpasses are not required. So the available space for data must grow by two - three orders of magnitude beyond the previous estimate, per each observing session, reaching the 100 TBytes per night.

We elaborate also on the two other modes of Quantum Astronomy we have previously discussed, namely image reconstruction and intensity interferometry. Regarding image reconstruction, the optical quality of ELTs will certainly be very good, and the telescopes equipped with sophisticated adaptive optics devices to counteract the effects of atmospheric turbulence. However, their mirrors will be so large to prevent a full correction for the atmospheric wavefront distortions, especially in the blue and visible regions. Our quantum instrument would allow to make the best use of the light speckles on the focal plane to recover the undistorted image at the diffraction limit. The exceptionally good dynamic range would even insure the ability to image extraterrestrial planets very close to their primary star.

Optical Intensity Interferometry, combining several distant telescopes, might well allow to reconstruct a micro-arcsecond sharp image. We could also conceive (this time really dreaming wildly), to distribute an extremely accurate time to very many telescopes on the Earth and in Space, a sort of 'distributed aperture' gigantic telescope requiring the help of the most powerful computer technology.

7. Conclusions

We have seen that, whatever the application, the necessary storage and computational power of Astronomy pushed to quantum limits is really large. At the end of the night of observing, the astronomer will be left with an enormous data base, for sure well organized and conveniently stored, on which sophisticated search algorithms will detect the elusive quantum properties. Possibly, one could follow precepts of data partitioning, following the topological properties of the space depicted by each event, that can be cast as the product of two Hypercubes, $I_1 \otimes I_2$, where the polarization vector $I_2 = (\mathbf{S})$ is independent from all the other quantities, as it commutes with each other physical quantity. As far as we know, one of the key factors that have a profound effect on the performance is space partitioning strategy, as discussed for instance in [16] for high-energy physics and astronomy applications. It might well be that in the next future quantum computing and quantum algorithms will become available, improving the computational and search capabilities (see the accompanying paper by Cariolaro and Occhipinti, [17]).

For completeness, we mention that there are other effects of quantum optics which will be exploited in astronomy. We refer to the Photon Orbital Angular Momentum (POAM), and the possibility to insert Optical Vorticity in the stellar light beam to realize a very efficient coronagraph (see [18] and references therein). However, the advantages offered by ELTs are less specific and the information content can be handled by conventional techniques, so that this topics will not be further treated here.

Acknowledgments

Thanks are due to Andrej Čadez, Tommaso Occhipinti and Fabrizio Tamburini for useful discussions.

References

[1] Dravins D, Barbieri C, Da Deppo V, Faria D, Fornasier S, Fosbury RAE, Lindegren L, Naletto G, Nilsson R, Occhipinti T, Tamburini F, Uthas H, Zampieri L (2005), *QuantEYE. Quantum Optics Instrumentation for Astronomy*, OWL Instrument Concept Study, ESO document OWL-CSR-ESO-00000-0162 (2005)

[2] Barbieri C, Dravins D, Occhipinti T, Tamburini F, Naletto G, Da Deppo V, Fornasier S, D'onofrio M, Fosbury RAE, Nilsson R, Uthas H (2007). *QuantEYE, a high speed photometer pushed to the Quantum limit for Extremely Large Telescopes.* Journal Of Modern Optics.**54(2-3)** special issue on "Single-Photon: Sources, Detectors, Applications and Measurement Methods", pp. 191-198 (2007).

[3] Dravins, D. 2007, *Photonic Astronomy and Quantum Optics*, to appear in D. Phelan, O. Ryan and A. Shearer, eds, 2007, *High Time Resolution Astrophysics,* Astrophysics and Space Science Library, Springer

[4] R.J. Glauber, 1963a, *The Quantum Theory of Optical Coherence*, Phys. Rev. vol. **130**, 2529

[5] R.J. Glauber, 1963b, *Coherent and Inchoerent States of the radiation Field*, Phys. Rev. vol. **131**, 2766

[6] R.J. Glauber, 1970, in *Quantum Optics*, S.M. Kay, A. Maitland eds.

[7] F.T. Arecchi, 1965, *Measurement of the statistical distribution of Gaussian and laser sources*, Phys.Rev.Letter **15**, 912-916

[8] R. Loudon, 2000, *The Quantum Theory of Light*, Oxford University Press

[9] R. Hanbury Brown, R. Twiss, 1956, *Correlation between photons in two coherent beams of light,* Nature **177**, 27-29
[10] R. Hanbury Brown, 1974, *Intensity Interferometer*, Wiley, John & Sons, Incorporated
[11] G. Scarcelli, V. Berardi, Y. Shih, 2006, *Can two photon corelation of chaotic light be considered as correlation of intensity fluctuations?* Phys. Rev. Letter **96**, 063602
[12] A. Ofir, E:N., Ribak 2006 a, *Off-Line, Multi-Detector Intensity Interferometers I: Theory,* MNRAS **368** (4), 1646-1651
[13] A. Ofir, E:N., Ribak 2006 b, *Off-Line, Multi-Detector Intensity Interferometers II: Implications and Applications*, MNRAS **368** (4), 1652-1656
[14] S. Cova, M. Ghioni, A. Lotito, I. Rech, F. Zappa, 2004, *Evolution and prospects for single-photon avalanche diodes and quenching circuits*, Journal of Modern Optics, 15 vol. **51**, no. 9–10, 1267–1288
[15] D. Phelan, O. Ryan and A. Shearer, eds, 2007, *High Time Resolution Astrophysics,* Astrophysics and Space Science Library, Springer
[16] J. Lukaszuk, R. Orlandic, 2004, *On accessing data in high-dimensional spaces: A comparative study of three space partitioning strategies*, The Journal of Systems and Software 73, 147–157
[17] G. Cariolaro, T. Occhipinti, 2007, *From the Qubit to the Quantum Search Algoritms*, this conference
[18] F. Tamburini; G. Anzolin; G. Umbriaco; A. Bianchini; C. Barbieri, 2007, *Overcoming the Rayleigh Criterion Limit with Optical Vortices*, Physical Review Letters, vol. 97, Issue 16, id. 163903

MINING THE STRUCTURE OF THE NEARBY UNIVERSE

RAFFAELE D'ABRUSCO [1,2], GIUSEPPE LONGO[1,3,4],

MASSIMO BRESCIA [4,3], ELISABETTA DE FILIPPIS[1],
MAURIZIO PAOLILLO[1,3,4], ANTONINO STAIANO[5], and
ROBERTO TAGLIAFERRI[6,3]

1 - Department of Physics, University Federico II, Napoli, Italy
2 - Institute of Astronomy, Cambridge UK
3 - INFN - Napoli Unit, via Cinthia 6, Napoli, Italy
4 - INAF - Napoli Unit, via Moiariello 16, Napoli, Italy
5 - University Parthenope, via Acton 8, Napoli, Italy
6 - DMI, University of Salerno, Fisciano, Italy
** E-mail: dabrusco@na.infn.it*

We discuss some preliminary results of our ongoing efforts to implement a new generation of data mining tools, largely based on neural networks and specifically tailored to work in a distributed computing environment. These tools have been applied to the Sloan Digital Sky Survey public dataset to produce a 3-D map of the galaxy distribution in the nearby universe.

Keywords: Massive data sets, neural networks, photometric redshifts

1. Introduction

Modern survey telescopes and projects, as well as the ongoing efforts to implement the Virtual Observatory, are producing an explosive growth in the quantity, quality and accessibility of multiwavelength and multiepoch archives of heterogeneous data. However, while on the one end it has become clear that the simultaneous analysis of hundreds of parameters for millions of objects can unveil previously unknown patterns and lead to a deeper understanding of the underlaying phenomena and trends, on the other end it has also become clear that traditional interactive data analysis and data visualization methods are far too inadequate to cope with data sets which are characterized by huge volumes and/or complexity (ten or hundreds of parameter or features per record). The field of Data Mining (hereafter DM) in distributed computing environments is therefore becoming of paramount importance for almost all fields of research.

This paper is structured as it follows. In the next section, as an introduction to the methods which have been implemented by our team and which are currently being ported under the U.K. Astrogrid platform, we shortly recall the distinction between supervised and unsupervised learning methods. We wish to stress that even though some of these tools have been implemented and tested in other data–rich fields of research, their application to the astronomical case is far from trivial due to the complexity of astronomical data which usually presents strong non linear correlations among parameters and are highly degenerate. Furthermore, most (if not all) astronomical data sets are usually plagued by a large fraction of missing data (id est data which is neither measured, nor detected, or just upper limits); a fact which renders most DM algorithms uneffective. In Section 3 we present an application to the evaluation of photometric redshifts for nearly 30 million galaxies in the nearby ($z < 0.5$) universe and in section 4 we present some preliminary results on the physical classification of galaxies in the nearby universe and draw some conclusions.

2. Supervised and Unsupervised Methods

Machine learning methods can be divided in two broad families: supervised and unsupervised. Supervised learning makes use of prior knowledge to group samples into different classes [1] . In unsupervised methods, instead, null or very little a priori knowledge is required and the patterns are classified using only their statistical properties and some *similarity measure* which can be quantified through a mathematical clustering objective function, based on a properly selected distance measure. Both approaches are used for a variety of applications and both have their advantages and their disadvantages, the choice depending mainly on the purpose of the investigation and on the structure of the dataset.

In the supervised case, the method classifies objects using a statistical significant subset of examples (*training set*) with attached labels representing the base of knowledge. These labels are used to learn a decision function which partitions the parameter space into disjoint regions, each associated with one of the classes. Typical decision function structures used in practice are neural networks, decision trees, and prototype-based classifiers. While being more accurate than the unsupervised ones, the supervised methods are more limited in scope and present several inconvenients. First of all the fact that the learning phase is usually slow and it is generally performed off-line. Moreover, in order to be effective, they require a relatively large number of labeled examples mapping with equal weights all regions of the parameter space, and such information is usually either lacking or very difficult to obtain.

While being much more general than the supervised ones, unsupervised methods are usually less accurate and require a much more accurate fine tuning to the specific needs of the problems under study. An additional family of methods is that of the 'hybrid or semisupervised' methods which learn on batches of data that consist of both labeled and unlabeled examples. On the one hand, appropriate use of unlabeled examples, in addition to labeled ones, can help to better learn the shapes of each of the classes (i.e., the class-conditional density functions). On the other hand, the use of some labeled examples can potentially help to guide unsupervised clustering methods toward solutions that capture the ground truth classes in the data. Finally, it must be stressed that all methods, supervised, unsupervised and hybrid ones, share the common problem that the number of clusters (classes) is or needs to be assigned "a priori".

3. Photometric Redshifts for the SDSS Galaxies

The 'photometric redshifts' technique was introduced in the eighties [2] , when it became clear that it could prove useful in two similar but methodologically very different fields of application: i) as a method to evaluate distances when spectroscopic estimates become impossible due to either poor signal-to-noise ratio or to instrumental systematics, or to the fact that the objects under study are beyond the spectroscopic limit [3] ; ii) as an economical way to obtain, at a relatively low price in terms of observing and computing time, redshift estimates for large samples of objects.

Photometric redshifts are of much lower accuracy then spectroscopic ones but even so, if available in large number and for statistically well controlled samples of objects, they still provide a powerful tool for a variety of applications among which we shall quote just a few: to study large scale structure [4] ; to constrain the cosmological constants and models [5–7] ; to map matter distribution using weak lensing [8].

3.1. *Dataset and base of knowledge*

In a recent paper [9], we have presented a method which, by making use of the new sets of tools implemented for the VO, such as TOPCAT, VO-Neural, etc. is capable to produce results having accuracy near to the theoretical limits of the photometric redshift technique [10]. All experiments described in what follows made use of the public archive of the Sloan Digital Sky Survey (hereafter SDSS - Data Release 4 [11, 12]) which is an ongoing survey to image approximately π sterad of the sky in five photometric bands (u, g, r, i, z) and it is also the only survey so far to be complemented by spectroscopic data for $\sim 10^6$ objects (cf. the SDSS webpages at http://www.sdss.org/ for further details). The SDSS data can be queried from the VO interface and visualized for pre-analysis and statistics using existing VO-tools such as TOPCAT.

We extracted from the Spectroscopic Subsample (SpS) of the SDSS-DR4 a *General Galaxy Sample* or GG, composed of 445,933 objects with $z < 0.5$, matching the following selection criteria: dereddened magnitude in r band, $r < 21$; $mode = 1$ which corresponds to primary objects only in the case of deblended sources. We also extracted a sample named Luminous Red Galaxies or LRG, accordingly to the recipy described in Padmahaban et al. (2005) [13] . Since it is well known that photometric redshift estimates depend on the morphological type, age, metallicity, dust, etc. it was to be expected that if some morphological parameters are taken into account besides than magnitudes or colors alone, estimates of photometric redshifts should become more accurate [14, 15] . Therefore, in order to be conservative and also because it is not always simple to understand which parameters might carry relevant information, for each object we extracted from the SDSS database not only the photometric data but also the additional parameters listed in Table 1. These parameters are of two types: those which we call 'features' (marked as F in Table 1), are parameters which potentially may carry some useful information capable to improve the accuracy of photometric redshifts, while those named 'labels' (marked as L) can be used to better understand the biases and the characteristics of the 'base of knowledge'.

Table 1. List of the parameters extracted from the SDSS database and used in the experiments. Column 1: running number for features only. Column 2: SDSS code. Column 3: short explanation. Column 4: type of parameter, either feature (F) or label (L).

N	Parameter		F/L
	objID	SDSS identification code	–
	ra	right ascention (J2000)	–
	dec	declination (J2000)	–
1	petroR50_i	50 % of Petr. rad. in 5 bands	F
2	petroR90_i	90 % of Petr. rad. in 5 bands	F
3	dered_i	dered. mag. in 5 bands	F
4	lnLDeV_r	morphology in r band	F
5	lnLExp_r	morphology in r band	F
6	lnLStar_r	degree of stellarity	F
z	spec. z		L
specClass	spec. class. index		L

3.2. *Features selection*

In order to evaluate the significance of the additional features, our first set of experiments was performed along the same line as described in [14] using a Multi Layer Perceptron [16] with 1 hidden layer and 24 neurons. In each experiment, the training, validation and test sets were constructed by randomly extracting from the overall dataset three subsets, respectively containing 60%, 20% and 20% of the total amount of galaxies. On the GG sample, we run a total of $N+1$ experiments. The first one was performed using all features, while the other N were performed taking away the $i-th$ feature with $i = 1, ..., N$. For each experiment, following [17], we used the test set to evaluate the robust variance σ_3 obtained by excluding all points whose dispersion is larger then 3σ. The final results are listed in Table 2.
As it can be seen, the most significant parameters are the magnitudes (or the colors), while other parameters affect only the third digit of the robust sigma and, due to the large increase in computing time during the training phase and to avoid loss of generality at higher redshifts, where additional features such as the Petrosian radii are either impossible to measure or affected by large errors, we preferred to drop all additional features and use only the magnitudes. We also notice that the redshift distribution of the objects in the spectroscopic subsample presents a clear

Table 2. Results of the feature significance estimation. Column 1: features used (cf. Table 1). Column 2: robust σ of the residuals.

parameters	σ_3
all parameters	0.0202
all parameters but 1	0.0209
all parameters but 2	0.0213
all parameters but 4 & 5	0.214
all parameters but 6	0.215
only magnitudes	0.0199

discontinuity at $z \simeq 0.25$ (86% of the objects have $z < 0.25$ and only 14% are at a higher redshift) and in practice no objects are present for $z > 0.5$. The dominance of LRGs at $z > 0.25$ implies that in this redshift range the base of knowledge offers a poor coverage of the parameter space. We therefore restricted ourselves to objects having $z < 0.5$ and adopted a two steps approach: first we trained a MLP [16] to recognize nearby (id est with $z < 0.25$) and distant ($z > 0.25$) objects, then we trained two separate networks to work in the two different redshift regimes. This approach ensures that the NNs achieve a good generalization capabilities in the nearby sample and leaves the biases mainly in the distant one. To perform the separation between nearby and distant objects, we extracted from the SDSS-4 SpS training, validation and test sets weighting, respectively, 60%, 20% and 20% of the total number of objects (449,370 galaxies). The resulting test set, therefore, consisted of 89,874 randomly extracted objects. Extensive testing (each experiment was done performing a separate random extraction of training, validation and test sets) on the network architecture lead to a MLP with 18 neurons in 1 hidden layer. This NN achieved the best performances after 110 epochs and the results are detailed, in the form of a confusion matrix, in Table 3.
As it can be seen, this first NN is capable to separate the two classes of objects with an efficiency of 97.52%, with slightly better performances in the nearby sample (98.59%) and slightly worse in the distant one (92.47%).

Table 3. Confusion matrix

	SDSS nearby	SDSS far
NN nearby	76498	1096
NN far	1135	11145

Once the first network has separated the nearby and distant objects, we can proceed to the derivation of the photometric redshifts working separately in the two regimes. For the nearby sample we trained the network using objects with spectroscopic redshift in the range $[0.0, 0.27]$ and then considered the results to be reliable in the range $[0.01, 0.25]$. In the distant sample, instead, we trained the network over the range $[0.23, 0.50]$ and then considered the results to be reliable in the range $[0.25, 0.48]$.

Also in this case, in order to select the optimal NN architecture, extensive testing was made varying the network parameters and for each test the training, validation and test sets were randomly extracted from the SpS. The results of the Bayesian learning of the NNs were found to depend on the number of neurons in the hidden layer; for the GG (LRG) sample the performances were best when this parameter was set to 24 for the nearby sample and for the distant one (24 and 25 respectively for the LRG sample). It needs to be stressed that, at this point of the procedure, some systematic trends are still present [10] , in the nearby sample for $z < 0.1$ and in the distant sample at $z \sim 0.4$. The first feature is due to the fact that at low redshifts faint and nearby galaxies cannot be easily disentangled by luminous and more distant objects having the same color. The second one is instead due to a degeneracy in the SDSS photometric system introduced by a small gap between the g and r bands. At $z \sim 0.4$, the Balmer break falls into this gap and its position becomes ill defined [13].

In order to minimize such systematic trends, but paying the price of a slight increase in the variance of the final catalogues we applied to both data sets an interpolative correction computed separatedly in the two redshift intervals. After this correction we obtain a final robust variance $\sigma_3 = 0.0197$ for the GG sample and 0.0164 for the LRG samples, computed in both cases over the whole redshift range. The resulting distributions for the two samples are shown in Fig. 1.

4. Steps Towards a Photometric Classification of Galaxies

In order to precisely define the scope of the present paragraph we must first constrain what we mean with the word "classification". In fact, from the data mining point of view, many if not most of the operations which astronomers perform on measured data can be interpreted, one way or the other, as classification tasks. Besides obvious examples such as the morphological classification of galaxies or the spectral classification of stars, other typical tasks which can be regarded as classification ones are, for instance, the Star/Galaxy separation in images, the identification of specific subclasses of objects selected accordingly to their photometric properties (e.g. quasar candidates or high redshift object candidates). In a very schematic simplification, any classification scheme can be reconduced at partitioning the parameter space into N disjoint regions populated by homogeneous sets of objects, and their union should cover entirely the parameter space. Also in this case, the partition (id est the classification) of the parameter space can be obtained following the supervised or the unsupervised approaches. In the supervised case, one has some extensive a priori information and optimizes the partition in order to reproduce as well as possible the a priori information. This would be the case, for instance, of someone trying to disentangle in a color-color diagram galaxies of different morphological types.

In the unsupervised case, the partition is performed with no a priori information using only the statistical properties of the data themselves. For instance, this would

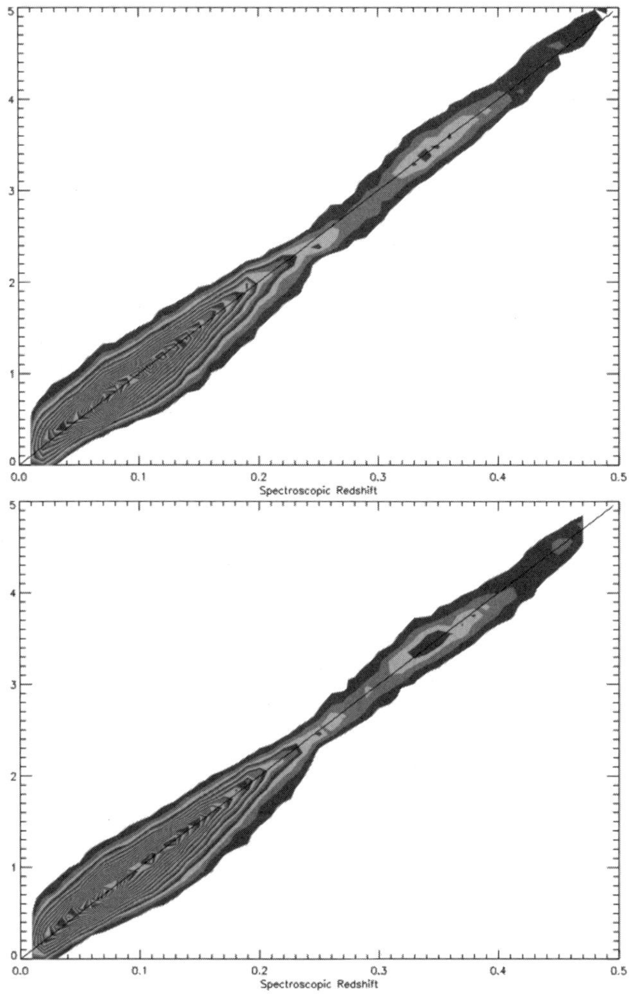

Fig. 1. Trend of spectroscopic versus photometric redshifts for the GC (upper panel) and the LRG (lower) samples.

be equivalent to find in the color-color diagram groups (hereafter clusters) of objects which accordingly to some metric criterium are close in the parameter space. It has to be stressed, however that when dealing with complex phenomenologies, no classification scheme will ever match all three above listed criteria and usually a compromise needs to be found.

In what follows we shall discuss the preliminary results of an attempt to cluster a subspace of the SDSS parameter space using an unsupervised clustering method based on the use of the Probabilistic Principal Surfaces (hereafter PPS [18]) together with an agglomerative method based on the Negative Entropy Clustering (hereafter NEC) concept.

We first applied the PPS algorithm to the sample of spectroscopically selected SDSS DR-4 objects using as parameters for the clustering the 4 colours obtained from model magnitudes (u-g,g-r,r-i,i-z) of SDSS archive. We fixed the number of latent variables and latent bases of the PPS to 614 and 51 respectively, so obtaining at the end of this step 614 clusters, each formed by objects which only respond to a certain latent variable. We choose a large number of latent variables in order to obtain an accurate separation of objects and to avoid that any group of distinct but near points in the parameter space could be projected into the same cluster by chance. The clusters found by the PPS are graphically represented by groups of points with the same colour (a different colour for each cluster) on the surface of a 2-d sphere embedded in the 3-dimensional latent space. These first order clusters were then fed to the NEC algorithm which determined the final number of clusters. The plateau analysis of the agglomerative process and the inspection of the dendrogram allowed to set the treshold to a value corresponding to 31 clusters. We present in Table 4 the ten most populous clusters together with the distribution of the *specClass* index within each cluster. The additional 21 clusters represent less than 10 % of the objects and are still under investigation. It needs to be emphasized that the clustering makes use of the photometric data only and the spectroscopic information is used only to validate them. As it can be seen, galaxies (SP2) clearly dominate clusters 1, 2 and 6.

Table 4. Distribution of objects in the most significant clusters found by our procedure. The columns correspond to different values of the *specClass* index.

Cl. n	SP0	SP1	SP2	SP3	SP4	SP6
1	69	145	9362	48	0	12
2	25	133	13370	10	0	12
3	149	132	63	64	0	5
4	44	3396	1530	189	67	1
5	202	85	447	2428	6	10
6	26	125	13728	12	0	12
7	0	0	0	0	0	484
8	1	1	1	0	0	329
9	541	1507	127	4750	18	1
10	89	474	2117	19	4	529

Whether this separation reflects or not some deeper differences among the three groups (such as, for instance, different morphologies), cannot be assessed on the grounds of presently available data. AGNs (SP3) dominate clusters 5 and 9 even though some contamination from galaxies exists. Late type stars (SP6) populate mainly clusters 7 and 8 and are strong contaminants of cluster 10 which also is dominated by galaxies. It needs to be stressed, however, that the use of the *specClass* as label must be taken with some caution.

Acknowledgments

This work took place in the framework of the European Project VO-Tech (Virtual Observatory Technological Infrastructures) and partially funded by the Italian PON-SCOPE. We wish to thank all our colleagues in the VO-Tech and the ASTROGRID projects for their herculean efforts and outstanding results.

References

[1] Su-In L., S. Batzoglou, Genom Biology 2003, 4: R76
[2] Butchins, S. A. 1981, A&A, 97, 407
[3] Bolzonella, M., Pelló R. & Maccagni D. 200, A&A, 395, 443
[4] Brodwin, M., Brown, M. J. I., Ashby, M. L. N., Bian, et al. 2006, astro-ph/0607450
[5] Blake, C., & Bridle, S. 2005, MNRAS, 363, 1329
[6] Budavári, T., et al. 2003, ApJ, 595, 59
[7] Tegmark, M. et al., 2006, astro-ph/0608632
[8] Edmondson E.M., Miller L. & Wolf C. 2006, astro-ph/0607302
[9] D'Abrusco R., et al. 2007, astro-ph/0703108
[10] Connolly, A. J., Csabai, I., Szalay, A. S., Koo, D. C., Kron, R. G., & Munn, J. A. 1995, AJ, 110, 2655
[11] Adelman-McCarthy, J.K., et al. 2006, ApJS, 162, 38
[12] Eisenstein, D. J., et al. 2001, AJ, 122, 2267
[13] Padmanabhan, N., et al. 2005, MNRAS, 359, 237
[14] Tagliaferri, R. et al. 2002, astro-ph/0203445
[15] Vanzella, E., et al. 2004, A&A, 423, 761
[16] Bishop C.M., 1995, Neural Networks for Pattern Recognition, Oxford University Press
[17] Csabai, I., et al. 2003, AJ, 125, 580
[18] A. Staiano, *Unsupervised Neural Networks for the Extraction of Scientific Information from Astronomical Data*, 2003, Ph.D. Thesis, University of Salerno

NUMERICAL CHARACTERIZATION OF THE OBSERVED POINT SPREAD FUNCTION OF THE VST WIDE-FIELD TELESCOPE

GIORGIO SEDMAK* and SIMONE CAROZZA

Universitá degli Studi di Trieste, Dipartimento di Fisica,
Via Valerio 2, Trieste 34127, Italia
**E-mail: sedmak@ts.infn.it*

GABRIELLA MARRA

INAF VSTceN,
Via Moiariello 16, Napoli 34131, Italia
E-mail: marra@na.astro.it

The numerical model of the observed Point Spread Function of the VST wide-field telescope is computed over the field of view by means of the convolution of the simulation of the atmospheric seeing and the ray-tracing, Fast-Fourier-transform-based Point Spread Function of the optical configuration of the telescope. The images obtained are compressed by means of two methods, a polynomial-modulated gaussian fit and a wavelet decomposition. The compressed images are mapped over the field of view by means of the interpolation of the fit or wavelet coefficients. The original and mapped Point Spread Function images are used for the evaluation and mapping over the field of view of various intensity and position error figures. The error maps confirm the high quality of the design of the VST telescope optics. Finally, the error maps can be input into the scientific processing of the astronomical observations.

Keywords: wide-field telescopes, Point Spread Function modeling, image compression, Point Spread Function mapping.

1. Introduction

The VST telescope [1] is a project of INAF Osservatorio Astronomico di Capodimonte, Naples, Italy, started in 1997 and developed in cooperation with ESO. It is now managed by the INAF VSTceN center at Osservatorio Astronomico di Capodimonte in cooperation with Italian industry. It is a ground-based, alt-azimuth telescope with a primary mirror of 2.61 meters aperture, designed for seeing limited performance over a field of view of (1 x 1) square arcdegrees. The optical scheme is a modified Ritchey-Cretien and includes two exchangeable wide-fields correctors, one standard and one equipped with an atmospheric dispersion compensation system which allows the telescope to operate at large zenith angles up to 55 arcdegrees. The telescope and its optical scheme are shown in Fig.1. The optical schemes of the two wide-field correctors are shown in Fig.2.

The VST telescope is equipped with a large mosaic CCD camera (OmegaCAM)

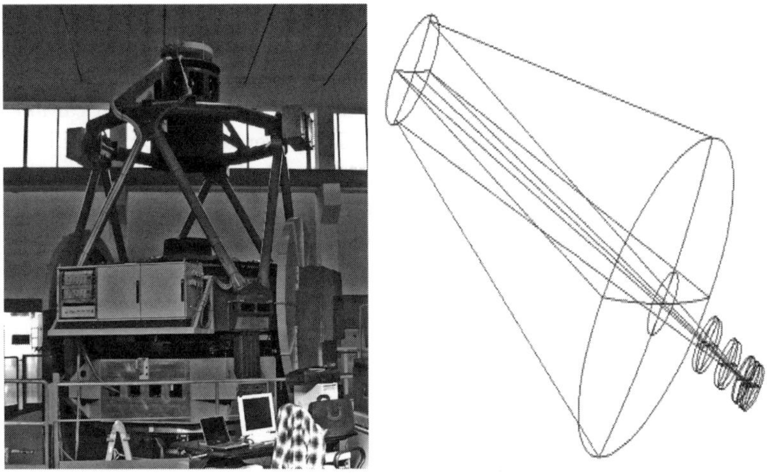

Fig. 1. The Italian VST wide-field telescope at the Mecsud integration area of Scafati (Salerno) in September 2006 (left) and its Zemax^{TM} ray-tracing optical scheme (right).

[2] with a format of (16K x 16K) pixels of size of (15 x 15) square microns and a scale of 0.21 arcsecond per pixel. The telescope and its camera will operate in the enclosure built by ESO at the ESO VLT Paranal Observatory in Chile. The VST telescope is presently in the disassembly and packing phase, while the active optics is under the final verification and test phase, carried out jointly with ESO. It is planned to ship the telescope from Italy and reassembly it in Chile since April 2007 in order to start the scientific operations possibly within the year 2007 or early in the year 2008.

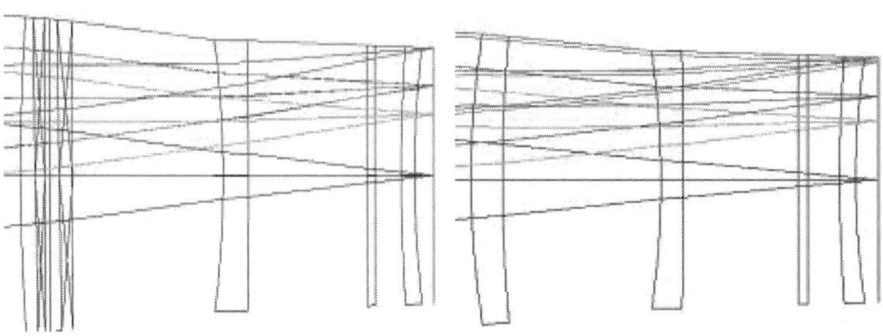

Fig. 2. The Zemax^{TM} ray-tracing optical scheme of the VST telescope wide-field corrector with atmospheric dispersion compensation (left) and for the standard wide-field corrector (right).

The scientific operations require the knowledge of the observed Point Spread Function of the telescope over the field of view. This can be done by mapping the Point Spread Function images evaluated from a suitable, but possibly small number of sparse, isolated single stars observed in the field of view. In order to study the implementation of such mapping, which is required to characterize the telescope performance and to carry out the scientific data processing, a detailed simulation was done by means of the convolution of the numerical models of the telescope Point Spread Function, computed using a ray-tracing code, and the numerical model of the atmospheric seeing of the ESO Paranal VLT Observatory, computed by means of a simulated atmosphere. The results of the simulations mapped over the field of view allow one to characterize the system performance, while the procedure can be used on observed astronomical data in order to check the actual system performance. One important application of the procedure consists in using the simulated characterization maps, calibrated by a few observed astronomical objects, in order to estimate the telescope Point Spread Function in regions of the field of view where there are insufficient observed data. This is the only viable solution to carry out the scientific data processing in such regions.

2. Simulation of the Atmospheric Seeing

The atmospheric seeing observed at ESO Paranal observatory shows a median value of 0.65 arcsecond with marginal values as small as 0.30 arcsecond. In this work the numerical simulation of the seeing is implemented by means of two-dimensional gaussian functions of standard deviation of 0.65 and 0.30 arcsecond respectively.

The accuracy of the gaussian model of the seeing was checked with respect to the sum of a large number of atmospheric speckles simulated by means of the McGlamery Fast-Fourier- Transform-based algorithm [3]. The results reported in Fig.3 show that the gaussian model approximates the physical model of the seeing with small errors in the percent range in the wings.

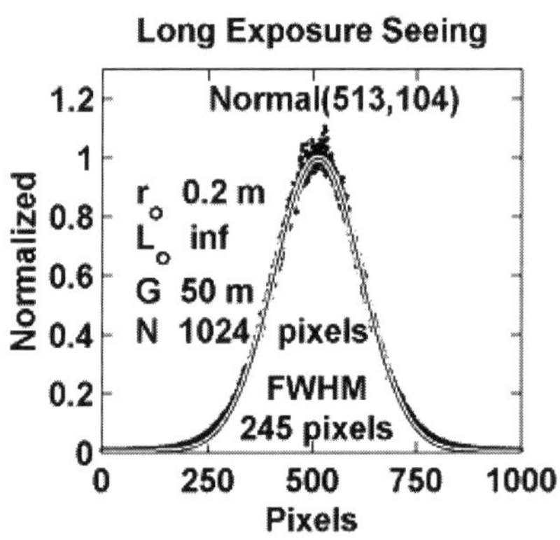

Fig. 3. Numerical model of the seeing obtained by summing simulated atmospheric speckles and the gaussian model approximation. The turbulence parameters correspond to a seeing of about 0.56 arcsecond in the visual range.

3. Simulation of the Observed Point Spread Function

The observed Point Spread Function of the VST telescope is simulated by means of the Fast-Fourier-Transform-based convolution of the Zemax^{TM} Fast-Fourier-transform-based Point Spread Function of the optics, computed on a grid of (9 x 9) positions equispaced in the (1 x 1) square arcdegrees field of view of the telescope, with the atmospheric seeing simulated by a two-dimensional gaussian. The 81 Point Spread Function images, computed by the Zemax^{TM} code, are located on the field of view of the telescope as shown in Fig.4. The Point Spread Function images are extracted from the original Zemax^{TM} (512 x 512) pixels support and rebinned to an (128 x 128) pixels support with a scale of 0.21 arcsecond per pixel, equal to the scale of the OmegaCAM VST CCD mosaic camera. The origin option selected for the Zemax^{TM} ray-tracing code is the chief ray of the optics the telescope. i.e. the ray passing through the center of the telescope pupil. The calculations are done for a polychromatic light in the wavelength range from 320 nm to 1014 nm for the standard wide-field corrector and from 365 nm to 1014 nm for the wide-field corrector with atmospheric dispersion compensation. The images are distorted by the optical aberrations proportionally to the distance (angular offset) from the optical axis and show an axial symmetry with respect to the telescope optical axis.

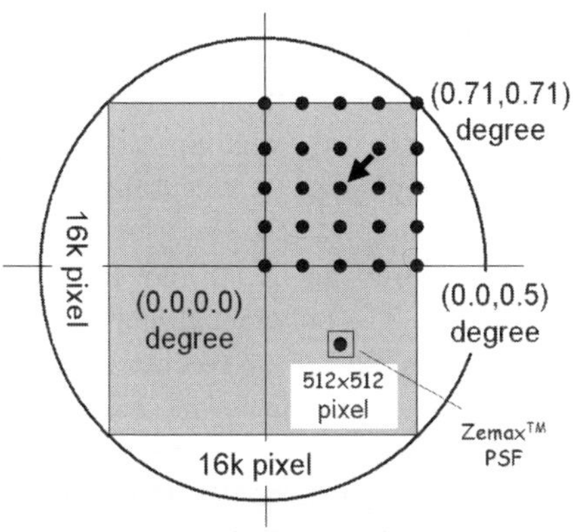

Fig. 4. The grid of Zemax^{TM} Point Spread Function arrays over the telescope field of view. The arrow marks the (0.3,0.3) arcdegrees offset data set used in this paper to show the results of the simulations.

The convolution to the seeing acts as a spatial low pass filter and the observed Point Spread Functions approximate the seeing, as it was expected for a seeing limited telescope design. One sample sequence of data, seeing, and convolved data is shown in Fig.5.

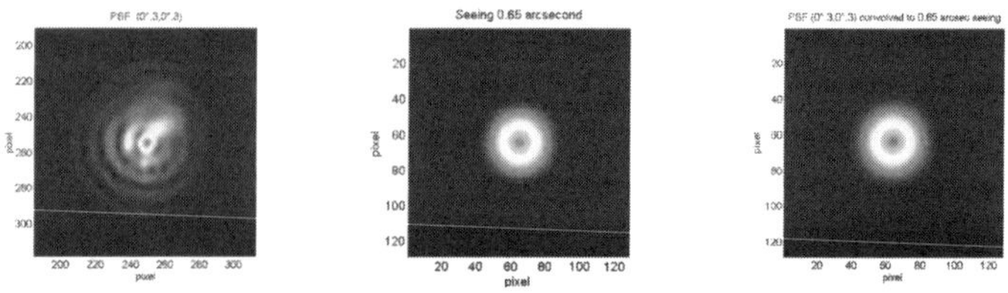

Fig. 5. One sample simulation of the VST telescope observed Point Spread Function. Left: the Zemax^{TM} Fast-Fourier-transform-based Point Spread Function of the optical configuration with atmospheric dispersion compensation, computed at the angular offset of (0.3,0.3) arcdegrees from the telescope axis. Center: the atmospheric seeing approximated by a gaussian of standard deviation of 0.65 arcsecond. Right: the observed Point Spread Function.

4. Mapping the Point Spread Function over the Field of View

The characterization of the observed Point Spread Function is done by mapping over the field of view of the telescope the selected figures of merit. The figures of merit used in this work are the maximum value of the observed Point Spread Function and the vector of the displacement of its centroid with respect to the position of the optical chief ray of the Zemax^{TM} ray-tracing scheme. Since in this work the Point Spread Function samples the field of view over a (9 x 9) grid, then an appropriate two-dimensional interpolation is required before mapping. The interpolation is implemented by a standard smoothed bi-cubic spline algorithm. Since the support of the telescope field of view is large, i.e. (16K x 16K) pixels, and the support of the data is not small, i.e. (512 x 512) pixels, it is necessary to compress the data of the observed Point Spread Functions in order to limit the computing time and memory space to acceptable values. Two methods were used in this work in order to compress the data for the interpolation, one based on the polynomial-modulated gaussian fit [4] and another based on the wavelet decomposition and compression. The polynomial-modulated gaussian fit allows the observed Point Spread Functions data to be reconstructed with about one percent rms fit error with typically 20 coefficients. The wavelet compression allows the data to be reconstructed with less than one percent rms error with typically 200 coefficients and a computing time much smaller than the time required by the fit. The results obtained for the (0.3,0.3) arcdegrees offset are shown in Fig.6.

5. Results

The procedure described above was used to compute the maximum value and the centroid of the observed Point Spread Functions gridded over the field of view of the VST telescope. The calculations used two values of the atmospheric seeing, the ESO Paranal median of 0.65 arcsecond and the best value of 0.30 arcsecond. The sampling of the field of view used a (9 x 9) grid over the (1 x 1) square arcdegrees area. The data compression method selected for the mapping over the field of view used a wavelet decomposition of type Matlab^{TM} SYM4 with 200 coefficients. The

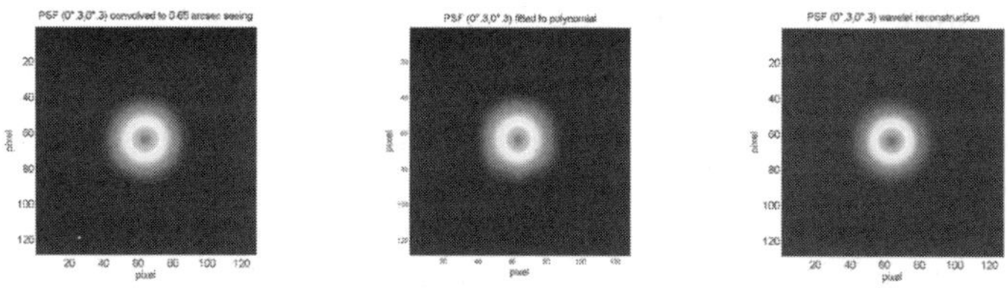

Fig. 6. The VST telescope observed Point Spread Function simulated at the angular offset of (0.3,0.3) arcdegrees from the telescope axis (left) with the polynomial-modulated gaussian fit reconstruction with 20 coefficients (center) and the wavelet reconstruction with 200 coefficients(right)

wavelet decomposition was used because it is faster and more robust than the fit modeling, while the computing time spent for interpolating the larger number of coefficients for the mapping is not very important due to the high speed of the smoothed bi-cubic spline interpolation algorithms available. All the calculations were done using a Matlab^{TM} and a standard C++ platform. Sample results are shown in Table 1 and Fig.7.

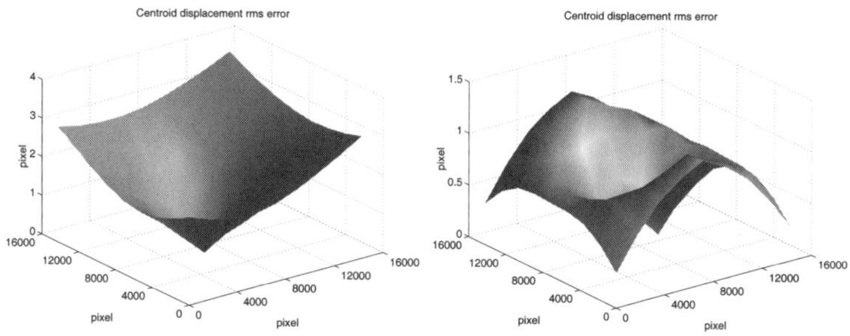

Fig. 7. The rms errors of the centroid displacement of the simulated observed Point Spread Function of the VST telescope mapped over the field of view of (16K x 16K) pixels of the camera for a seeing of 0.65 arcsecond. Left: standard two lens wide-field corrector. Right: wide-field corrector with atmospheric dispersion compensation at zenith angle of 55 arcdegrees.

Table 1. Sample figures of merit of the simulated observed Point Spread Function of the VST telescope as a function of the x-axis angular offset. Seeing 0.65 arcsecond. Field of view of the individual Point Spread Function images (2.52 x 2.52) square arcseconds on a support rebinned to (128 x 128) pixels.

Point Spread Function position (arcdeegree off axis on x-axis)	Maximum value ($\times 10^{-3}$)	Modulus of centroid displacement (pixel of 0.21 arcsecond/pixel)
0.0	2.114	0.36
0.2	2.111	0.63
0.3	2.105	1.00
0.4	2.072	1.24
0.5	2.034	1.26

6. Conclusions

The results obtained by means of numerical simulations confirm the high quality of the design of the VST telescope optics, which meets the specs of being seeing limited over the field of view of (1 x 1) square arcdegree for the ESO Paranal median seeing. One further finding is that the one percent error scale size of the observed Point Spread Function approximates the grid size of the field of view used in this work, i.e.about 10 percent of the detector area. This allows one to use easily the maps of the compressed, simulated Point Spread Functions for mapping the sky regions where there are only a few observed objects suitable for the calibration.

The simulated results presented in this work need to be confirmed by means of true astronomical observations. However, the simulated error maps can be used by now for the implementation of the scientific data processing procedures planned for the VST telescope.

Acknowledgments

This work has been carried out under COFIN contract number 2004020323-002 funded by MIUR PRIN 2004 national program..

References

[1] Capaccioli M., Mancini D., Sedmak G., *VST: a dedicated wide field imaging facility at Paranal*, Survey and other telescope technologies and discoveries, A.Tyson and S.Wolff Eds., SPIE, **4836-09**, 43-52, 2002.

[2] Kuijken K., Bender R., Cappellaro E., Muschielok B., Baruffolo A., et al., *OmegaCAM: the 16K x 16K CCD camera for the VLT survey telescope*, The ESO Messenger, **110**, 15, 2002. ,

[3] Sedmak G., *Implementation of Fast-Fourier-transform based simulations of extra-large atmospheric phase and scintillation screens*, Applied Optics, **43**, 21, 2004.

[4] Alard C., Lupton R.H., *A method for optimal image subtraction*, Ap.J **503**, 325, 1998.

PART B

Biology, Biochemistry and Bioinformatics

FROM GENOMES TO PROTEIN MODELS AND BACK*

ANNA TRAMONTANO†

Department of Biochemical Sciences, University "La Sapienza",
P.le A. Moro 5, Rome, 00185, Italy
E-mail: anna.tramontano@uniroma1.it

ALEJANDRO GIORGETTI

Department of Biochemical Sciences, University "La Sapienza",
P.le A. Moro 5, Rome, 00185, Italy

MASSIMILIANO ORSINI

Bioinformatics, CRS4, Parco Scientifico e Tecnologico Polaris,
Edificio 3, Pula (CA), 09010, Italy

DOMENICO RAIMONDO

Department of Biochemical Sciences, University "La Sapienza",
P.le A. Moro 5, Rome, 00185, Italy

The alternative splicing mechanism allows genes to generate more than one product. When the splicing events occur within protein coding regions they can modify the biological function of the protein. Alternative splicing has been suggested as one way for explaining the discrepancy between the number of human genes and functional complexity. We analysed the putative structure of the alternatively spliced gene products annotated in the ENCODE pilot project and discovered that many of the potential alternative gene products will be unlikely to produce stable functional proteins.

1. Introduction

1.1. *The ENCODE project*

The ENCODE (ENCyclopedia Of DNA Elements) Project [1] aims to identify all functional elements in the human genome sequence. The pilot phase of the Project is organized as an international consortium of computational and laboratory-based scientists working to develop and apply high-throughput approaches for detecting all sequence elements that confer biological function. In this phase of the project, over 30 labs provided experimental results focused on the origin of replication,

*This work is supported by the BioSapiens Network of Excellence, European. Commission Grant LSHG-CT-2003-503265.

†Work partially supported by Istituto Pasteur Fondazione Cenci Bolognetti.

DNase I hypersensitivity, chromatin immunoprecipitation, promoter function, gene structure, pseudogenes, non-protein-coding RNAs and transcribed RNAs.

One important result of the project has been the availability of a reference set of manually annotated splice variants by the GENCODE consortium [2]. The annotation by the GENCODE consortium is an extension of the manually curated annotation by the Havana team at The Sanger Institute.

1.2. *The BioSapiens consortium*

The objective of the BIOSAPIENS Network of Excellence is to provide a large-scale, concerted effort to annotate genome data by laboratories distributed around Europe, using both informatics tools and input from experimentalists. The Network has created a European Virtual Institute for Genome Annotation, bringing together many of the best laboratories in Europe. The BioSapiens partners were called upon and asked to focus on information, from a protein annotation perspective, on the GENCODE gene set. Special attention was given to the potential aspect of alternative splicing and the putative effect it has on function by altering domain, structure, localisation, post-translational modification etc.

1.3. *Protein structure prediction*

The knowledge of the amino acid sequence of a gene product is still not sufficient to understand its function at the molecular level: the function of a protein is determined by and large by its three-dimensional structure.

In general, and with very rare exceptions, the three dimensional structure of a protein is determined solely by its amino acid sequence, so that the knowledge of the latter should in principle be sufficient to predict the former. But, in practice, things are very different. The deciphering of the code relating protein sequence to protein structure has eluded years of investigations and keeps representing a major problem in modern biology [3]. We can predict the structure of a protein from its amino acid sequence in some cases, not in all cases.

In a famous pioneer experiment, Anfinsen [4] showed that a protein (in its particular case Ribonuclease S) once denatured – that is unfolded – in vitro, would go back to the native conformation "by itself" when the denaturing agents were removed from the test tube. This implies that the information about the three-dimensional structure of a protein is contained in its amino acid sequence, but also that the native three-dimensional structure is a well defined state, and this cannot be but the state at minimum energy among those achievable by the protein.

Both enthalpic and entropic forces are responsible for the folding process: a protein has to reach a minimum free energy state through a kinetically practicable path. A protein has a polar backbone and both hydrophobic and hydrophilic side chain groups. Consequently, when it is in a random coil state in a polar solvent, the main chain polar atoms will form hydrogen bonds with the solvent, but the hydrophobic side chains will not, and this will cause the polar solvent to get "ordered" around them, just as an oil drop in water, causing an unfavourable loss of entropy.

In a folded protein most hydrophobic side chains will be shielded from the solvent, thus gaining entropy, but some of the polar groups will be buried too and this is energetically unfavourable. To reduce the enthalpy loss, the hydrogen bond potential of the polar atoms has to be saturated.

This implies that the stability of a protein is due to the difference between large entropic and enthalpic terms. It is unfortunate (from a computational point of view) that the difference between these large numbers is just a small number of the order of a few Kcal/mol. In other words, the stability of a protein at room temperature is marginal. Therefore, if we want to distinguish between the native protein structure and all its other possible conformations, we need to calculate entropic and enthalpic forces with sufficiently high precision to get an accurate estimate of the small residual difference.

However, our calculations of the energy terms involved are necessarily approximate due to the complexity of the system and they are affected by an error that, as of today, is too high to allow us to discriminate between the native structure and other close and reasonable conformations. Nevertheless, energy based methods are often used to explore the conformation of proteins around their native structure and have demonstrated their usefulness especially in helping the resolution of structures by experimental methods [5–7].

Another problem is, of course, that even if we could calculate the energy with infinite precision, the exploration of all possible conformations would be unfeasible. Not even the protein itself can do it, as enunciated by the famous "Levinthal paradox" [8], of which there are many descriptions but that we will state as follows:

Suppose that we have a protein of 100 amino acid (proteins are on average larger than this), that its structure is only determined by the two main chain angles phi and psi (in reality there are also side chain angles and the omega angle is not always perfectly planar), that these two angles could only assume two values each (in fact they can assume more!), then there would be 4^{100} possible conformations equal to more than 10^{60} possibilities. Even if the exploration of one conformation would only take one femtosecond, then it would take about 10^{40} years to fold a protein! Proteins obviously do not explore all possible conformations to find their native structure and there are recent theories that can solve the Levinthal's paradox, but the number of conformations involved is anyway too large to be treated exhaustively even with modern computers.

The discussion so far implies that we cannot, as of today and probably for some years to come, simulate the folding process of a protein. The only reason we can continue to discuss the problem of structure prediction is indeed that proteins are a product of evolution. The evolutionary mechanisms imply that proteins mostly evolved via small sequence variation, usually single amino acid substitutions, insertions and deletions. Therefore the sequences of proteins that are "sufficiently" closely evolutionary related in evolution preserve detectable similarities. An unfolded polypeptide is risky for the cell, because it exposes hydrophobic groups that favour aggregation with other proteins. Consequently, we can assume that each of the evolutionary steps has produced a structure compatible with the function of the protein. Note that function is usually brought about by few key amino acids, but is dependent on their correct positioning in the active site, that is on the correct folding of the polypeptide.

All the above implies that evolutionary related proteins not only have similar sequences but also similar structures [9]. In other words, if two proteins have a sequence sufficiently similar to guarantee that they are evolutionary related, we might be reasonably certain that they also have a similar structure.

We will not discuss what "sufficiently similar" means, and just mention that there are statistical methods to calculate the likelihood that a given sequence similarity is not due to chance alone. As a rule of thumb, if two sequences of length above 100 amino acids have a sequence identity above 30%, it is highly likely that they are evolutionary related and, consequently, that they have a similar structure.

1.4. *The prediction of the structures of the GENCODE gene products*

This observation forms the basis of a protein structure prediction method called "comparative modelling" or "modelling by homology" that we applied to the GENCODE gene product set.

We analysed the putative structure of all the gene products of known structure or for which a reliable three-dimensional model could be built and analysed the results asking the question of whether these products could give raise to a functional element [10].

2. Results

There are 2,608 annotated transcripts for 487 distinct loci in GENCODE, 1,097 transcripts from 434 loci are predicted to be protein coding with, on average 2.5 protein coding variants per locus (see Fig. 1). A large proportion of the splice isoforms in the data set have identical protein sequences. These coding sequence-identical variants are alternatively spliced in the 5 and 3 untranslated regions [10].

It is important to know which of these genes can produce functional proteins and which, instead, could be misfolded and quickly removed by the cell degradation machinery. If the proteins are translated and fold properly, one could try and predict which functional role they might play in the cell.

Structures are known for proteins from 42 different loci (almost 10% of the total). More than half of the sequences in the set could be modelled by homology and, in 85 cases, we could map the changes resulting from splicing events at the protein level onto the structures or the models.

The most striking results of our analysis is that, for 49 of these 85 alternative transcripts, the resulting protein structure is likely to be substantially altered in relation to that of the principal sequence [10].

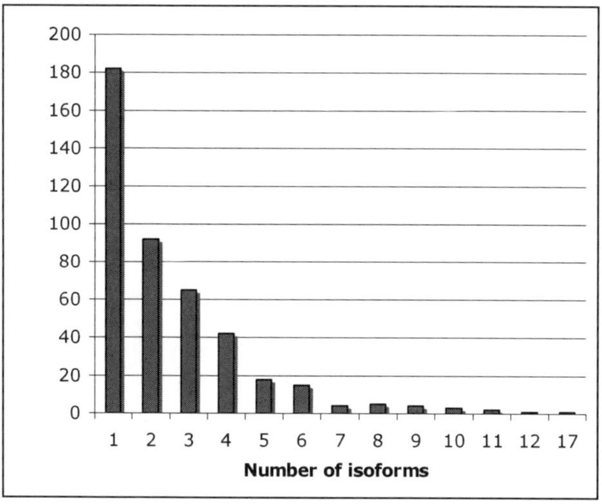

Fig. 1. Distribution of isoforms in the ENCODE gene set. The number of isoforms per locus varies considerably. There are 182 loci that have only one isoform, while there is one locus that has 17 different splice variants (RP1-309K20.2).

Here we illustrate this finding with two examples. The gene product of locus RP4-61404.1 (ITGB4BP) can be modelled since its sequence share 75% sequence identity with a yeast protein of know structure (PDB code 1G62) [11, 12] Isoform 005 of this locus has an internal substitution of 85 residues (Figure 2) which disrupts the interactions of a very crucial and central part of the structure.

There are also cases where variants can be expected to be functional, as is the case for IL-4, locus AC004039.4. This gene codes for a cytokine. The GENCODE set contains a single alternative splice variant (isoform 002, IL-4d2) that has a deletion of the second exon, a total of 16 residues. The structure of human IL-4 is known [13]. It is a four-helix bundle with long connecting loops between helices held together by three cysteine bridges. The region coded for by the missing exon is in the first of the long loops (see Fig. 3), and the resulting new structure is compatible with a stable functional form of the proteins, if some local structural rearrangements take place.

3. Discussion

Proteins generally evolve by stepwise single base-pair mutations. In contrast alternative splicing usually involves large changes and, as shown here, alternative splicing is not necessarily a mechanism for subtle changes in structure and function. The rearrangements that we observed in many of the alternative isoforms are likely to disrupt the structure and function of the encoded protein. It seems therefore that alternative splicing can lead to many evolutionary dead ends We cannot exclude

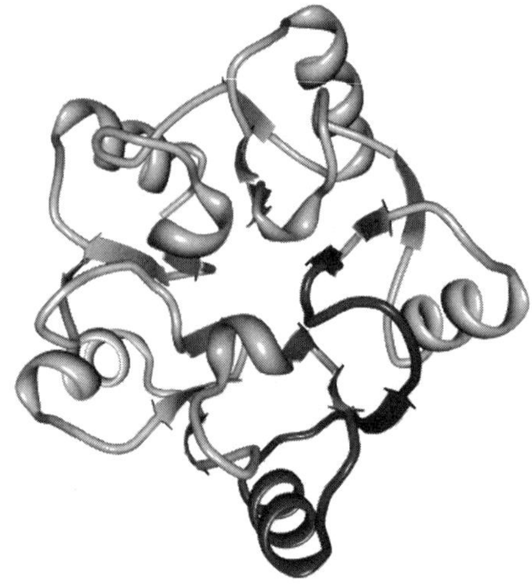

Fig. 2. The effect of alternative splicing mapped on a model for ITGB4BP. The darker region corresponds to the alternate exonic sequence of isoform 5.

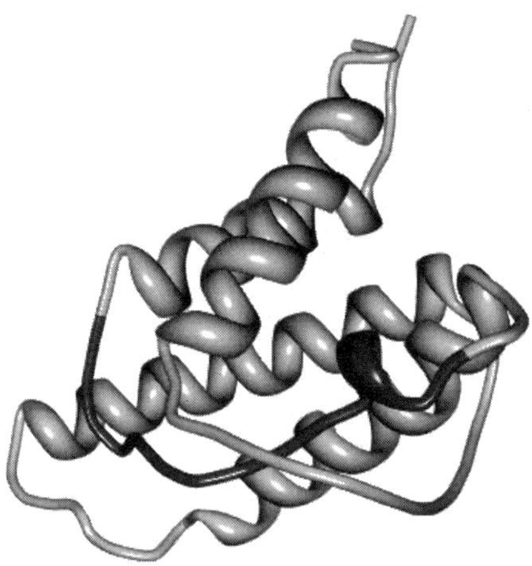

Fig. 3. The effect of alternative splicing mapped on the structure of IL-4. The darker region corresponds to the missing exon in isoform 2.

the possibility that the expression of alternative transcripts has implications for the control of gene expression. It is also likely that some alternative isoforms may have developed a function that is useful for the cell by folding in a different way or that they can do so by interacting with other gene products.

It remains, though, that functional splicing isoforms are not very common. It is possible that many splice variants are expressed only as a result of some disease event.

Our results leave open two important questions, namely which is the role of the putative non-functional variants and, equally importantly, how can we reconcile the number of functions of a human organism with its apparently very low coding content.

Acknowledgments

This work was supported by the BioSapiens project funded by the European Commission within its FP6 Programme, under the thematic area "Life sciences, genomics and biotechnology for health," contract number LSHG-CT-2003-503265.

References

[1] The ENCODE Project Consortium *Science* **306**, 636 (2004)
[2] J. Harrow, F. Denoeud, A. Frankish, A. Reymond, C-K Chen, J. Chrast, J. Lagarde, JGR Gilbert, R. Storey, D. Swarbreck, et al . *Genome Biol* **7**, S4(2006)
[3] A. Tramontano Protein Structure Prediction Wiley Inc. Weinheim, Germany (2006)
[4] C. B. Anfinsen Principles that Govern the Folding of Protein Chains. *Science* **181**, 223 (1973)
[5] M. Levitt. The birth of computational structural biology, *Nature Struct. Biol.*, **8**, 392 (2001).
[6] W.F. van Gunsteren and H.J.C. Berendsen In: "Molecular Dynamics and Protein Structure", J. Hermans ed., Polycrystal Book Service (1985)
[7] W.F. van Gunsteren, R. Boelens, R. Kaptein, R.M. Scheek and E.R.P. Zuiderwe In: "Molecular Dynamics and Protein Structure", J. Hermans ed., Polycrystal Book Service, (1985)
[8] C. Levinthal, in Mossbauer Spectroscopy in Biological Systems, P. Debrunner, J.C.M. Tsibris and E. Munck eds. (1969)
[9] C. Chothia, AM. Lesk. *EMBO J.* **5**, 823 (1986)
[10] M. L. Tress et al.. *Proc Nat Acad Sci USA* **104**, 5495 (2007)
[11] H.M. Berman, J. Westbrook, Z. Feng, G. Gilliland, T.N. Bhat, H. Weissig, I.N. Shindyalov, P.E. Bourne. *Nucleic Acids Research,* **28** pp. 235 (2000)
[12] C.M. Groft, R. Beckmann, A. Sali, S.K. Burley. *Nat.Struct.Biol.* **7,** 1156 (2000)
[13] L.J. Smith, C. Redfield, J. Boyd, G.M. Lawrence, R.G. Edwards, R.A. Smith, C.M. Dobson. *J.Mol.Biol.* **224**, 899 (1992)

EXPLORING BIOMOLECULAR RECOGNITION BY MODELING AND SIMULATION

REBECCA WADE

EML Research, Schloss-Wolfsbrunnenweg 33, 69118 Heidelberg, Germany
E-mail: rebecca.wade@eml-r.villa-bosch.de

Biomolecular recognition is complex. The balance between the different molecular properties that contribute to molecular recognition, such as shape, electrostatics, dynamics and entropy, varies from case to case. This, along with the extent of experimental characterization, influences the choice of appropriate computational approaches to study biomolecular interactions. I will present computational studies in which we aim to make concerted use of bioinformatics, biochemical network modeling and molecular simulation techniques to study protein-protein and protein-small molecule interactions and to facilitate computer-aided drug design.

Keywords: Biomolecular Interactions: simulation, modeling

Biomolecular interactions are essential to biological function. An understanding of the determinants of biomolecular recognition and binding specificity, thermodynamics and kinetics is not only necessary for studying the function of biomolecules but also for designing molecules for pharmaceutical and biotechnology purposes. Given the diversity of biomolecules, a battery of computational techniques is necessary for modelling and simulating biomolecular interactions. Here, I outline some of the ways in which computational approaches can be used to study different aspects of biomolecular recognition, with examples from our own studies.

Consider the process of a macromolecule, e.g. a protein, binding to a small molecule or another macromolecule in an aqueous environment. In a first step, the two molecules need to be transported or diffuse into each others proximity to form an encounter complex. This step will be particularly influenced by the electrostatic interactions between the molecules [1]. In a second step, the molecules may adapt their conformation (through induced fit) and short-range hydrogen bonds and hydrophobic interactions will be formed resulting in a fully bound complex. The time and spatial scales of these two steps are generally quite different and therefore different computational techniques and a multiscale framework are required to simulate them.

The first step, the diffusional association, can be studied by Brownian dynamics (BD) simulation [2]. In these simulations, the solvent is treated as a continuum exerting frictional and stochastic effects on the diffusional motion of the two solutes. The solute molecules may be treated as atomic detail rigid bodies or at a less detailed level by coarse graining. The electrostatic interactions between solutes are typically described by the Poisson-Boltzmann equation. One application of BD simulations is to compute bimolecular diffusional association rate constants. Such calculations have shown how electrostatic interactions between the biomolecules can not only enhance the association rate constants, e.g. in diffusion-controlled enzymes1, but also depress the association rate constants [3].

The second step generally requires a more detailed model. The individual water molecules make important contributions to binding. Water is displaced upon binding but can also remain trapped at the interface and mediate intermolecular interactions. Thus water molecules can affect binding specificity and have entropic and enthalpic effects. The protein motions are important to permit favourable short-range interactions, van der Waals and hydrogen-bonding, as well as Coulombic interactions, to be made. Classical atomic-detail molecular dynamics (MD) techniques can be used for simulating this step. With standard MD, the motions of a moderately sized protein can be simulated for times on the order of 10 ns. This is not adequate to cover all the types of motion that govern molecular recognition and binding, e.g. protein loop motions may gate ligand binding and occur on timescales up to ms. Therefore enhanced sampling techniques are often required in using MD techniques to study receptor-ligand binding and unbinding processes. One example is the use of Random Acceleration Molecular Dynamics (RAMD) to explore routes by which substrates can access and products can egress from the buried active site of cytochrome P450 enzymes [4, 5]. These simulations show how the wide variations in the mechanism of ligand passage to and from the active site are related to the enzyme specificity, suggesting that not only the interactions at the binding site but also those influencing the binding/unbinding process contribute to specificity.

Simulations of the binding process can give insights into binding mechanisms and allow computation of thermodynamic and kinetic quantities, but are computationally demanding. The findings of such studies can provide a basis for developing computationally efficient tools aimed at studying binding specificity and molecular design. For example, electrostatic interactions between molecules are important for binding and catalytic properties. Therefore a comparison of molecular electrostatic potentials can provide information about the functional properties of the molecules. We have developed the PIPSA (Protein Interaction Property Similarity Analysis) approach [6] (available at `http://projects.villa-bosch.de/mcm`) as a systematic and objective way to compare molecular interaction fields, such as molecular electrostatic potential, for a set of proteins with similar structures. Despite sharing the same three-dimensional structural fold, structurally related proteins from a given protein family may show significant variation in their molecular recognition properties, and this can be examined by comparing their molecular electrostatic potentials. The value of the potential is calculated at points on a grid around the superimposed proteins. The similarity/dissimilarity between the electrostatic potentials of the proteins can be quantified by computing similarity/distance indices with PIPSA [6]. PIPSA has been used to assist functional classification of large protein families, e.g. in comparative analysis of over a thousand proteins of the ubiquitination and related pathways [7, 8], see `http://www.ubiquitin-resource.org`. A PIPSA study of electrostatic potentials in the region of the active sites of orthologous enzymes shows how the electrostatic potential can be related to kinetic parameters of the enzymes and therefore used to validate and estimate enzymatic kinetic parameters, e.g. in the context of mathematical modelling of biochemical metabolic pathways [9].

Comparison of macromolecular molecular interaction fields [10] can also aid the design of target specific ligands such as drugs that e.g. inhibit a viral protein but not the structurally similar human homologue. More detailed structure-based ligand design requires modelling of receptor-ligand complexes and prediction of their binding affinities. Comparative approaches can again be applied, not only to interaction fields but also to receptor-ligand energies. This can be done with COMBINE (Comparative Binding Energy) analysis which permits the target-specific derivation of 3D-quantitative structure-activity relationships (QSARs) for series of protein-ligand complexes (for review, see ref.11). The COMBINE models can highlight the amino acid residue interactions (electrostatic and van der Waals) most important for variations in receptor-ligand binding affinity in a set of related complexes and this information can be exploited in molecular design for ligand optimization [?].

The complexity of biomolecular recognition means that many complementary approaches are required to model and simulate biomolecular interactions. This article provides a brief overview of some of our work on this topic; it necessarily covers only a small subset of the approaches being used to study biomolecular recognition.

Acknowledgments

The contributions of all present and past group members and the financial support of the group by the Klaus Tschira Foundation are gratefully acknowledged.

References

[1] Wade R.C., Gabdoulline R.R., Ldemann S.K., Lounnas V. Electrostatic steering and ionic tethering in enzyme-ligand binding: insights from simulations. Proc Natl Acad Sci USA (1998) 95, 5942-5949.

[2] Madura, J. D., Briggs, J. M., Wade, R. C. & Gabdoulline, R. R. in Encyclopedia of Computational Chemistry, eds. von Rague Schleyer, P., Allinger, N. L., Clark, T., Gasteiger, J., Kollman, P. A. & Schaefer, H. F. (Wiley, Chichester, U.K.), (1998) 1, 141-154.

[3] Gabdoulline, R.R., Kummer, U., Olsen, L.F. and R.C. Wade Concerted Simulations Reveal How Peroxidase Compound III Formation Results in Cellular Oscillations Biophys. J. (2003), 85, 1421-1428.

[4] P. J. Winn, S. K. Ldemann, R. Gauges, V. Lounnas and R. C. Wade Comparison of the dynamics of substrate access channels in three cytochrome P450s reveals different

opening mechanisms and a new functional role for a buried arginine Proc. Natl. Acad. Sci. USA, (2002), 99, 5361-5366.

[5] K. Schleinkofer, Sudarko, P. J. Winn, S. K. Ldemann and R. C. Wade. Do mammalian cytochrome P450s show multiple ligand access pathways and ligand channelling? EMBO Reports (2005) 6, 584-589.

[6] Wade,R.C., Gabdoulline, R.R. and De Rienzo, F. Protein Interaction Property Similarity Analysis Intl. J. Quant. Chem. (2001) 83, 122-127.

[7] P. J. Winn, T. L. Religa, J. D. Battey, A. Banerjee and R. C. Wade Determinants of Functionality in the Ubiquitin Conjugating Enzyme Family. Structure (2004) 12, 1563-74.

[8] P. J. Winn, M. Zahran, J. N.D. Battey, Y. Zhou, R. C. Wade, A. Banerjee Structural and electrostatic properties of ubiquitination and related pathways. Frontiers in Bioscience (2007) 12, 3419-3430.

[9] M. Stein, R.R. Gabdoulline, R.C. Wade Bridging from molecular simulation to biochemical networks Curr. Op. Struct. Biol. (2007) in press.

[10] R. C. Wade Calculation and Application of Molecular Interaction Fields. In 'Molecular Interaction Fields. Applications in Drug Discovery and ADME Prediction', Ed. Cruciani, G., Wiley-VCH, Weinheim. (ISBN 3-527-31087-8) (2005) Ch. 2, pp27-42.

[11] R. C. Wade, S. Henrich and T. Wang Using 3D protein structures to derive 3D-QSARs Drug Discovery Today: Technologies (2004) 1(3), 241-246.

[12] K. Schleinkofer, U. Wiedemann, L. Otte, T. Wang, G. Krause, H. Oschkinat and R. C. Wade Comparative Structural and Energetic Analysis of WW Domain-Peptide Interactions J. Mol. Biol. (2004) 344, 865-881.

FROM ALLERGEN BACK TO ANTIGEN: A RATIONAL APPROACH TO NEW FORMS OF IMMUNOTHERAPY

PAOLO COLOMBO*, ANTONINO TRAPANI, DOMENICO GERACI

Istituto di Biomedicina e di Immunologia Molecolare "Alberto Monroy"
Consiglio Nazionale delle Ricerche, Via Ugo La Malfa 153, 90146 Palermo, Italy
** E-mail: paolo.colombo@ibim.cnr.it*

MASSIMILIANO GOLINO, FABRIZIO GIANGUZZA

Dipartimento di Biologia Cellulare e dello Sviluppo,Parco D'Orleans, Palermo, Italy

ANGELA BONURA

Istituto di Biomedicina e di Immunologia Molecolare "Alberto Monroy"
Consiglio Nazionale delle Ricerche, Via Ugo La Malfa 153, 90146 Palermo, Italy

Mapping an epitope on a protein by gene fragmentation and/or point mutations is often expensive and time consuming. Analysis of a 3D model can be utilized to detect the amino acids residues which are exposed to the solvent surface and thus represent potential epitope residues. Parj1 and Parj2 are the two major allergens of the *Parietaria judaica* pollen belonging to the Lipid Transfer Protein family. Using their three-dimensional structures as a guide, a head to tail dimer expressing disulphide bond variants of the major allergens was generated by means of DNA recombinant technology. The hybrid was expressed in *E.coli* and its immunological activity studied *in vivo* and *in vitro*. Our results demonstrate that a hybrid polypeptide expressing disulphide bond variants of the major allergens of the Parietaria pollen displayed reduced allergenicity and enhanced T cell reactivity for induction of protective antibodies able to block human IgE induced during the natural course of sensitization against the *Parietaria* pollen.

Keywords: Immunotherapy; Modeling Allergens; Mutational Analysis

1. Introduction

Allergy is an immunological disorder induced by inhalation of pollen grains and leading to hypersensitivity of the respiratory tract including several manifestations like allergic rhinitis, rhino-conjunctivitis, urticaria and asthma [1]. The allergic reaction is directed against various environmental proteins (known as allergens) innocuous for the majority of the population. The symptoms of the IgE-mediated allergic reactions can be transiently ameliorated by pharmacological treatments which do not interfere with the ethiogenesis and progression of the disease. The only treatment able to modify the natural outcome of the disease restoring a normal immunity against allergens is specific immunotherapy (SIT) [2]. So far, immunotherapy is performed by s.c. injection or mucosal administration of crude extracts from natural sources without taking care of the individual sensitisation profile of the patients and, possibly, inducing new IgE specificity [3]. The use of recombinant purified allergens can overcome many of these problems and present the advantage to make possible to produce genetically modified molecules [4]. A new strategy may be represented by the use of high doses of hypoallergenic recombinant molecules with no risk of side effect. *Parietaria judaica* (Pj) pollen allergens are proteins from dicotyledonous weeds of the Urticaceae family. Immunologically, this family represents a relevant species since its pollen is one of the main outdoor sources of allergens in the Mediterranean area [5]. In particular, almost 30% of all the allergic subjects in the Southern part of Italy present a Skin Prick Test (SPT) reactivity towards the *Parietaria* pollen and more than 50% of these subjects have experienced bronchial asthma with high levels of bronchial hyper-responsiveness [6]. The prevalence of this weed as a major cause of allergic disease in the Mediterranean countries underlines the importance of designing new and improved formulations of immune-therapeutics to ameliorate *Parietaria*-induced allergic disorders.

The composition of the allergenic extracts of the Pj pollen has been studied in details and by molecular cloning, the two major allergens (Parj1 and Parj2) have been sequenced and characterized [7–9]. Parj1 and Parj2 allergens are two small polypeptides of 14.400 and 11.344 Da respectively, with a few common features rising from their common evolutionary origin [10]. Sequence analysis showed a high level of homology with a family of plant proteins named nsLTP for their capability of transporting lipids through membrane *in vitro*. 3D modelling by homology [11] and enzymatic digestion [12] have shown that both allergens attain a three-dimensional structure consistent with that of the ns-LTP composed of four α-helices and one β-sheet stabilized by four disulphide bridges according to the invariant order Cys4-Cys52, Cys14-Cys29, Cys30-Cys75, Cys50-Cys91. In a previously published paper by our group, we described the identification of an IgE binding region in the first 30 aa of the major *Parietaria judaica* pollen allergens. In particular, either deletion or Serine substitutions of the Cysteine residues at position 14 and 29 suggested that these amino acids were essential for the IgE binding. In the same way, dot blot assays in the presence of 2-β-mercaptoethanol showed that the reduction of the disulphide bonds to sulphydryl groups caused loss of IgE binding. Furthermore, to better define the structure of the epitope, we performed a site-specific mutational analysis of AA residues laying between Cysteine 14 and Cysteine 29 that allowed us to define that the K21, K23, E24 and K27 amino acids are essential for the IgE antibody recognition [11]. For these reasons, we decided to target disulfide bridges developing a family of full-length three-dimensional mutants of the rParj1 allergen. *In vivo* and *in vitro* analysis showed that IgE binding recognition is dependent on the three-dimensional structure of rParj1 [13]. This structural analysis represents the rational approach to this study. In this report, we describe the design of a head to tail dimer expressing disulphide bond variants of the major *Parietaria* allergens (PjEDcys) generated by site-directed mutagenesis of cysteine residues 4,29 and 30. This derivative was expressed in *E.coli* and its immunological properties studied by independent strategies.

2. Experimental Results

In this report, we describe the design of a head to tail dimer expressing disulphide bond variants of the major *Parietaria* allergens (PjEDcys) generated by site-directed mutagenesis of cysteine residues 4,29 and 30. This derivative was expressed in *E.coli* and its immunological properties studied by independent strategies. Immunoblot experiments performed with sera from Pj allergic patients showed that the PjEDcys derivative displayed a reduced capability of binding human Pj-specific IgE. Furthermore, when basophils from Pj allergic patients were stimulated *in vitro* with the hybrid and the wild-type recombinant allergens, we detected a strong reduction in the PjEDcys allergenicity. A few studies performed with purified allergens have demonstrated that during the course of immunotherapy a new allergen specific immune response may arise with an increase in nasal IFN-γ producing cells [14, 15] and a rise in allergen specific IgG suggesting that the vaccination induces a new immune response coexisting with the allergic response [16]. For these reasons, we decide to test the immunogenicity of the hypoallergenic derivative looking at the pattern of antibody production in a mouse model of sensitisation. Despite of the changes introduced by mutagenesis, mice immunized with the PjEDcys polypeptide mounted a strong IgG response towards the w.t. rParj1 and rParj2 molecules suggesting that an immunization protocol performed with the recombinant hybrid molecule can induce antibodies capable of binding natural allergens preventing the IgE-allergen reaction.

3. Conclusions

These results demonstrated that a rational approach to site-specific mutagenesis can help to generate new immuno-therapeutics which may represent improved tools for the therapy of allergic disorders.

References

[1] D'Amato, G., S. Dal Bo, and S. Bonini.. Ann Allergy 433 (1992)
[2] Bousquet, J., R. Lockey, and H. J. Malling.. J Allergy Clin Immunol 558 (1998)
[3] Moverare, R., L. Elfman, E. Vesterinen, T. Metso, and T. Haahtela. Allergy 423 (2002)
[4] Valenta, R. Nat Rev Immunol 446 (2002)
[5] D'Amato, G.. Clin Exp Allergy 628 (2000)
[6] D'Amato, G., F. T. Spieksma, G. Liccardi, S. Jager, M. Russo, K. Kontou-Fili, H. Nikkels, B. Wuthrich, and S. Bonini. Allergy 567 (1998)
[7] Costa, M. A., P. Colombo, V. Izzo, H. Kennedy, S. Venturella, R. Cocchiara, G. Mistrello, P. Falagiani, and D. Geraci. FEBS Lett 182 (1994)
[8] Duro, G., P. Colombo, M. A. Costa, V. Izzo, R. Porcasi, R. Di Fiore, G. Locorotondo, R. Cocchiara, and D. Geraci.. Int Arch Allergy Immunol 348 (1997)
[9] Duro, G., P. Colombo, M. A. Costa, V. Izzo, R. Porcasi, R. Di Fiore, G. Locorotondo, M.G Mirisola, R. Cocchiara, and D. Geraci, FEBS Lett 295 (1996)

[10] Colombo, P., A. Bonura, M. Costa, V. Izzo, R. Passantino, G. Locorotondo, S. Amoroso, and D. Geraci, Int Arch Allergy Immunol 173 (2003)
[11] Colombo, P., D. Kennedy, T. Ramsdale, M. A. Costa, G. Duro, V. Izzo, S. Salvadori, R. Guerrini, R. Cocchiara, M. G. Mirisola, S. Wood, and D. Geraci. J Immunol 2780 (1998)
[12] Amoresano A., P. P., Duro G., Colombo P., Costa M.A., Izzo V., Lamba D., Geraci D., Biol Chem. 1165 (2003)
[13] Bonura, A., S. Amoroso, G. Locorotondo, G. Di Felice, R. Tinghino, D. Geraci, and P. Colombo. Int Arch Allergy Immunol 126 (2001)
[14] Wachholz, P. A., and S. R. Durham.. Clin Exp Allergy 1171 (2003)
[15] Durham, S. R., S. Ying, V. A. Varney, M. R. Jacobson, R. M. Sudderick, I. S. Mackay, A. B. Kay, and Q. A. Hamid. J Allergy Clin Immunol 1356 (1996)
[16] Ball, T., W. R. Sperr, P. Valent, J. Lidholm, S. Spitzauer, C. Ebner, D. Kraft, and R. Valenta. Eur J Immunol 2026 (1999).

SULFONYLUREAS AND GLINIDES AS NEW PPARγ AGONISTS: VIRTUAL SCREENING AND BIOLOGICAL ASSAYS* †

MARCO SCARSI[1,2@], MICHAEL PODVINEC[1,2‡]

ADRIAN ROTH[1], HUBERT HUG[3,4], SANDER KERSTEN[5], HUGO ALBRECHT[6], TORSTEN SCHWEDE[1,2], URS A. MEYER[1], CHRISTOPH RÜCKER[1]

1- Biozentrum, University of Basel, Klingelbergstr. 50-70, CH-4056 Basel, Switzerland
2- Swiss Institute of Bioinformatics, Klingelbergstr. 50-70, CH-4056 Basel, Switzerland
3- TheraSTrat AG, Gewerbestr. 25, CH-4123 Allschwil, Switzerland
4- DSM Nutritional Products, Human Nutrition and Health, Wurmisweg 576, CH-4303 Kaiseraugst, Switzerland
5- Nutrition, Metabolism and Genomics Group, Wageningen University, PO BOX 8129, 6700 EV Wageningen, The Netherlands
6- BioFocus DPI, Gewerbestr. 16, CH-4123 Allschwil, Switzerland
@ E-mail: marco.scarsi@unibas.ch

This work combines the predictive power of computational drug discovery with experimental validation by means of biological assays. In this way, a new mode of action for type 2 diabetes drugs has been unvealed. Most drugs currently employed in the treatment of type 2 diabetes either target the sulfonylurea receptor stimulating insulin release (sulfonylureas, glinides), or target PPARγ improving insulin resistance (thiazolidinediones). Our work shows that sulfonylureas and glinides bind to PPARγ and exhibit PPARγ agonistic activity. This result was predicted in silico by virtual screening and confirmed in vitro by three biological assays. This dual mode of action of sulfonylureas and glinides may open new perspectives for the molecular pharmacology of antidiabetic drugs, since it provides evidence that drugs can be designed which target both the sulfonylurea receptor and PPARγ. Targeting both receptors could in principle allow to increase pancreatic insulin secretion, as well as to improve insulin resistance.

Keywords: Molecular Pharmacology; Computational Methods

1. Introduction

The peroxisome proliferator activated receptors (PPARs) are ligand-dependent transcription factors involved in the regulation of lipid and glucose metabolism [1]. Three subclasses of PPARs are known, called PPARα, PPARγ, and PPARδ. Of

*This research was funded by the Swiss Commission for Technical Innovation (KTI/CTI, Grant 6570.2 MTS-LS)

‡Marco Scarsi and Michael Podvinec contributed equally to this work

these, PPARγ is mostly expressed in adipose tissue, where it is essential in adipocyte differentiation and controls the storage of fatty acids, increasing triglyceride synthesis and storage within adipocytes. Additionally, there is strong evidence that PPARγ regulates glucose homeostasis [1]. Activation of PPARγ improves insulin resistance, and therefore PPARγ is an established molecular target for the treatment of type 2 diabetes [2].

For PPARγ, several unsaturated fatty acids have been proposed as natural ligands. A few synthetic PPARγ agonists are approved drugs, *e.g.*, rosiglitazone, pioglitazone, which are members of the glitazone (thiazolidinedione) class [1], or are under development as antidiabetics, *e.g.*, tesaglitazar [3] and muraglitazar [4]. All PPARγ agonists in clinical use or development, and in fact most known PPARγ agonists are either thiazolidinediones or carboxylic acids [5]. Many drug therapies targeting PPARγ have their disadvantages, e.g. the liver toxicity of glitazones [6], weight gain, fluid retention, enhanced adipogenesis, and cardiac hypertrophy [7]. Therefore demand is increasing for new PPARγ ligands, and compound classes other than carboxylic acids or thiazolidinediones could be of special interest.

The goal of the present study was to identify new PPARγ agonists among known drugs and biologically active compounds, by combining virtual screening with experimental verification in biological assays. This strategy provides a detailed model of ligand-receptor complexes, together with an experimental confirmation of ligand-receptor binding and the consequent biological activity. We followed a two step approach. First, a virtual screening search on two large databases of drugs and biologically active compounds allowed us to identify a few glinides and sulfonylureas as promising PPARγ ligands, and prompted us to concentrate on these two drug classes, screening several more members thereof. Most of these compounds showed good affinities to PPARγ *in silico*. In a second step, we found that selected sulfonylureas and glinides bind to PPARγ and enhance PPARγ-mediated gene expression *in vitro*.

Sulfonylureas and glinides are hypoglycemic drugs in clinical use for the treatment of type 2 diabetes, by virtue of their insulin secretagogue properties. These compounds bind to the sulfonylurea receptor SUR1 on the membrane of β-cells, triggering the closure of the nearby potassium channel, which in turns leads the β-cell to increase insulin secretion [8]. Our discovery that some insulin secretagogue drugs activate PPARγ has attractive implications for the pharmacological treatment of type 2 diabetes. Moreover, sulfonylureas are a new class of PPARγ agonists.

2. Materials and Methods

Virtual screening database: The TheraSTrat AG inhouse database [9], contains most marketed drugs and many of their metabolites (approximately 8000 compounds). The freely available Chembank database contains about 6000 bioactive compounds [10].

Ligand docking: Each compound was docked into the PPARγ binding site using the AutoDock 3.0.5 software [11]. AutoDock finds several low-energy arrangements ("poses") of a given flexible ligand into a given receptor assumed to be rigid. For each pose, a pK_i value is calculated. The PPARγ 3D structure was obtained from PDB entry 1FM9. This is a 2.1 Å resolution crystal structure of the heterodimer of the human RXRα and PPARγ ligand binding domains, bound to 9*cis*-retinoic acid and farglitazar, respectively, together with coactivator peptides [12]. The PPARγ-farglitazar complex was imported in MOE [13] where hydrogens were added and energy-minimized. The resulting structure without farglitazar was imported in AutoDock, where the protonation state of acidic and basic groups was adjusted (His323 and His449 were protonated), and partial charges were assigned. The protonation state of ligands to be docked was adjusted to the species assumed predominant at physiological pH. In particular, carboxylic acid, thiazolidinedione, sulfonylurea and *N*-acylsulfonamide moieties were deprotonated. Partial charges were assigned according to the MMFF94x force field [13]. For method verification and calibration we docked to PPARγ ga set of 121 carboxylic acids that are PPARγ agonists with known experimental binding affinities, a collection detailed in our earlier work [14]. For the compounds whose best pose showed both carboxylate oxygen atoms within 2 Å of the corresponding atoms of farglitazar in the X-ray structure (83% of the total), the pK_i calculated by AutoDock and the experimental pK_i correlated with $r^2 = 0.6$ (slope and intercept of the linear regression were 0.9 and 3.5, respectively). Since in this test AutoDock consistently overestimated experimental pK_i values, all calculated pK_i values given in the following are linearly rescaled using the above numbers for slope and intercept. Among the many poses returned by AutoDock for each compound, we selected as best the one assigned the highest pK_i value, provided it had a hydrogen bond to Tyr473 and at least two further hydrogen bonds to His323, His449, or Ser289. These constraints are justified by the facts that in all PPARγ-ligand complexes with known X-ray structures such hydrogen bonds seem to be essential for binding [12, 15–18], and that the hydrogen bond to Tyr473 was proposed to play a vital role for PPARγ co-activator recruitment [15].

Ligand binding assay. The Green PolarScreen PPAR Competitor Assay (Invitrogen, Carlsbad, USA) was used according to the manufacturer's instructions. IC_{50} values (concentrations at which 50% of the fluorescent ligand is displaced) and pIC_{50} values (negative decadic logarithm of IC_{50}) were determined for test compounds by measuring fluorescence polarization values for a series of concentrations.

Transactivation assay. CV-1 cells were cultured in DMEM (4500mg/l glucose), supplemented with 10% FBS and 50 U Penicillin/Streptomycin. Three days before transfection, cells were sterol-depleted by exchanging the culture medium to DMEM/F12 without Phenol Red, supplemented with 10% charcoal-stripped FBS and 50 U Penicillin/Streptomycin. Cells were plated in a 96-well dish with a density of 500'000 cells/ml (100 μl per well). DNA transfection was carried out in Opti-Mem I without Phenol Red using Lipofectamine. Each well received 8ng expression vector, 20ng reporter vector and 60ng β-galactosidase vector. 24h after transfection drugs dissolved in DMSO were added in DMEM/F12 without Phenol Red, supplemented with 10% charcoal-stripped delipidated FBS (Sigma-Aldrich, Buchs, Switzerland), and 50 U Penicillin/Streptomycin. 16h after addition of drugs, cells were lysed in CAT lysis buffer (Promega, Catalys AG, Wallisellen, Switzerland). Supernatants were analyzed for luciferase activity by addition of luciferase reagent (Promega, Catalys AG). Background normalization was carried out by measuring β-galactosidase activity as previously described [19]. EC_{50} values (concentrations at which 50% of the maximal gene expression is induced) and pEC_{50} values (negative decadic logarithm of EC_{50}) were determined for test compounds by measuring the increase of PPARγ target gene expression induced at different concentrations. Experiments were performed in quadruplicate, and error bars represent standard deviations.

Measurement of PPARγ target gene expression. 3T3-L1 fibroblasts were amplified in DMEM/10% calf serum and subsequently seeded into 6-well plates. Two days after the cells reached confluency, the medium was changed to DMEM/10% fetal calf serum containing 0.5 M isobutylmethylxanthine, 2 μg/mL insulin (Actrapid, Novo Nordisk), and 0.5 μM dexamethasone, to which the experimental compounds were added. Two days after induction of the cells the medium was changed to DMEM/10% fetal calf serum containing 2 μg/mL insulin, to which the experimental compounds were freshly added. Compounds were tested at the following concentrations: rosiglitazone (1 μM), pioglitazone (10 μM), gliquidone (10 μM), glipizide (100 μM, 200 μM), nateglinide (50 μM, 200 μM), repaglinide (50 μM, 100 μM, 200 μM). The cells were harvested in Trizol (Invitrogen, Breda, the Netherlands) 5 days after induction, and RNA was isolated by the standard procedure. 1 μg of RNA was used for cDNA synthesis using iScript (Biorad, Veenendaal, the Netherlands). cDNA was amplified with Platinum Taq polymerase using SYBR green on a Biorad MyiQ cycler. Specificity of the amplification was verified by melt curve analysis and evaluation of the amplification efficiency. Subsequently, expression of the genes of interest was normalized using cyclophilin as housekeeping gene.

3. Results

In a first step, all compounds in the TheraSTrat and Chembank databases were docked to PPARγ. Among them repaglinide (a carboxylic acid belonging to the glinide group of drugs), sulfadimidine (an acidified sulfonamide), and glimepiride (a sulfonylurea) were thereby assigned relatively high binding affinities [20]. In a second step of virtual screening we focused on these compound classes, docking several more members thereof.

Several glinides and sulfonylureas are docked to PPARγ with a high binding affinity.

Figures 1A, 1B, and 1C depict repaglinide, nateglinide and mitiglinide, respectively, docked into PPARγ and superimposed to the farglitazar complex X-ray structure (for structures see Figure 2). The predicted bound conformations to PPARγ are similar to that of farglitazar. The carboxylate group of the glinides superimposes well with that of farglitazar, forming hydrogen bonds with residues His323, His449, Ser289, and Tyr473 of PPARγ, as farglitazar does. Repaglinide forms several hydrophobic contacts in the large apolar cavity (bottom left of Figure 1A) that in the case of farglitazar is occupied by its long hydrophobic tail. Nateglinide and mitiglinide fit well the smaller hydrophobic cavity (bottom right of Figure 1B and 1C) that in the farglitazar complex is occupied by the benzoylphenyl group.

Repaglinide, mitiglinide, nateglinide, and meglitinide are predicted to bind PPARγ with pK_i values of 7.2, 6.3, 5.9, and 5.0, respectively. The smaller molecules mitiglinide, nateglinide, and meglitinide (molecular weight < 350) bind at a lower affinity compared to the larger repaglinide (molecular weight 453), as the latter forms more favorable contacts between the hydrophobic cavities of the PPARγ binding site and its large hydrophobic moieties.

Figures 1D, 1E, and 1F show gliquidone, glimepiride, and glipizide, respectively, docked into PPARγ and superimposed to the farglitazar complex X-ray structure (for structures see Figure 2). The predicted binding mode of gliquidone and glimepiride in the polar part of the binding site exhibits interesting similarities to that of farglitazar. In analogy to the carboxylate oxygens of farglitazar, the sulfonamide oxygen atoms point toward the pocket built by the side chains of His323, His449, Ser289, and Tyr473 and form hydrogen bonds to the H donor atoms in these residues. Notably, in the case of gliquidone and glimepiride there are two alternatives for H bond formation to His449, either to a sulfonyl oxygen or to the urea oxygen atom. Glimepiride deviates considerably in the lower part of the binding cavity from the bound conformation of farglitazar. This is not unexpected, since this part of the binding cavity is wide, allowing some conformational freedom for the ligand [16, 21]. Glipizide exhibits a slightly different binding mode compared to gliquidone and glimepiride, in that the sulfonyl group does not superpose to the carboxylate of farglitazar. In this case two hydrogen bonds are formed by the urea oxygen atom with Tyr473 and with His449, while the deprotonated sulfonamide nitrogen forms a hydrogen bond with Ser289. Here the deprotonated amide seems to be almost as good a mimic of farglitazar's carboxylate as is the sulfonyl group in the other examples.

Many sulfonylureas are predicted to bind PPARγ with pK_i values ranging from 3.7 to 8.8. As in the case of the glinides, smaller sulfonylureas such as tolbutamide and chlorpropramide (molecular weight < 300) are assigned lower binding affinities (pK_i $\sim$4) than larger ones such as glimepiride, glipizide, and glisamuride (molecular weight > 400, pK_i $\sim$6-8). The larger molecules form a higher number of favorable contacts to the hydrophobic walls of the receptor's binding cavity.

Sulfonylureas and glinides bind to PPARγ

The predicted PPARγ ligands gliquidone, glipizide, glimepiride, repaglinide, nateglinide, and mitiglinide, as well as two known ligands, linoleic acid, an endogenous agonist, and pioglitazone, a synthetic drug used in the treatment of type 2 diabetes were tested in a PPARγ competitor binding assay. Gliquidone, glimepiride, repaglinide, nateglinide, pioglitazone and linoleic acid bind to PPARγ and completely displace the reference ligand at different concentrations. The pIC_{50} values resulting from this experiment are 5.1 for gliquidone, 3.9 for glimepiride, 2.8 for repaglinide, 3.5 for nateglinide, 6.5 for pioglitazone, and 6.6 for linoleic acid. Glipizide and mitiglinide partially displace the reference ligand at concentrations between 500 and 2000 μM (maximal concentration measured), but IC_{50} values could not be determined.

Sulfonylureas and glinides activate PPARγ in a cell-based transactivation assay.

The eight compounds measured in the binding assay were also tested in a cell-based transactivation assay for PPARγ agonistic activity. All tested compounds activate PPARγ, albeit at various concentrations. Figure 2 reports the increase in gene expression induced by these compounds.

Among the sulfonylureas tested, gliquidone is the most potent PPARγ agonist (pEC_{50} 5.0), followed by glipizide (pEC_{50} 4.6) and glimepiride (pEC_{50} 4.0). Among the tested glinides, repaglinide shows the highest potency (pEC_{50} 4.8), followed by nateglinide (pEC_{50} 4.0) and mitiglinide (pEC_{50} 3.7). As to the standard agonists, pioglitazone (pEC_{50} 6.0) was found far more active than linoleic acid (pEC_{50} 3.2).

Ranking the compounds by decreasing potency, pioglitazone is followed by the sulfonylureas (gliquidone, glipizide, glimepiride), the glinides (repaglinide, nateglinide, mitiglinide), and by linoleic acid. Gliquidone approaches pioglitazone in terms of potency, reaching similar agonistic activity at a concentration one order of magnitude higher.

Sulfonylureas and glinides enhance PPARγ-induced target gene expression.

The effects of gliquidone, glipizide, nateglinide, mitiglinide, pioglitazone, and rosiglitazone on the expression of PPARγ target genes were measured in 3T3-L1 fibroblasts. For the sulfonylureas and glinides, concentrations bracketing EC_{50} values from the activation study were chosen (see materials and methods). Three bona fide target genes of PPARγ [22] were selected for analysis by quantitative RT-PCR: adiponectin, aP2 and GLUT4. Gliquidone, nateglinide and glipizide significantly enhanced the expression of these genes. For these three compounds, maximal induction was observed at the lowest measured concentration (10 μM for gliquidone, 50 μM for nateglinide, and 100 μM for glipizide). In contrast, repaglinide did not show any induction at concentrations ranging from 50 μM to 200 μM. Figure 3 shows the results for the selected sulfonylureas and glinides, together with pioglitazone as positive control. The induction of gene expression is reported relative to that observed in the presence of 1 μM of rosiglitazone, a strong thiazolidinedione PPARγ agonist. Gliquidone is as potent as pioglitazone and at 10 μM causes about 80% of the induction observed in the presence of 1 μM of rosiglitazone. Nateglide and glipizide show somewhat lower activities (between 30% and 70% compared to 1 μM of rosiglitazone) at higher concentrations compared to gliquidone.

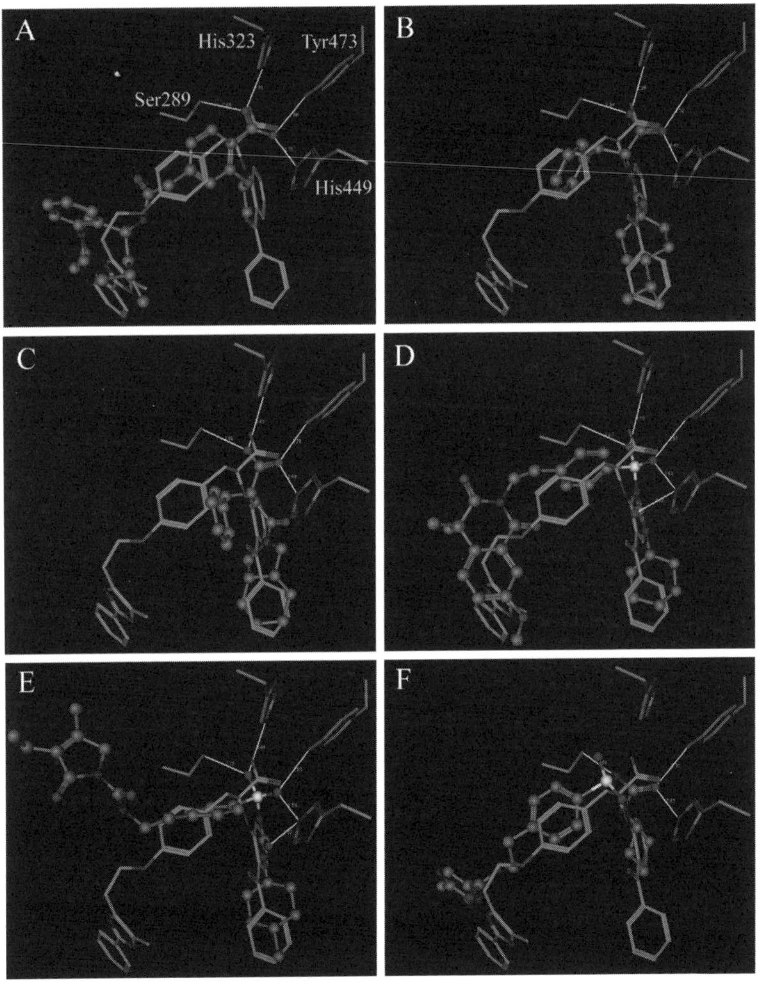

Fig. 1. 3D structures of three glinides and three sulfonylureas docked to PPARγ. Repaglinide (A), nateglinide (B), mitiglinide (C), gliquidone (D), glimepiride (E), and glipizide (F) (grey ball and stick models) docked to PPARγ and superimposed to the farglitazar complex X-ray structure (green stick model). Of PPARγ, only the side chains of His323, His449, Ser289, and Tyr473 are shown (grey stick models).

4. Discussion

The major results of this work are that several sulfonylurea and glinide drugs bind to and activate PPARγ *in vitro*, and that a detailed 3D binding mode underlying this activation is proposed. Experimental evidence for direct binding to PPARγ has been provided in a competitor binding assay, while PPARγ agonistic activity was measured both in a transactivation assay and by observing target gene levels

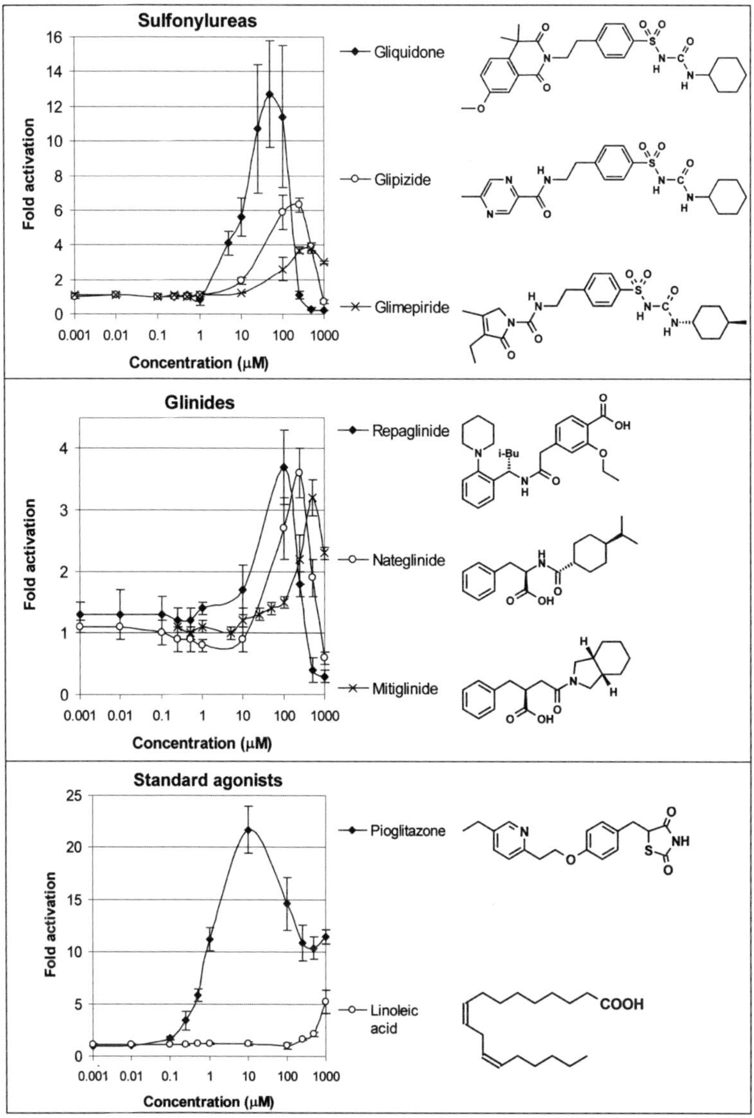

Fig. 2. Induction of PPARγ-mediated gene expression. The effects of three sulfonylureas (gliquidone, glipizide, and glimepiride, upper graph), three glinides (repaglinide, nateglinide, and mitiglinide, middle graph), and two standard agonists (pioglitazone and linoleic acid, bottom graph) on PPARγ-dependent transactivation are shown. Reporter gene activity is expressed as fold increase relative to the activity in the absence of compounds (see methods). Values are mean ± S.D. (n=4).

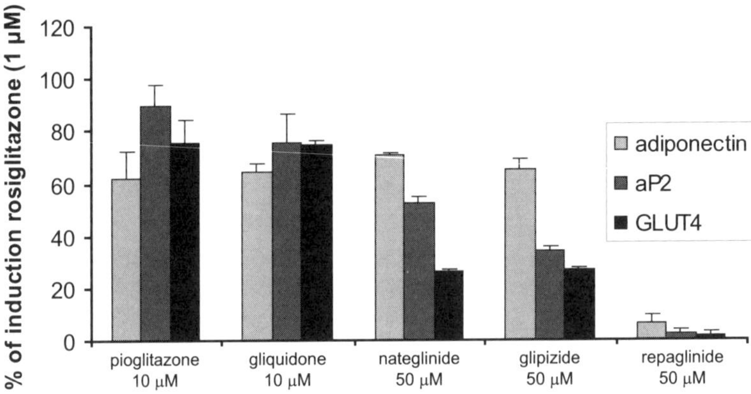

Fig. 3. Effect of selected compounds on the expression of three PPARγ target genes (adiponectin, aP2 and GLUT4). For each compound, the lowest effective dose is shown. Data are normalized and shown relative to the induction observed with 1μM rosiglitazone. Values are mean ± S.D. (n=3).

in 3T3-L1 cells. In all these experiments gliquidone showed the strongest PPARγ agonistic activity among the measured sulfonylureas and glinides.

While this study was underway, two sulfonylureas, glimepiride and glibenclamide, were reported to activate PPARγ [23, 24]. Our work provides strong evidence that additional sulfonylureas, as well as glinides (which equally target the sulfonylurea receptor), can bind and activate PPARγ, and allows the interpretation of binding data on the basis of docking results.

Sulfonylureas and glinides are standard treatments for type 2 diabetes. So far, members of these classes were presumed to act by a mechanism independent of PPARγ. According to this mechanism, they bind to the sulfonylurea receptor SUR1 in pancreatic islet β cells, closing K^+ channels, and leading to increased insulin production [8]. In contrast, here we provide evidence that binding to and activating PPARγ may be a new mode of action for at least some of these drugs, resulting in enhanced insulin sensitivity in peripheral tissue. This discovery opens new pharmacological perspectives for drugs targeting both SUR1 and PPARγ.

For this hypothesis to be useful from a clinical point of view, it is important that the minimal drug concentrations required for PPARγ activity are reached under pharmacological treatment According to availabele data, there is evidence that gliquidone, glipizide, and nateglinide may activate PPARγ at pharmacologically relevant concentrations, while glimepiride, repaglinide, and mitiglinide only activate PPARγ at concentrations higher than the those reached under clinical circumstances [25–34].

References

[1] Willson, T.M., Lambert, M.H., Kliewer, S.A. (2001) *Annu Rev Biochem* **70**, 341

[2] Staels, B., and Fruchart, J. C. (2005) *Diabetes* **54**(8), 2460

[3] Ericsson, H., Hamren, B., Bergstrand, S., Elebring, M., Fryklund, L., Heijer, M., and Ohman, K. P. (2004) *Drug Metab Dispos* **32**(9), 923

[4] Cox, S. L. (2005) *Drugs Today (Barcelona)* **41**(9), 579

[5] Martin, J. A., Brooks, D. A., Prieto, L., Gonzalez, R., Torrado, A., Rojo, I., Lopez de Uralde, B., Lamas, C., Ferritto, R., Dolores Martin-Ortega, M., Agejas, J., Parra, F., Rizzo, J. R., Rhodes, G. A., Robey, R. L., Alt, C. A., Wendel, S. R., Zhang, T. Y., Reifel-Miller, A., Montrose-Rafizadeh, C., Brozinick, J. T., Hawkins, E., Misener, E. A., Briere, D. A., Ardecky, R., Fraser, J. D., and Warshawsky, A. M. (2005) *Bioorg Med Chem Lett* **15**(1), 51

[6] Hug, H., Dannecker, R., Schindler, R., Bagatto, D., Stephan, A., Wess, R. A., and Gut, J. (2004) *Drug Discov Today* **9**(22), 948

[7] Picard, F., and Auwerx, J. (2002) *Annu Rev Nutr* **22**, 167

[8] Farret, A., Lugo-Garcia, L., Galtier, F., Gross, R., and Petit, P. (2005) *Fundam Clin Pharmacol* **19**(6), 647

[9] Gut, J., and Bagatto, D. (2005) *Expert Opin Drug Metab Toxicol* **1**(3), 537

[10] ChemBank. (2004) http://chembank.broad.harvard.edu/data_downloads/ bioactives-2004_05_01-sdf.zip.

[11] Morris, G. M., Goodsell, D. S., Halliday, R. S., Huey, R., Hart, W. E., Belew, R. K., and Olson, A. J. (1998) *J Comput Chem* **19**(14), 1639

[12] Gampe, R. T., Jr., Montana, V. G., Lambert, M. H., Miller, A. B., Bledsoe, R. K., Milburn, M. V., Kliewer, S. A., Willson, T. M., and Xu, H. E. (2000) *Mol Cell* **5**(3), 545

[13] MOE. (2006) Molecular Operating Environment, version 2006.07. Chemical Computing Group Inc.: 1255 University Street, Montreal, Quebec, Canada

[14] Rucker,C., Scarsi,M., Meringer,M. (2006) *Bioorg Med Chem* **14**(15), 5178

[15] Cronet, P., Petersen, J. F., Folmer, R., Blomberg, N., Sjoblom, K., Karlsson, U., Lindstedt, E. L., and Bamberg, K. (2001) *Structure* **9**(8), 699

[16] Nolte, R. T., Wisely, G. B., Westin, S., Cobb, J. E., Lambert, M. H., Kurokawa, R., Rosenfeld, M. G., Willson, T. M., Glass, C. K., and Milburn, M. V. (1998) *Nature* **395**(6698), 137

[17] Sauerberg, P., Pettersson, I., Jeppesen, L., Bury, P. S., Mogensen, J. P., Wassermann, K., Brand, C. L., Sturis, J., Woldike, H. F., Fleckner, J., Andersen, A. S., Mortensen, S. B., Svensson, L. A., Rasmussen, H. B., Lehmann, S. V., Polivka, Z., Sindelar, K., Panajotova, V., Ynddal, L., and Wulff, E. M. (2002) *J Med Chem* **45**(4), 789

[18] Xu, H.E., Lambert, M.H., Montana, V.G., Plunket, K.D., Moore, L.B., Collins, J.L., Oplinger, J.A., Kliewer, S.A., Gampe, R.T., Jr., McKee, D.D., Moore, J.T., Willson, T.M. (2001) *Proc Natl Acad Sci U S A* **98**(24), 13919

[19] Iniguez-Lluhi, J. A., Lou, D. Y., and Yamamoto, K. R. (1997) *J Biol Chem* **272**(7), 4149

[20] Scarsi, M., Rücker, C., Dannecker, R., Hug, H., Gut, J., and Meyer, U. A. (2005) *Poster presented at the "Third international symposium on PPARs efficacy and safety", Monte Carlo (Monaco) March 19-23, 2005* (http://www.lorenzinifoundation.org/ppars2005/abstract_book.pdf page 39)

[21] Xu, H. E., Lambert, M. H., Montana, V. G., Parks, D. J., Blanchard, S. G., Brown, P. J., Sternbach, D. D., Lehmann, J. M., Wisely, G. B., Willson, T. M., Kliewer, S. A., and Milburn, M. V. (1999) *Mol Cell* **3**(3), 397

[22] Knouff, C., and Auwerx, J. (2004) *Endocr Rev* **25**(6), 899-918

[23] Fukuen, S., Iwaki, M., Yasui, A., Makishima, M., Matsuda, M., and Shimomura, I. (2005) *J Biol Chem* **280**(25), 23653

[24] Inukai, K., Watanabe, M., Nakashima, Y., Takata, N., Isoyama, A., Sawa, T., Kurihara, S., Awata, T., and Katayama, S. (2005) *Biochem Biophys Res Commun* **328**(2), 484
[25] Anonymous. (2001)
[26] Hazama, Y., Matsuhisa, M., Ohtoshi, K., Gorogawa, S., Kato, K., Kawamori, D., Yoshiuchi, K., Nakamura, Y., Shiraiwa, T., Kaneto, H., Yamasaki, Y., and Hori, M. (2006) *Diabetes Res Clin Pract* **71**(3), 251
[27] Jaber, L. A., Ducharme, M. P., Edwards, D. J., Slaughter, R. L., and Grunberger, G. (1996) *Pharmacotherapy* **16**(5), 760
[28] Jönsson,A., Chan,J.C., Rydberg,T., Vaaler,S., Hallengren,B., Cockram,C.S., Critchley,J.A., and Melander,A. (2000) *Eur J Clin Pharmacol* **56**(9-10), 711
[29] Luzio, S. D., Anderson, D. M., and Owens, D. R. (2001) *J Clin Endocrinol Metab* **86**(10), 4874
[30] McLeod, J. F. (2004) *Clin Pharmacokinet* **43**(2), 97
[31] Novartis (2005) Starlix Prescribing Information, http://www.starlix.info/ starlix/content/pages/basic.php
[32] Pfizer (2000) Glucotrol Full U.S. Prescribing Information, www.pfizer.com/ pfizer/download/uspi_glucotrol.pdf
[33] von Nicolai, H., Brickl, R., Eschey, H., Greischel, A., Heinzel, G., König, E., Limmer, J., and Rupprecht, E. (1997) *Arzneimittelforschung* **47**(3), 247
[34] Weaver, M. L., Orwig, B. A., Rodriguez, L. C., Graham, E. D., Chin, J. A., Shapiro, M. J., McLeod, J. F., and Mangold, J. B. (2001) *Drug Metab Dispos* **29**(4 Pt 1), 415

A MULTI-LAYER MODEL TO STUDY GENOME-SCALE POSITIONS OF NUCLEOSOMES

VITO DI GESÚ, GIOSUÉ LO BOSCO, LUCA PINELLO

Dipartimento di Matematica ed Applicazioni, Università di Palermo
Via Archirafi 34, I-90123 Palermo, Italy
E-mail:digesu,lobosco@math.unipa.it,lucapinello@gmail.com

DAVIDE CORONA, MARIANNA COLLESANO

Istituto telethon Dulbecco c/o DiSBi del Policlinico, Università di Palermo
Via del Vespro 129, I-90127 Palermo, Italy

GUO-CHENG YUAN

Department of Biostatistics, Harvard School of Public Health
677 Huntington Ave, MA-02115 Boston, USA

The positioning of nucleosomes along chromatin has been implicated in the regulation of gene expression in eukaryotic cells, because packaging DNA into nucleosomes affects sequence accessibility. In this paper we propose a new model (called MLM) for the identification of nucleosomes and linker regions across DNA, consisting in a thresholding technique based on cut-set conditions. For this purpose we have defined a method to generate synthetic microarray data fully inspired from the approach that has been used by Yuan et al. [3] . Results have shown a good recognition rate on synthetic data , moreover, the MLM shows a good agreement with the recently published method based on Hidden Markov Model when tested on the *Saccharomyces cerevisiae* chromosomes microarray data.

Keywords: Multi-Layers methods, Nucleosomes positioning, Microarray data analysis, BioInformatics.

1. Introduction

Eukaryotic DNA is packaged into a highly compact and dynamic structure called chromatin. While this packaging provides the cell with the obvious benefit of organizing a large and complex genome in the nucleus, it can also block the access of transcription factors and other proteins to DNA. Biochemical and genetic analysis have shown that covalent modifiers of chromatin and ATP-dependent multisubunit complexes, known as chromatin remodelling factors, directly alter chromatin structure to regulate gene expression. It is becoming clear that alterations in the spectrum of chromatin modifications underlie many human diseases including cancer [1] . The recognition that alterations in chromatin structure can result in a variety of diseases, including cancer, highlights the need to achieve a better understanding of the molecular processes modulating chromatin dynamics.

Cell cycle control relies on a fine balance between transcriptional activation and repression. The role played by ATP-dependent remodellers, such as the SWI/SNF complex, in chromatin opening, transcriptional activation and cell cycle control has just started to be revealed. However, the contribution to cell cycle played by other ATP-dependent remodellers linked to chromatin condensation and transcription repression still remain elusive. The ATP-dependent remodeller ISWI is a subunit of several evolutionary conserved eukaryotic chromatin remodelling complexes [2] . ISWI-containing complexes play a critical role in making chromatin dynamic and have been implicated in transcriptional repression, DNA replication and chromosome condensation [2] .

ISWI in vitro is an ATP-dependent nucleosome spacing factor and in vivo its nucelosome spacing activity is necessary to maintain chromosome organization. To determine the potential contribution of nucleosome spacing defects to the cell cycle and differentiation problems observed in ISWI mutant cells we propose to conduct genome-scale identification of nucleosome positions in wild type and ISWI mutant cells.

To measure nucleosome positions on a genomic scale, a DNA microarray method has been recently used to identify nucleosomal and linker DNA sequences on the basis of susceptibility of linker DNA to micrococcal nuclease [3] . Nucleosomal DNA is labelled with Cy3 fluorescent dye (green), and mixed with total genomic, labelled with Cy5-fluorescent dye (red). This mixture is then hybridized to DNA microarrays printed with overlapping 50-mer oligonucleotide probes tiled every 20 base pairs across DNA chromosomal regions of interest. A signal of green/red ratio values for spots along the chromosome shows nucleosomes as peaks about 140 base pairs long, or six to eight microarray spots, surrounded by lower ratio values corresponding to linker regions (regions where no nucleosomes are present, see Fig. 1). In this work we introduce and validate a new method for nucleosome identification starting from a signal coming from a microarray designed as described above.

Note that, because of noise in the data signal and large-scale trends in mean hybridization values, a naive threshold-based approach for determining nucleosome positions was highly inaccurate (see Fig. 1 for an example of input signal).

By the way, methods based on probabilistic networks [5] are suitable to face sequences of noisy observed data; however, they suffers of the high computational complexity and results are biased from the memory steps parameter.

In particular, the specific problem of nucleosome positioning has been faced recently by Yuan et al. [3] by using Hidden Markov Models (HMM) [6].

Generally, the analysis of stochastic signals aims to extract "significant" patterns from noise background and to study their spatial relations (periodicity, long term variation, burst,etc). The problem is more complex whenever the noise background has an underlying structure.

Here, we propose a Multi-Layers Model of analysis (MLM), that belongs to a class of methods successfully used in the analysis of very noisy data because of their ability to recover statistical properties of a signal, starting from several views of the input data-set. Main advantages of the MLM approach are computational and a global view of the whole phenomena.

We choose to test the MLM on synthetic and real data. In the latter case, data comes from microarray data of the *Saccharomyces cerevisiae* where their complexity makes hard the manual discrimination of regions representing nucleosomes and linkers. Thus, we compared the MLM with the recently published method based on Hidden Markov Model [3] on the above mentioned data. This comparison plays a significant role to evaluate the accuracy of the recognized signal components; moreover, the correspondence in the outputs of the two methods increases the validity of their classification. Results have shown an average recognition of 76% on synthetic data for the nucleosome and linker regions and an agreement of 71% with the Hidden Markov Model on *Saccharomyces cerevisiae.*
Section 2 shows the proposed Multi-Layers analysis model, experimental results for the validation and assessment of the Multi-Layers Model are given in Section 3. Final remarks are given in Section 4.

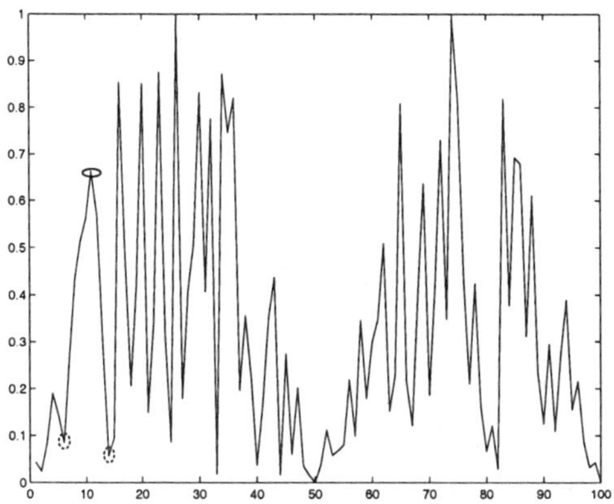

Fig. 1. *Input Signal:* The input Signal coming from the microarray, each value in the x axes represents a spot on the microarray and its y value is the log ratio Green/Red. Nucleosomes correspond to peaks of about 140 base pairs long, or six to eight microarray spots (marked by black circle), surrounded by lower ratio values corresponding to linker regions (marked by dashed circles). Noise in the data signal and large-scale trends in mean hybridization values makes inaccurate the determination of nucleosome positions by a naive threshold-based approach.

2. Multi-Layers Model

In this Section we provide an outline of the MLM method for the analysis of monodimensional signal. The proposed analysis procedure is carried out as follows:
- Preprocessing. Starting from the input signal, $\mathbf{S}$ we convolute it by an hamming

window so defined :

$$w[j+1] = \beta - (1-\beta)\ cos(\frac{2\pi j}{n-1})$$

Note that, the input signal **S** is composed of T fragments $S_1, \cdots, S_T$ which are not necessary contiguous. This happens, for example, in the case of the *Saccharomyces cerevisiae* data-set which is the one we have analyzed. Such kind of smoothing is thus applied to each fragment S_t.

- Model construction. Each fragment S_t is processed in order to find $L(S_t)$ local maxima $M_t^{(l)}$ for $l = 1, \cdots, L(S_t)$. By looking around each $M_t^{(l)}$ in the signal, we extract the sub-fragment F_t^l in such way:

$$F_t^l = S_t([M_t^{(l)} - os, M_t^{(l)} + os])$$

where os is a window value that is set depending on the microarray design (in our case os=4). Note that F_t^l is the sub-fragments of S_t of size $2 * os$ centered on the maximum $M_t^{(l)}$. Not all the sub-fragment are used in the next steps of the process, in fact we preserve only the one which satisfies the following condition:

$$\begin{cases} F_t^l(j+1) - F_t^l(j) > 0 & j = 1, \cdots, os \\ F_t^l(j+1) - F_t^l(j) < 0 & j = os+1, \cdots, 2 * os \end{cases} \tag{1}$$

After this selection process we have $G(S_t)$ sub-fragments, and we build the model of the *well positioned nucleosome*:

$$\overline{F}(j) = \frac{1}{T}\sum_{t=1}^{T} \frac{1}{G(S_t)} \sum_{k=1}^{G(S_t)} F_t^k(j) \quad j = 1, \cdots, 2 * os + 1$$

That is, for each j, the mean value of all the sub-fragments satisfying Equation 1.

-Interval identification. Giving the convolved signal **X** of **S**, a set of intervals, $\mathbf{R} = \{\mathbf{R_k} \mid 1 \le k \le K\}$ is retrieved where:

$$\mathbf{R}_k = \{(\lambda_k^i, b_k^i, e_k^i) \mid \mathbf{X}(b_k^i) = \mathbf{X}(e_k^i) = t_k\}$$

in the previous definition, t_k with $k = 1, 2, ..., K$ are the threshold values and λ_k^i is the label of the interval $[b_k^i, e_k^i]$ at threshold t_k.

- Interval merging. In this step intervals are merged in a bottom up way following this rule :

if $\forall(\lambda_{k+1}^j, b_{k+1}^j, e_{k+1}^j) \in \mathbf{R}_{k+1}$

$\exists(\lambda_k^i, b_k^i, e_k^i) \in \mathbf{R}_k$ such that $Card(\{\lambda_{k+1}^j \mid [b_{k+1}^j, e_{k+1}^j] \subseteq [b_k^i, e_k^i]\}) = 1$

then $\lambda_{k+1}^j = \lambda_k^i$.

In practice this rule is used to assign to a label λ_{k+1}^j at threshold $k+1$ the value of the label λ_k^i at threshold k if $[b_{k+1}^j, e_{k+1}^j]$ is the unique subset of $[b_k^i, e_k^i]$.

- Pattern detection. In this step we find the *interesting intervals* which are the ones whose labels belong to the set **P** so defined :

$$\mathbf{P} = \bigcup_{k=1}^{K-m} \bigcap_{n=k}^{k+m} \Lambda_n$$

where $\Lambda_n = \{\lambda_n^j \mid (\lambda_n^j, b_n^j, e_n^j) \in \mathbf{R}_n\}$ are the labels of all the intervals at threshold level t_n. The intersection $\bigcap_{n=k}^{k+m} \Lambda_n$ give us the labels of the intervals that remain for exactly m threshold values starting from threshold value t_k, finally, the union results in all the labels of the intervals which remain for at least m threshold values. The value m is said the *minimum number of permanences*.

- Feature extraction. Each $p_i \in \mathbf{P}$ defines a pattern of feature vector $\mathbf{f_i}$

$$\mathbf{f_i} = (f_j^{(i)} f_{j+1}^{(i)} \cdots f_{j+l}^{(i)})$$

where j is the minimum threshold value where the interval labelled p_i has been found, $l \geq m$ and $f_{j+n}^{(i)} = \begin{pmatrix} b_{j+n}^i \\ e_{j+n}^i \end{pmatrix}$. Informally, for each interesting interval labelled p_i we consider as feature vector the limits of each interval labelled p_i from the lower threshold j to the upper threshold l where it appears. The representation in a multi-dimensional feature space is used to characterize different types of patterns.

- Pattern similarity. At this purpose we can define a distance between patterns in such way:

$$d(p_r, p_s) = (1-\alpha)(A_r - A_s) + \alpha \sum_{i \in I} (a_r^i - a_s^i) \qquad (2)$$

where A_r and A_s are the surfaces of the two polygons bounded by the set of vertexes $V = \bigcup_{i \in I} \{b_s^i, e_s^i, b_r^i, e_r^i\}$, and $a_r^i = e_r^i - b_r^i$, $a_s^i = e_s^i - b_s^i$. This distance will be used by a classifier in order to distinguish the type of pattern.

An example of signal and the relative patterns is given in Fig. 2.

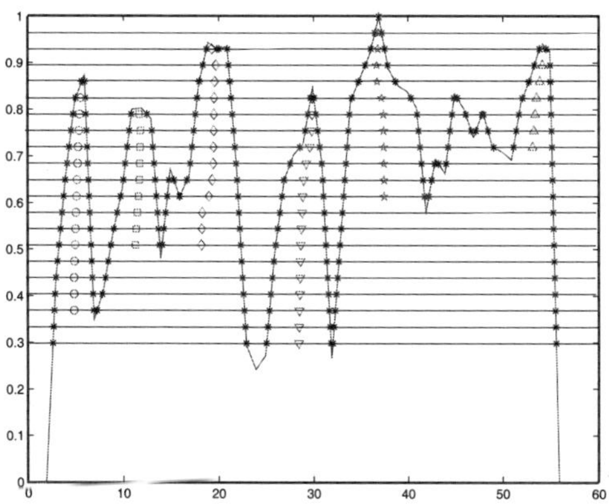

Fig. 2. *Pattern identification and extraction:* Given the input signal x, we intersect it by horizontal lines each one representing a threshold value t_k. In this example we retrieve 5 patterns identified by round, square, diamond, triangle up, star, triangle down. Each pattern identifier is replicated for each of its feature values and pointed in the middle of it. In this example the minimum number of permanences is $m = 5$.

3. *MLM* validation

At this stage of the work, we have focused our attention on the classification of two types of patterns : Nucleosome (N) and Linker regions (L). It means that the MLM is able to find where a nucleosome or a linker zone is present. For each pattern p_s, we evaluate the distance $d(p_s, \overline{F})$ (d is defined in Equation 1, $\overline{F}$ is the model), the rule to classify p_s is :

$$c(p_s) = \begin{cases} 1 & d(p_s, \overline{F}) \leq \phi \\ 0 & \text{otherwise} \end{cases} \tag{3}$$

where 1 means nucleosomal region, 0 linker. The value of ϕ in our experiments has been estimated by the subfragments F_t^l and is defined below in the Experimental Results. Afterwards, for each nucleosomal region p_i, $i = 1, \cdots, N$, we consider the feature vector $\mathbf{f_i} = (f_j^{(i)} f_{j+1}^{(i)} \cdots f_{j+l}^{(i)})$, and we calculate the *center of the nucleosomal region* C_i :

$$C_i = \frac{1}{l} \sum_{k=j}^{j+l} \frac{e_k^i - b_k^i}{2}$$

If the nucleosomal region p_i covers at most $2 * os + 1$ spots, we set its range to $[C_{i-4}, C_{i+4}]$, otherwise to $[b_j^i, e_j^i]$, which corresponds to the range of the largest interval labeled p_i. In the case this new ranges overlap, we merge them considering their union. The output of the classifier will be a string of $0's$ and $1's$ of the same length of the signal, in particular the $1's$ are replicated inside each range of the nucleosomal region.

3.1. *Synthetic data generation*

In order to validate the MLM method we have set up a generator of synthetic signals. This signal should be the one coming from a microarray where each spot represents a probe i of 50 base pairs overlapping 20 base pairs with probe $i+1$. This means that, we move from left to right across a chromosome by a window (probe) of width 50 base pairs, such that two consecutive window (probes) have an overlap of 20 base pairs.

We start from a representation I of the chromosome in base pairs, it is a discrete signal in the range $\{0, 1\}$, where the 1's indicates the presence of a nucleosome (see Fig. 3). Note that, we know that the width of a nucleosome is about 140 base pairs long. Thus, we consider m replicates of I, $I_1, \cdots, I_m$, where each starting point x_i of a nucleosome is perturbated to $x_i + \epsilon$ with ϵ uniformly distributed in the range $[-d, d]$ (d represents the maximum decentralization). Then, three different distribution are considered : N_{gn} which is normal, is the distribution of green values for nucleosomes, U_{gl} which is uniform, models the green values in a linker zone, N_r which is also normal, models the values of red in the whole genome. Note that, since the nucleosomes are marked with green, we expect more green than red around a nucleosome, while the opposite for the linker regions. The resulting simulated signal X is :

$$X(i) = \{log_2 \frac{1}{n} \sum_{j=1}^{m} \frac{I_j(k) * n_\epsilon + l_\epsilon}{G_\epsilon} \mid (r-o)i - r + 2o \le k \le (r-o)i + o\} \qquad (4)$$

with $n_\epsilon \sim N_{gn}$, $l_\epsilon \sim U_{gl}$ and $G_\epsilon \sim N_r$. The values r and o are the ovlerlap and resolution values which represents the number of base pairs of a probe (50) and the overlap base pairs of two probes respectively (20).

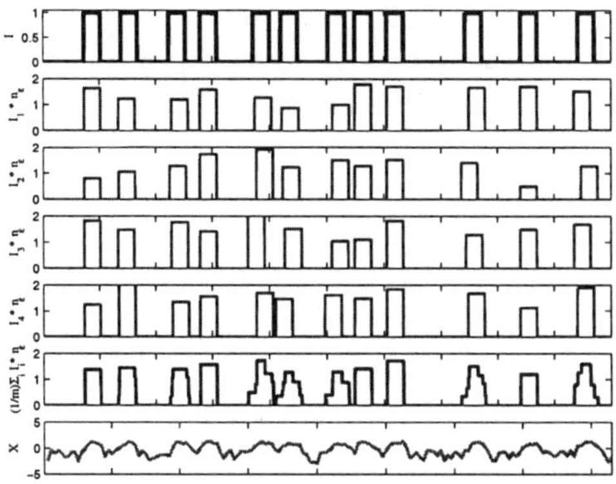

Fig. 3. *Synthetic signal generation:* We start from the signal I (the first up plot), then we perturbate each position of nucleosomes in each replicate I_i multiplying also for n_ϵ (from second to fourth plot). Thus all the replicates are added together (the fifth plot) and finally the signal X (the last plot) is obtained according to Equation 4.

3.2. *Experimental results*

The accuracy of the classification has been measured in terms of correspondence between binary values. In the case of the synthetic signal, the output of the classifier has been compared with I, in the case of the real data set it has been compared with the output of the Hidden Markov model opportunely converted into a binary string.

In all the experiments, the same value $\phi = mean(d(F_t^l, \overline{F})) - 3std(d(F_t^l, \overline{F}))$ has been considered, where F_t^l are all the sub-fragments used on the construction of the model $\overline{F}$. Moreover, the offset os has been set to $os = 4$ because, by biological consideration, we know a nucleosome is 8 probes long. The MLM depend also on the values of K (number of thresholds), and m (the minimum number of permanences). We have set the values of K, m and α as the values which have shown the best performances all over the ROC curves $R_{km\alpha}$ carried out for $k, m = 20, \cdots, 100$ (by step of 10) and $\alpha = 0, \cdots, 1$ (by step of 0.1). For the synthetic data, the resulting values have been $K = 100$, $m = 20$ and $\alpha = .25$. Note that the ROC curves $R_{km\alpha}$ has been carried out using a training set of 100 signal of length 10000 (in base pairs). Finally, we have generated 10 signal of 200000 base pairs reaching on average 76% of correct classification on nucleosome and linker regions. In Fig. 4 the accuracy is shown for each of the ten experiments. Moreover, we have also computed

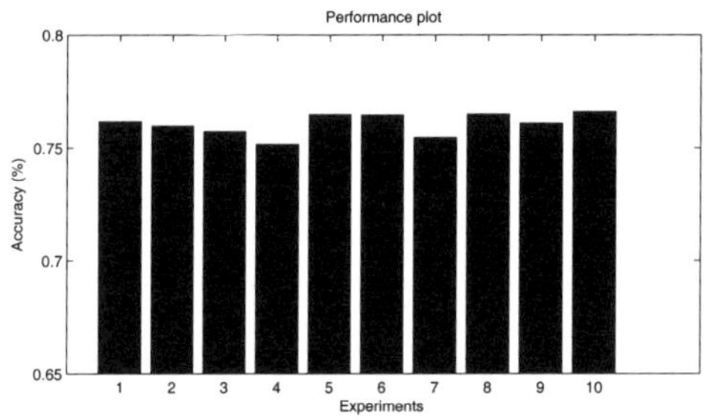

Fig. 4. *Accuracy of the classification:* Correct classified nucleosomes and linkers for 10 experiments using signal of length 200000 base pairs.

the classification for *Saccharomyces cerevisiae* chromosomes microarray data [8]. In this case, we have compared the accordance of the MLM approach with the Hidden Markov Model that is recently used to analyze the above data set (for the details about this model see [3]). The input signal is composed by 215 contiguous fragments for a total of 24167 base pairs. In such experiment, we have considered $K = 20$, $m = 3$, $alpha = .5$ as resulting values of the ROC study. The confusion matrix which show the accordance of the two methods is in Table 2.

Table 1. Performances of the MLM on the *Saccharomyces cerevisiae* data set.

Method	*Linkers by* MLM %	*Nucleosomes by* MLM %
Linkers by HMM	0.64	0.36
Nucleosomes by HMM	0.21	0.79

4. Final remarks

In this paper we have presented a new method able to find nucleosome and linker regions along the chromosome by using microarray data. For this purpose we have defined a method to generate synthetic microarray data fully inspired from the microarray technique that has been used by Yuan et al. [3] . Thus we have tested our model on both synthetic and real data, reaching, in the first case an accuracy of 76% and in the second case an accordance of 71% with a previously used method. Future work will be devoted to improve and refine the MLM in order to discriminate well positioned from decentralized nucleosomes (the two kind of different nucleosome regions). In the future we intend to apply MLM to complex biological data as, for example, the drosophila chromosome.

References

[1] S. Jacobson, L. Pillus, "Modifying chromatin and concepts of cancer", Curr. Opin. Genet. Dev., Vol.9, pp.175–184, 1999.

[2] D. F. V. Corona, and J. W. Tamkun, "Multiple roles for ISWI in transcription, chromosome organization and DNA replication", Biochim Biophys Acta, Vol. 1677(1-3), pp.113–119, 2004.

[3] Y. Guo-Cheng Yuan et al., "Genome-Scale Identification of Nucleosome Positions in S. cerevisiae", Science, Vol.309, pp. 626–630, 2005.

[4] A. L. Delcher, S. Kasif, H. R. Goldberg, W. H. Hsu, "Protein secondary structure modelling with probabilistic networks", Proc. of Int. Conf. on Intelligent Systems and Molecular Biology, pp. 109-117, 1993.

[5] F. V. Jensen, "Bayesian Networks and Decision Graphs", Springer, 2001.

[6] Y. Ephraim and N. Merhav, "Hidden Markov processes", IEEE Trans. Inform. Theory, Vol.48, pp. 1518–1569, June 2002.

[7] J. Swets and R. Pickett, "Evaluation of Diagnostic Systems: Methods from Signal Detection Theory", Academic Press, New York, 1992.

[8] http://www.sciencemag.org/content/vol0/issue2005/images/data/1112178/DC1/Yuan.SOM.data.zip.

BIOINFOGRID: BIOINFORMATICS SIMULATION AND MODELING BASED ON GRID*

LUCIANO MILANESI

Institute of Biomedical Technology, CNR,
Via Fratelli Cervi, 93, Segrate, 20090, Italy
E-mail: Luciano.Milanesi@itb.cnr.it

Genomics sequencing projects and new technologies applied to molecular genetics analysis are producing huge amounts of raw data. In future the trend of the biomedical scientific research will be based on computing Grids for data crunching applications, data Grids for distributed storage of large amounts of accessible data and the provision of tools to all users. Biomedical research laboratories are moving towards an environment, created through the sharing of resources, in which heterogeneous and dispersed health data, such as molecular data (e.g. genomics, proteomics), cellular data (e.g. pathways), tissue data, population data (e.g. Genotyping, SNP, Epidemiology), as well the data generated by large scale analysis (eg. Simulation data, Modelling). In this paper some applications developed in the framework of the European Project "Bioinformatics Grid Application for life science - BioinfoGRID" will be described in order to show the potentiality of the GRID to carry out large scale analysis and research worldwide.

Keywords: Bio-Informatics, Modelling, GRID platform

1. The GRID Bioinformatics Platform

1.1. *Introduction*

Starting from the genomic and proteomic sequence data, a complex computational infrastructure has been established with the objective to develop a GRID based system to characterize the functional interactions of protein involved in genetic diseases.

The current interest in the genome-wide analysis of cells at the level of transcription ('transcriptome') and translation ('proteome'), the third level of analysis is the 'metabolome'. A new level of experiments are required to obtain an overall picture of when, where, and how genes are expressed.

More specifically the GRID platform will make large scale research in the fields of Genomics, Proteomics, Transcriptomics and applications in Drug Discovery much easier, reducing data calculation times thanks to the distribution of the calculation at any one time on thousands of computers across Europe and the rest of the world.

*This work is supported by BioinfoGRID (n. 026808), EGEE and the "LITBIO" and "ITALBIONET" MIUR-FIRB projects.

The massive potential of Grid technology will be indispensable when dealing with both the complexity of models and the enormous quantity of data, for example, in searching the human genome or when carry out simulations of molecular dynamics for the study of new drugs.

Grid networking promises to be a very important step forward in the Information Technology field. Grid technology will make a global network made up of hundreds of thousands of interconnected computers possible, allowing the shared use of calculating power, data storage and structured compression of data.

This goes beyond the simple communication between computers and aims instead to transform the global network of computers into a vast joint computational resource. Grid technology is a very important step forward in sharing of information over the internet.

Bioinformatics applications in biotechnology which specifically focuses on the discovery of new therapeutic proteins is currently becoming an ideal research area where computer scientists apply and further develop new intelligent computing methods in both experimental and theoretical cases. These progresses made it essential to devise technological platforms able to ensure appropriate support to research activities performed in the areas of Bioinformatics and Medical Informatics.

Calculation time barriers, geographical boundaries and new data mining tools implemented in HPC and GRID technologies [1] can be successfully used to deal with the increasing amount, complexity and heterogeneity of biological and biomedical data. These potentialities of the GRID have been explored by the BioinfoGRID [2] project which combines Bioinformatics services and applications for molecular biology users utilising the Grid Infrastructure created by the European EGEE II Project [3].

1.2. *BioinfoGRID project*

The BIOINFOGRID project proposes to combine Bioinformatics services and applications for molecular biology users with the Grid Infrastructure created by EGEE (6th Framework Program).

The BioinfoGRID project aims to carry out Bioinformatics research and to develop new applications in the sector using a network of services based on futuristic Grid networking technology that represents the natural evolution of the Web.

The BIOINFOGRID initiative plans to evaluate genomics, transcriptomics, proteomics and molecular dynamics applications studies based on GRID technology.

More specifically the BIOINFOGRID project will make research in the fields of live science much easier, reducing data calculation times thanks to the distribution of the calculation at any one time on thousands of computers across the world (Figure 1).

A series of programs have been implemented to automate the analysis, prediction and annotation processes of genomic DNA. To support of this type of analysis, several algorithms have been used to recognize biological signals involved in the

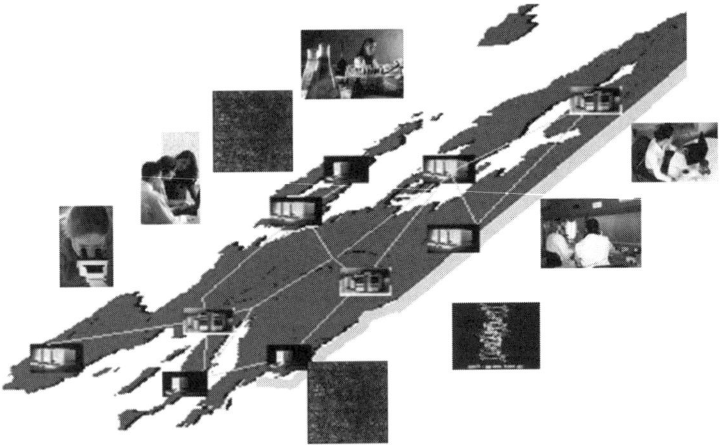

Fig. 1. Researchers can perform their Bioinformatics activities regardless of geographical location. Data analysis specific for bioinformatics allows the GRID user to store and search genetics data, with direct access to the data files stored on Data Storage elements on GRID servers.

identification of genes and proteins. The system implemented can be used to analyze the content of the large number of genomic sequences. For this reason, the system realized is capable of using a computational architecture specifically designed for intensive computing based on GRID technologies developed by EGEE European project.

Some of the applications and goals addressed by the BioinfoGRID project are:

(1) GRID based analysis of cDNA data.
(2) GRID based analysis of sequence similarity with publicly available databases.
(3) GRID based analysis of rule-based multiple alignments.
(4) GRID based analysis for protein functional domain analysis.
(5) GRID based analysis Transcriptomics and Phylogenetics data.
(6) GRID based for microarray analysis.
(7) GRID based access to the biological data files stored on Data Storage element on GRID servers.
(8) GRID based Database applications to manage and access biological experimental data using the GRID.
(9) GRID based analysis in order to cluster the gene products by their functionality using Gene Ontology.
(10) GRID based analysis Molecular Dynamics Applications
(11) GRID based analysis challenge of the Wide In Silico Docking.
(12) To expand Grid awareness inside the bioinformatics community in conjunction with the European Grid Infrastructure Projects which is achieved by the promotion of dissemination and tutoring events where Grid experts can discuss the available Grid services and the user requirements with real life examples.
(13) The evaluation and adaptation of high-level user interfaces which will be com-

mon to all the different BioinfoGRID applications.

(14) To exploit, in a more user-friendly approach, the Grid services provided by European Grid Infrastructures.

(15) The organization of Bioinformatics Portals to aid in the simplification of the services invocation or the jobs submission to the Grid and Workflows which enables the dynamically establishment of complex biologic analysis

2. Methodology

The following grid based applications demonstrate the potential of access to very large computing resources and are essential for high throughput large scale deployment analysis addressing bioinformatics challenges in life science.

2.1. *Technology description*

Bioinformatics usually needs to perform very complex workflow analysis. Some applications can perform and scale very well in a Grid environment while other are better suited for a dedicated cluster of computers especially in the case of software licensing or specialized complex software procedures [4].

For any Grid computing effort, middleware is a crucial component. The European EGEE project, has developed and re-engineered most of this middleware stack into a new middleware solution, gLite, now being deployed on the pre-production service. The gLite stack [5] combines low level core middleware with a range of higher level services. Distributed under a business friendly open source license, gLite integrates components from the best of current middleware projects such as Condor and the Globus Toolkit as well as components developed for the LCG project.

The product is a best-of-breed, low level middleware solution, compatible with schedulers such as PBS, Condor and LSF, built with interoperability in mind and providing foundation services that facilitate the building of Grid applications from all fields. The Grid Infrastructure adopted in this work is a network of computing and storage resources connected together with the gLite middleware that provides a framework for the management of grid nodes and a complete command line API for access the distributed resources (job and data). In the Figure 2 the main components of the EGEE Grid and their interactions are shown.

A job, described in the JDL format, is submitted by the user from a *User Interface* (UI) to the *Resource Broker* (RB) that processes the job in order to find a *Computing Element* (CE) matching user requirements. When the Computing Element receive a job from the Resource Broker, executes it in a cluster of *Worker Nodes* (WN). Security is handled through *Personal Certificates* that allow users to be recognized in the infrastructure and to execute their jobs for a limited authentication time. Usually users belong to a *Virtual Organization* (VO) that represents a logical group of users that share common research interests and projects

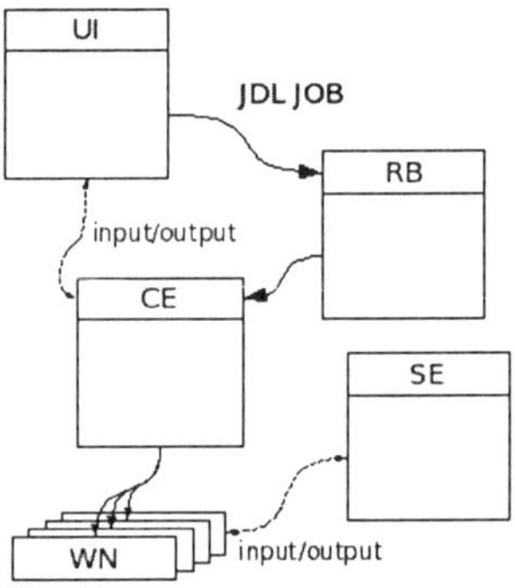

Fig. 2. Main components of the EGEE Grid Infrastructure.

2.2. *GRID database applications example*

BLAST [6] is probably the most used application in bioinformatics teams. BLAST complexity tends to be a concern when the query sequence sets and reference databases are large. In the framework of the BioifoGRID project, we have developed BioinfoGridBlast (BGBlast): an approach for handling the computational complexity of large BLAST executions by porting BLAST to the Grid platform, leveraging the power of the thousands of CPUs which compose the EGEE infrastructure.

BGBlast is an innovative porting of BLAST onto the Grid providing the following capabilities shown in Figure 3:

(1) Automatic Database Updater (ADU): ensures the users always work with the latest version of every Blast Reference Database (BRD), and this without needing human staff to manually monitor the release of newer versions of BRDs or manually performing database updates over the Grid
(2) Adaptive Replication (AR) for the BLAST Reference Databases: ensures that the most used BLAST databases are replicated more times than less used databases. The optimal number of replicas for each BRD is calculated dynamically based on the relative usage of the specific database in recent times. This keeps a constant optimization of Grid queue times vs Grid storage costs.
(3) Version Regression for BLAST Reference Databases: allows the user to specify an older version of a certain BRD to be used for the computation. This allows the user to reproduce exactly computations obtained in the past, something that might be needed to confirm results that were obtained. The storage of older version of BRDs is performed with a delta-encoding efficient in both space (storage costs) and time (a short download time and a short time to patch a BRD for regressing it to an earlier version).

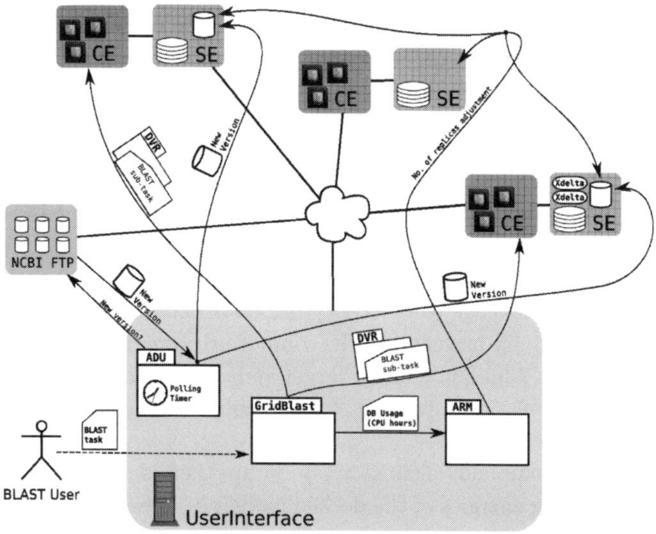

Fig. 3. Interactions between parts of BioinfoGridBlast and Grid elements. The user launching a BLAST task on the Grid. The BLAST task is split by GridBlast in jobs of equal size ("sub-tasks") and sent to Computing Elements (CEs) for execution. The code for performing the DVR is sent to the CEs together with the Grid jobs. The ADU uses a timer-based polling to detect and fetch new versions of a database from the FTP site of origin, then it updates the databases located on the Storage Elements (SEs). The ARM receives notification of CPU time spent on a database by the GridBlast core, then it adjusts the number of replicas of the database on the SEs.

2.3. *GRID large scale in silico docking application on avian flu*

A collaboration of Asian, Africa and European laboratories has analyzed 300,000 possible drug components against the avian flu virus H5N1 using the EGEE Grid infrastructure [3].

The goal was to find potential compounds that can inhibit the activities of an enzyme on the surface of the influenza virus, the so-called neuraminidase, subtype N1. To study the impact of small scale mutations on drug resistance, a large set of compounds was screened against the same neuraminidase target and slightly different structures with mutations.

The main goal from the in silico screening is to predict which compounds and chemical fragments are most effective for blocking the active neuraminidases in case of mutations.

In silico virtual screening requires intensive computing, of the order of a few TFlops during one day to compute 1 million docking probabilities or for the molecular modelling of 1000 ligands on one target protein. In April and May 2006, the data challenge of the EGEE project led by Academia Sinica in Taiwan, CNRS-IN2P3 in France and CNR-ITB in Italy in collaboration with the European SSA BioinfoGRID project, was set up to identify new drugs for the potential variants of the Influenza A virus [7].

The computational tools for the work are based on molecular docking engines, such as AutoDock [8], to carry out a quick conformation search of small compounds in the binding sites, fast calculation of binding energies of possible binding poses, prompt selection for the probable binding modes, and precise ranking and filtering for good binders. Input for the avian flu data challenge consisted of 8 protein targets predicted from the neuraminidases subtype 1 (N1) to simulate the possible mutations of the H5N1 virus and 308,585 chemical compounds selected from ZINC and a chemical combinatorial library.

During the data challenge, the activity has distributed 54,000 jobs on 60 Grid CEs. The 6-weeks activity has covered the computing power of about 88 CPU years and has docked about 2 million pairs of target and chemical compounds. More than 60 000 output files with a data volume of 600 Gigabytes were created and stored in a relational database.

Potential drug compounds against avian flu are then identified and ranked according to the binding energies of the docked models [9]. Besides the biological goal of reducing the time and cost of the initial investment on structure-based drug design, there were two Grid technology objectives for this activity: one was to improve the performance of the in silico high-throughput screening (HTS) environment; the other to test another environment which enables users to have efficient and interactive control of the massive molecular dockings on the Grid.

2.4. *GRID based microarray expression profiling analysis*

DNA microarray technology allows the monitoring of expression levels for thousands of genes simultaneously. This allows monitoring of collective cell behaviour and cell cycles which is an important fact when annotating transcription binding sites. Genes with similar temporal and spatial gene expression patterns are most likely to be subject to a common regulatory logic [10].

The molecular signatures are based on expression profiles, and these profiles can then be used to automatically separate normal cells or tissues into their correct category. The role of clustering these expression profiles is significant towards the attempt to infer functional characteristics of a novel gene [11]. If it is grouped together with genes of known function, the newly discovered gene can be expected to have similar characteristics to them.

In the frame of the BioinfoGRID project we used the EGEE grid infrastructure to execute workflows exploiting the capabilities of a grid environment specific to clustering algorithms on gene expression data from samples taken from individuals with breast cancers of different types, West et al. [12]. For this purpose the K-Means algorithm was submitted in GRID and in a single PC. When the numbers of iterations were increased by increasing the length of a single computation, the execution time on the Grid increased as well but much smaller than the increase on a single PC. This study shows that when the number of clusters increases the

Grid's performance is not affected as the only thing that had changed is the number of resources used.

2.5. *GRID based system biology model simulator*

A system biology model simulator has been realized in order to calculate the model based on the Cell Cycle Database [13], an integrative resource which collects the main information regarding genes and proteins involved in budding yeast S.cerevisiae and mammalian cell cycle process.

The simulation software chosen for our system is XPP [14], a computational device frequently used in systems biology numerical calculations. XPPAUT allows the solution of differential equations using many different options for the numerical algorithm. It is widely used for the modelling of different biological pathways.

The models are essentially based on differential equations and they can describe abundances, kinetics and binding affinities of pathway components and their interactions [15]. In this work we consider models based on a system of nonlinear ordinary differential equations (ODE system) in which each X_i state variable (usually species concentrations) can be described by the Eq.(1):

$$\frac{dX_i}{dt} = F_i(X_1, X_2, ..., X_n; p_1, p_2, ..., p_m); i = 1, ..., n \tag{1}$$

where the function F_i is the rate of change or the rate law of the state variable X_i and pi are parameters of the function F_i. The time course of each state variable is obtained by solving the ODE system which requires a set of initial conditions $X_i(t{=}0)$. For each set of ODE system simulation which must be calculated, a grid job is submitted: it means that according to the user parameters selected a JDL script is dynamically generated with the information about the input and the related job requirements. The jobs are routed by the *Resource Broker* to the best *Computing Element* that is available at the moment (see Fig.2). The simulation of the cell cycle pathway is used for better understanding of cell cycle control in normal and transformed mammalian cells to help in discovery of anticancer drugs. The grid-oriented approach to solve ODE systems describing cell cycle models can be used to make the numerical simulations of the biological process easier and more accurate.

3. Conclusion

In this paper some of the on-going activities in the fields of functional genomics, transcriptomics, proteomics and virtual screening on Avian Flu based on GRID technology have been presented.

A range of different Bioinformatics and System Biology applications have been used GRID developed by the EGEE and BioinfoGRID project. An increasing number of applications in the fields of Bioinformatics, Biology, Computational Chemistry, Medicine and Biotechnology, are using GRID computing. On the base of this technology several experts of various disciplines are collaborating to search for solutions to highly complex problems in the fields of genome analysis, drug discovery and protein functional analysis.

Acknowledgments

The author would like to thank WISDOM, BioinfoGRID, EGEE and the Laboratory for Bioinformatics Technology team members.

References

[1] Gagliardi, F., Jones, B., Grey, F., Bégin, M.E., Heikkurinen, M. Building an infrastructure for scientific Grid computing: status and goals of the EGEE project, *Philosophical Transactions: Mathematical, Physical and Engineering Sciences*, 363: 1729-1742 (2005).
[2] BioinfoGRID: Bioinformatics applications in GRID, http://www.bioinfogrid.eu
[3] EGEE: Enabling Grids for E-science in Europe, http://public.eu-egee.org
[4] K.A. Karasavvas, R. Baldock, A. Burger: A criticality-based framework for task composition in multi-agent bioinformatics integration systems. *Bioinformatics*, Vol. 21, pp. 3155-3163(2005)
[5] gLite: Lightweight Middleware for Grid Computing, http://glite.web.cern.ch/glite
[6] S.F. Altschul, W. Gish, W. Miller, E.W. Myers and D.J. Lipman: Basic Local Alignment Search Tool, *J. Mol. Biol.* 215:403410 (2005)
[7] Irwin K. S. Li, Y. Guan, J. Wang, G. J. D. Smith, K. M. Xu, L. Duan, A. P. Rahardjo, P. Puthavathana, C. Buranathai, T. D. Nguyen, A. T. S. Estoepangestie, A. Chaisingh, P. Auewarakul, H. T. Long, N. T. H. Hanh, R. J. Webby, L. L. M. Poon, H. Chen, K. F. Shortridge, K. Y. Yuen, R. G. Webster and J. S. M. Peiris. Genesis of a highly pathogenic and potentially pandemic H5N1 influenza virus in eastern Asia. *Nature* 430:209-213 (2004)
[8] G. M. Morris, D. S. Goodsell, R. S. Halliday, R. Huey, W. E. Hart, R. K. Belew and A. J. Olson. Automated Docking Using a Lamarckian Genetic Algorithm and Empirical Binding Free Energy Function. *J. Computational Chemistry,* 19:1639 (1998)
[9] H. C. Lee, J. Salzemann, N. Jacq, HY. Chen, LYung Ho, I. Merelli, L. Milanesi, V. Breton, S. C. Lin, YT. Wu (2007). Grid-enabled High-throughput in silico Screening against Influenza A Neuraminidase. *IEEE transaction in Nanobioscience.* dec;5(4):288-95 (2006)
[10] B.P. Berman, Y. Nibu, B. D. Pfeiffer, P. Tomancak, S. E. Celniker, M. Levine, G. M. Rubin, and M. B. Eisen. Exploiting transcription factor binding site clustering to identify cis-regulatory modules involved in pattern formation in the Drosophila genome *PNAS* January 22, vol. 99 |no. 2 757-762 (2002)
[11] De Hoon, M., J., L., Imoto, S., Nolan, J., Miyano, S., Open Source Clustering Software, *Bioinformatics*, Vol.20, no. 9, pp. 1453-1454 (2004)
[12] West,M., Blanchette, C., Dressman, H., Huang, E., Ishida, S., Spang, R., Zuzan, H., Marks, J., Neris, J. Predicting the clinical status of human breast cancer using gene expression profiles. PNAS 98: 1146211467 (2001)
[13] Cell Cycle Database [http://www.itb.cnr.it/cellcycle]
[14] M. Hucka et al, The Systems Biology Markup Language (SBML): A Medium for Representation and Exchange of Biochemical Network Models, *Bioinformatics*, **19(4)**, 524-531(2003)
[15] T. Ideker, D. Lauffenberger, Building with a scaffold: emerging strategies for high- to low-level cellular modelling, *Trends in Biotechnology*, **21(6)**,255-62(2003)

GEOMETRICAL AND TOPOLOGICAL MODELLING OF SUPERCOILING IN SUPRAMOLECULAR STRUCTURES

LUCIANO BOI

HSS, Centre de Mathmatiques, and
LUTH, Observatoire de Paris-Meudon
54, Bd Raspail, 75006 Paris (France)

Both microscopic and macroscopic worlds are plentiful in knotted and linked structures. But even at the mesoscopic level, works in molecular topology have shown that Nature builds DNA molecules, which form actual knots and links between themselves. Qualitative similarities may be observed between forms of the microscopic world, as well as between those of the macroscopic world; and by extracting the structures, it is possible to study their dynamics thanks to topological methods and techniques. The aim of this paper is to show that knots and links are ubiquitous scale-independent objects carrying a tremendous amount of precious information on the emergence of new forms and structures especially in living matter. In fact, knotting and unknotting are "universal" principles underlying these forms and structures. This study is focused at showing that differential geometry and topological knot's theory can be used notably (a) to describe 3-dimensional structures of DNA and protein-DNA complexes, and (b) to explain how geometrical supercoiling occur in the most fundamental supra-molecular structures and why it is essential for their biological activity.

1. The topological compaction of the double-helix molecule into the chromatin and the role of supercoiling

One of the most striking phenomena which reveal the deep connection between topological problems and biological processes is the compaction of chromatin into the chromosome of cells, whose full explanation is one of most challenging task of biology today. Here we are faced with a genuine problem of differential topology. What kind of deformations the double-strands linear DNA molecule undergoes in order that it condenses into an extremely compact form, corresponding to the metaphase of the chromosome? The answer to this question is actually far of being clear or complete. However, some facts have been elucidated very recently.

The key distinguishing characteristic of the eukaryotic genome is its tight packaging into chromatin, a hierarchically organized complex of DNA and histone and non-histone proteins. How genome operates in the chromatin context is a central question in the molecular genetics of eukaryotes. The chromatin packaging consists of different levels of organization. Every level of chromatin organization, from nucleosome to higher-order structure up to its intranuclear localization, can contribute to the regulation of gene expression, as well as affect other functions of the genome, such as replication and repair. Concerning gene expression, chromatin is important not only because of the accessibility problem it poses for the transcription apparatus, but also due to the phenomenon of chromatin memory, that is, the apparent ability of alternative chromatin states to be maintained through many cell divisions. This phenomenon is believed to be involved in the mechanism of epigenetic inheritance, an important concept of developmental biology.

Supercoiling is one of the three fundamental aspects of DNA compaction; the other two are flexibility and intrinsic DNA curvature. For example, the problem of DNA compaction in *E. coli* can be putted in the following words: the DNA must be compacted more than a thousand fold in the cell, yet it still needs to be available to be transcribed. In order this compaction to be achieved, it is requested some kind of anisotropic flexibility or "bendability" of DNA, which is very much sequence-specific, and is different from the static "rigidity" of DNA. Whereas *persistence length* of DNA is relatively non-specific, and just has to do with its overall "rigidity", anisotropic flexibility is a measure of a particular sequence to be deformed by a protein (or some other external forces). Some sequences are both isotropically flexible and "bendable"-for example, the TATA motifs. Perhaps one of the best examples of this is the binding site for the Integration Host Factor (IHF)-there are certain base pairs that are highly distorted upon binding of this protein. It is quite impressive that this protein induces a bend of 180 degrees into a DNA helix. In other words, the curvature, say K, at each sequence of the two strands of DNA helix must be very sharp in order the DNA double helix may assume its extremely compact form. So the relationship between (intrinsic) curvature and conformational (or topological) flexibility appear to be crucial in the understanding of the biological activity of cells.

Indeed, when one consider that the DNA must be compacted more than a thousand fold in the cell, it is probably not surprising that almost any protein that binds to DNA will bend it. Moreover, since the total curvature K of entire DNA double-helix segment depends on the torsional stress which applies to DNA strands, and these strands form hence a twisted curve, i.e., a curve of double curvature in the three-dimensional space of the cell nucleus, DNA must coil many times in a very ordered way inside the nucleus (otherwise, if the chromosome of a human cell were in the form of a random coil, they would not fit inside the nucleus). The DNA double helix coils first by overwinding or underwinding of the duplex. The supercoiled form of a circular DNA molecule is much more compact than the other possible conformations, i.e., nicked and linear. In its supercoiled form, DNA molecule minimize to the highest the space volume which its occupies in the nucleus. Supercoils condense DNA and promote the disentanglement of topological domains.

Today we know that DNA is topologically polymorphic. The overwound or underwound double-helix can assume exotic forms known as *plectonemes*, like the braided structures of a tangled telephone cord, or *solenoids*, similar to the winding of a magnetic coil. (i) Plectonemically supercoiled DNA is unrestrained and frequently branched, while toroidal supercoils is restrained by proteins and it is more compact. (ii) DNA can be either positive or negatively supercoiled. In particular, eukaryotic DNA is negatively supercoiled in and around genes, and it istransiently negatively supercoiled behind RNA polymerase during transcription. (iii) Negative supercoiling favors DNA-histone association and the formation of nucleosomes, the first step in packaging DNA. Because the solenoidal DNA wrapping around a nucleosome core creates about two negative supercoils, it is understandable that DNA that fulfills this topological prerequisite will more easily form nucleosome. (iv) These tertiary structures have an important effect on the molecule's secondary structure and eventually its functions. For example, supercoiling induced destabilization of certain DNA sequences and allows the extrusion of cruciform or even the transcriptional activation of eukaryotic promoters. Another essential process, DNA transcription, can both generate and be regulated by supercoiling. During replication, the chromosomes need to be partitioned and the two strands of DNA must be continuously unlinked during replication. The topoisomerases that accomplish this might instead be expected to entangle and knot chromosomes because of the huge DNA concentration in vivo. There are actually several factors that solve this problem and contribute to the orderly unlinking of DNA. A major contributor to chromosome partitioning is the condensation of daughter DNA upon itself soon after replication. DNA condensation is due primarily to supercoiling. Another factor promoting chromosome partitioning is that the type-II topoisomerases of all organisms do not just speed up the approach to topological equilibrium, but actually change the equilibrium position. They actively remove all DNA entanglements. This requires that topoisomerases sense the global conformation of DNA even though they interact with DNA only locally. In fact, topoisomerases achieve this because, by positioning themselves at sharp bends in DNA, they carry out net disentanglement of DNA. An equal partner to the topoisomerases in chromosome segregation is the helicases. They seem to convert the energy of ATP hydrolysis into unwinding DNA. All the enzymes that play critical roles in DNA unlinking and chromosome segregation, topoisomerases, helicases, and condensins, are motor proteins. They use the energy of ATP hydrolysis to move large pieces of DNA over long distances.

The previous discussion can be summed up by saying that supercoiling has three essential roles. (i) First, (–) supercoiling promotes the unwinding of DNA and thereby the myriad processes that depends on helix opening. ii) The second essential role of supercoiling is in DNA replication. For replication to be completed, the linking number of the DNA, Lk, must be reduced from its vast (+) value to exactly zero. In bacteria, DNA gyrase introduces (–) supercoils and thereby removes parental Lk. (iii) The third essential role of supercoiling is conformational. DNA manifests the difference between the relaxed and naturally occurring values of Lk by winding up into supercoils. These supercoils condense DNA and promote the disentanglement of topological domains. This can be accomplished equally well by (–) or (+) supercoiling. Let us still underline two important facts. First, the promotion of decatenation by supercoiling has also been directly demonstrated *in vivo*. Second, the volume occupied by a supercoiled molecule is much more smaller than that of a relaxed DNA. This difference in volume is due mostly to the formation of superhelical branches. Indeed, supercoiled DNA branches and bends itself into a ball. The decrease in chromosomal volume by supercoiling decrease the probability that the septum will pass through the chromosome during cell division.

2. A mathematical model for explaining the interphase folding of chromatin fiber

Of paramount importance to the understanding of gene expression and biological regulation, is the mechanism which drives and controls the packaging of DNA and its organization within the chromatin structure. The lowest level of organization is the nucleosome, in which two superhelical turns of DNA (a total of 165 base pairs) are wound around the outside of a histone octamer. Nucleosomes are connected to one another by short stretches of linker DNA. During chromatin assembly on nascent DNA, acetylated histones H3 and H4 are sequestered by the DNA first, histones H2A and H2B follow, and, finally, H1 binds, stabilizing chromatin folding within the irregular 30 nm fiber. At the next level of organization the string of nucleosomes is folded into a fiber about 30 nm in diameter, and these fibers are then further folded into higher-order structures. More precisely, during the progressive assembly of chromatin, DNA is compacted, nucleosome formation leads to a sevenfold compaction of DNA, and the subsequent formation of the 30 nm fiber contributes a further sevenfold compaction. These four successive steps of compactions represent the major topological constraints of DNA in eukaryotic nucleus. At levels of structure beyond the nucleosome the fundamental mechanisms of folding are still unknown. We know that the 11nm nucleosome units coil into a 30 nm solenoid structure which is stabilized by H1 histone. The solenoid forms loops that attach to a scaffold of non-histone protein, which leads to the chromatin supercoiling during condensation within metaphase chromatids. This intermediary and possibly crucial level of compaction of complexes DNA-proteins into the final form, a mitotic chromosome, is very scarcely understood.

Among the different hypothetical models that have been proposed over the last years for the folding of the chromatin fiber during interphase, the so-called radial-loop model seems to us the most suitable for explaining the formation of the 30-nm solenoid structure. Let us suggest, specifically, a theoretical model by applying methods and techniques from the classification theory of compact connected 2-manifolds, which has been one of the most important and far-reaching mathematical results of the 20th century. We start with the theorem

Theorem 2.1. *Let M be a closed simply-connected orientable manifold. M can be expressed as a union*

$$M = D \cup D' \cup \cup_{i=1}^{n} S_i$$

of polyhedral 2-cells with disjoint interiors, such that (1) for each i, each of the sets $S_i \cap D$ and $S_i \cap D'$ is the union of two disjoint arcs and (2) $D \cap D'$ is the union of

2n disjoint arcs.

The sets S_i will be called strips, and M' will be called a 2-cell with strips. Evidently, such an M' can always be imbedded in R^3, and thus can be described by a figure such as Fig 1 . Under the conditions of theorem 1 boundary $\partial M'$ must be a 1-sphere, but aside from this, the strips S_i may be attached to $\partial M'$ at any set of disjoint arcs. If $S_i \cup D$ is an annulus, then S_i will be called *annular* (relative to D, of course) and if $S_i \cup D$ is a Möbius band, then S_i we be called *twisted.* Thus, in figure 3, S_3 and S_6 are twisted, and the rest of the strips are not. Note that in investigating the topology of M', we do not care whether the sets $S_i \cup D$ are knotted. Note also that indicating "multiples twists" would contribute nothing to the generality of the figure. For example, a double twist gives an annulus, and a triple twist gives a Möbius band.

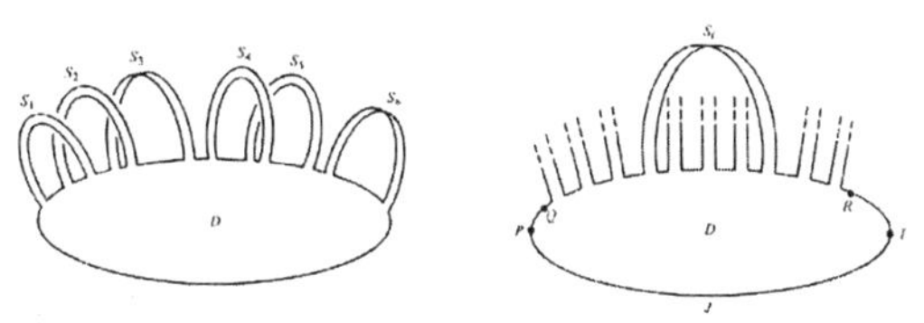

(a) (b)

Fig. 1.

We shall now simplify this representation of M in various ways. (1) Suppose that S_i is a twisted strip, so that $S_i \cup D$ is a Möbius band. Let $J = \partial(S_i \cup D)$, so that J is a polygon. As in Fig 2, let P, Q, R, and T be points of J, not lying in any set S_j; let PT be the arc in J, between P and T, that intersects ∂S_i suppose that P, Q, R and T appear in the stated order on J; and suppose that the arcs $PQ \subset PT$ and $RT \subset PT$ intersect no set S_j. We assert that there is a piecelinear homeomorphism (PLH):

$$h : M \Leftrightarrow M, J \Leftrightarrow J, D \cup S_i \Leftrightarrow D \cup S_i$$

$$P \to P, T \to T, QT \Leftrightarrow RT$$

such that $h|(J - PT)$ is the identity. Consider the 2-cell $D, h(D')$, S_i, and $h(S_j)(j \neq i)$. These have all the properties stated above for D, D', S_i, and S_j $(j \neq i)$. The operation which replaces the old system of 2-cells by the new will be called *operation* α. We now renumber the 2-cells S_i in such a way that $S_1, S_2, \ldots, S_k$ are annular, and $S_{k+1}, S_{k+2}, \ldots, S_n$ are twisted.

Lemma 2.1. *In the conclusion of Theorem 2.1, we can choose the 2-cells in such* *a way that (a) the intersections $S_i \cap D$ $(i > k)$ lie in disjoint arcs in ∂D and (b) $\cup_{i>k}(S_i \cap D)$ lies in an arc in ∂D which intersects no annular strip S_j.*

(2) If we have no annular strips, then we proceed to step (3) below. If we have an annular strip S_i, then there must be another annular strip S_j which is "linked with S_i on ∂D", as indicated in figure 3. (If not, $\partial M' = \partial D'$ would not be connected.) The set $D \cup S_i \cup S_j$ is then a handle.

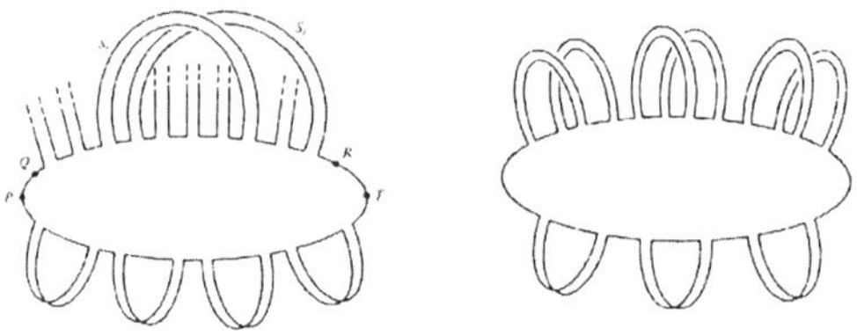

(a) (b)

Fig. 2.

Recall that by a *handle* we mean a space obtained by deleting from a torus the interior of a 2-cell. A *2-sphere with n holes* is a space obtained by deleting from a 2-sphere the interiors of n disjoint 2-cells. If a handle is attached to the boundary of each of the holes, the resulting space is a 2-sphere with n handles. A *projective plane* is a space defined in such a way that each pair of antipodal points of the circle are supposed to be identified. A *sphere with n cross-caps* is a space obtained by starting with a sphere with n holes and then attaching a Möbius band to the boundary of each of the holes.

By operation β, closely analogous to α, we slide the strip $S_r (r \leq k, r \neq i, j)$ along the arc PT, so as to get a situation in which $(S_i \cup S_j) \cap D$ lies in an arc in ∂D which intersects no set $S_r (r \neq i, j)$. We do this for each such handle. The figure now looks like figure 4.

(3) Let $m = n - k$ be the number of the twisted strips S_i, and suppose that $m > 2$. Consider the first three of the twisted strips (starting in some direction from the annular strips) as shown in figure 5. By two operations of the type α, we slide PQ along $\partial(D \cup S_s)$ so as to move it onto $P'Q' \subset IntAB \subset \partial D$; and we slide RT along $\partial(D \cup S_s)$ onto $R'T' \subset IntAB$. The figure now looks like figure 8. It is easy to check that the new strips $S'_r, \cup S'_t$ is a handle.

By another application of β, we move $S_s \cap D$ in figure 7. Thus we have introduced a new handle into the figure, and reduced the number of twisted strips by 2. Thus we may assume, in theorem 2.1, that the number m of twisted strips is ≤ 2. M is orientable if and only if $m = 0$ at the final stage. To each linked pair of annular strips, and to each twisted strip, we add a 2-cell lying in D, as indicated by the dotted arcs in figure 6. This gives a set $\{H_i\}$ of handles $(h \geq 0)$ and a set $\{B_j\}$ of m Mbius bands $(0 \leq m \leq 2)$.

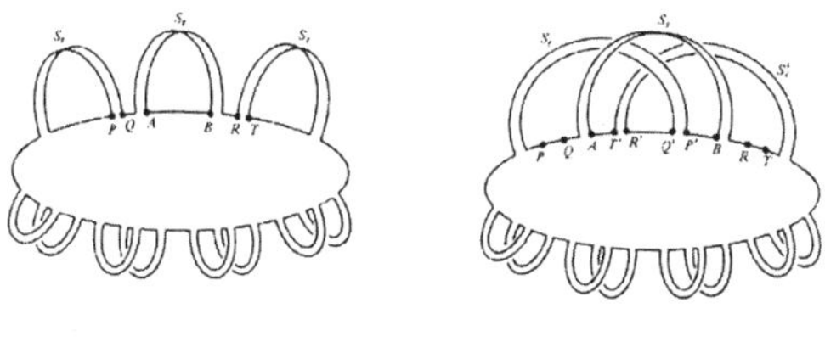

(a) (b)

Fig. 3.

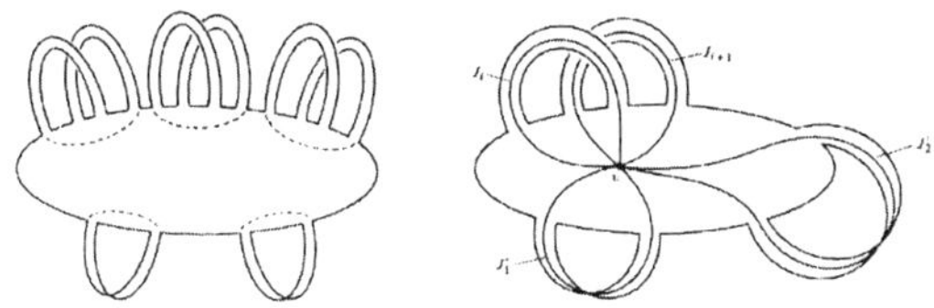

(a) (b)

Fig. 4.

Consider the set $N = Cl[M - \cup H_i \cup \cup B_j]$. N is the union of two 2-cells D_1 and D_2, with $D_1 \subset D$ and ∂D_2, it follows that N is a sphere with holes. Thus we have proved the following

Theorem 2.2. *Let M be a compact connected 2-manifold. Then M is a 2-sphere with h handles and m cross-caps* $(h \geq 0, 0 \leq m \leq 2)$.

We now define a new open cell decomposition of M, as follows. As indicated in figure 8, we choose a point v of $IntD$, and we define a collection $\{J_i, J'_j\}$ of polyhedral 1-spheres (one J_i for each annular strip, and one J'_j for each twisted strip) such that each of them "runs from v through the corresponding strip, and then returns to v," and such that each two of the sets in $\{J_i, J'_j\}$ intersect at v and nowhere else. This gives an open cell-decomposition C' of M, with one vertex v, $2h$ edges $J_i - \{v\}$, m edges $J'_j - \{v\}$, and one 2-face $C^2 = M - [\cup J_j \cup \cup J'_j]$.

Thus we have:

Theorem 2.3. *Let M be a 2*-sphere with h handles and m cross-caps$(h = 0, 0 \leq m \leq 2)$. *Then* $\chi(M) = 2 - (2h + m)$.

Proof. $V - E + F = 1 - (2h + m) + 1$. □

3. Biological justifications for the model above

One way to show the relevance of the topological model we sketched above is to investigate the spatial organization and functional compartmentalization of chromosomes, and the nucleus itself, within the quest to understand how the expression of complex genomes is regulated. Inside the higher eukaryotic chromosome, DNA is folded through DNA-protein interactions into multiple levels of organization. At the highest level, these yield a compaction ratio of more than 20 000: 1 in terms of the ratio of linear B-form DNA to the length of the fully compacted metaphase chromosome. While the extent of compaction within mitotic chromosomes is well known, less appreciated is the fact that compaction remains extremely high within interphase nuclei. The bulk of genomic DNA in interphase is likely to be packaged within large-scale structures well above the level of the 30 nm chromatin fiber (for further details see Widom et al. [1]).

For technical reasons, most research into chromosome structure has focused on the structure of maximally condensed, metaphase chromosomes. An experimental approach based largely on unfolding chromosome structure through extraction of chromosomal proteins has led to a radial loop model of chromosome structure. In this model, structural proteins, which are resistant to high salt and detergent extraction, anchor the bases of 30 nm chromatin fiber loops (~20-200 kb long) to a chromosome "scaffold", which itself may be helically coiled. Specific SAR/MAR DNA sequences (scaffold attachment regions or matrix attachment regions) are hypothesized to form the bases of these loops, attached to specific proteins which are predicted to make up the chromosome scaffold. Specific sequences are found remaining at the axial core in extracted human metaphase chromosomes, but it is not clear whether the same SAR/MAR sequences are attached to an underlying scaffold in both mitotic and interphase chromosomes. Experiments using fluorescence in situ hybridization (FISH) on cell nuclei has led to a giant-loop, random walk model for interphase chromosomes, based on statistical analysis of the mean separation between two chromosomes sites, as a function of genomic distance.

Ideally, any model of large-scale chromatin folding would unify mitotic and interphase chromosome structure and predict the structural transitions accompanying cell-cycle-driven chromosome condensation/decondensation. The radial-loop, helical-coil model of mitotic chromosome structure has been extended to interphase chromosomes (figure 5(a)). However, this has required postulating a particular loop geometry that might, under special circumstances, give rise to a fiber with an elliptical 60-90 nm cross-section [2] . An alternative model proposes a successive, helical coiling of 10 nm chromatin fiber into 30-50 nm tubes, and of these into 200 nm diameter tubes, which coil into *c.*600 nm metaphase chromatids [3] . Finally, a folded chromonema model is based on in vivo light microscopy combined with TEM ultrastructural analysis of folding intermediates during the transition into and out of mitosis. In this model, 10 and 30 nm chromatin fiber fold to form a *c.*100 nm diameter chromonema fiber, which then folds into a 200 – 300 nm diameter prophase chromatid, which itself coils to form the metaphase chromosome (figure 5(b)) (Belmont et al [4]) . It is still unclear how these structural models of mitotic and interphase chromosome structure integrates with the underlying biochemistry responsible for chromosome condensation. The two chief protein components of the mitotic chromosome scaffold, topoisomerase IIα and SCII, have more clearly identified DNA topological activities than structural roles. SCII is a component of the mitotic condensing complex, which recently has been demonstrated to have the ability to introduce positive supercoils into DNA in the presence of topoisomerase II in a stoichiometric manner. SCII also shows a non-ATP-dependent enhancement of re-annealing of complementarity DNA strands (see section 1 for more details on this topic).

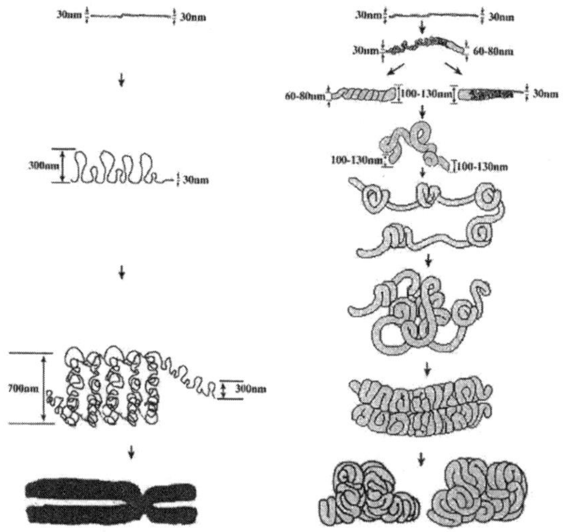

(a) (b)

Fig. 5. *(a) Radial-loop model for mitotic chromosome structure. A looping of the 30 nm fiber gives rise to a 300 nm structure in which 50-100kb looped DNA attaches at the base of the loop to a chromosomal scaffold. This structure coils helically to form the metaphase chromosome. (b) Chromosome model of interphase chromatin structure. Progressive levels of coiling of the 30 nm fiber into 60-80 nm and 100-130 nm fibers are depicted. Chromonema fibers kink and coil to form regions of more dispersed or compact chromatin. Extended chromonema fibers predominate in G_1 while more compact structures become abundant during cell-cycle progression. Chromonema folding culminates with the formation of the G_2 chromatid, which coils to form the compact metaphase chromosome.*

More specifically, the geometrical model we suggested in the previous section might fit well with the three-dimensional packing process of chromatin, first, into a 300 nm extended scaffold-associated form, followed by a 700 nm condensed scaffold-associated form. In fact, the condensation of metaphase chromosome results from several orders of folding and coiling of 30 nm chromatin fibers. For example, electron micrographs of histone-depleted metaphase chromosome from HeLa cells reveal long loops of DNA anchored to a chromosome scaffold composed of non-histone proteins.

This scaffold has the shape of the metaphase chromosome and persists even when the DNA is digested by nucleases. As depicted schematically in figure 6, megabase long loops of the 30 nm chromatin fiber are thought to associate with the flexible chromosome scaffold, yielding an extended form characteristic of chromosome during interphase. Coiling of the scaffold into a helix and further packing of this helical structure produces the highly condensed structure characteristic of metaphase chromosome.

Furthermore, in situ hybridization experiment with several different fluorescent-labeled probes to DNA interphase cells support the loop model shown in figure ??. In these experiment, some probe sequences separated by millions of base pairs in linear DNA appeared reproducibly very close to each other in interphase nuclei from different cells (figure 6). These closely spaced probe sites are postulated to lie close to specific sequences in the DNA, called scaffold-associated regions (SARs) or matrix-attachment regions (MARs), that are bound to the chromosome scaffold. SARs have been mapped by digesting histone-depleted chromosome with restriction enzymes and then recovering the fragments that are bound to scaffold proteins. In general, SARs are found between transcription units. In other words, genes are located primarily within chromatin loops, which are attached at their bases to a chromosome scaffold. Experiments with transgenic mice indicate that is some cases SARs are required for transcription of neighboring genes. In Drosophila, some SARs can insulate transcription units from each other, so that proteins regulating transcription of one gene do not influence the transcription of a neighboring gene separated by a SAR. Individual interphase chromosomes, which are less condensed than metaphase chromosomes, cannot be resolved by standard microscopy or electron microscopy. Nonetheless, the chromatin of interphase cells is associated with extended scaffolds and is further organized into specific domains.

4. Conclusions and prospects

All that strongly suggest that the secrets of life and what allows the biological growth of all organisms maybe lies in topology, namely in the fact that forms posses the capacity to convert structures and functions one into another. In fact, the topological compaction of our genome, the basic building block of which is the nucleosome (a protein-DNA structure), provides a whole repertoire of information in addition to that furnished by the genetic code. This mitotically stable information is not inherited genetically and is termed epigenetic. Epigenetic phenomena are propagated alternative states of gene expressions, and alternative states of protein folding, and they are closely linked with histone- and chromatin modifications. Still more than genetics events, one could say that the comprehension of epigenetic processes essentially requires a better understanding of the role played by topological transformations.

One of the challenges in chromatin research is to understand how levels of chromosome organization beyond the 30-nm chromatin filaments condense to form the

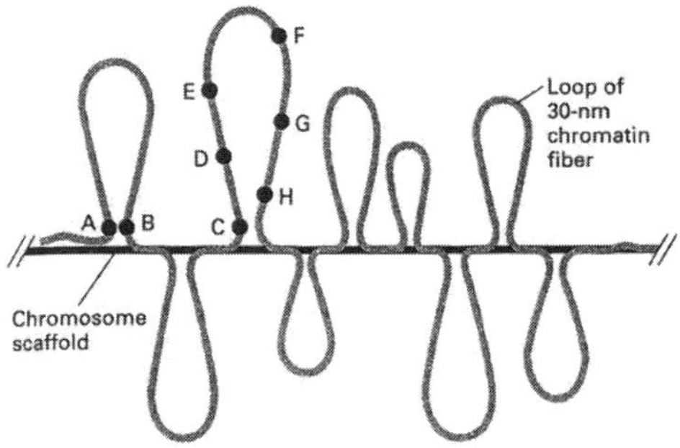

Fig. 6. *Experimental demonstration of chromatin loops in interphase chromosomes. In situ hybridization of interphase cells was carried out with several different fluorescent-labeled probes specific for sequences separated by known distances in linear, cloned DNA. Lettered circles represent probes. Measurement of the distances between different hybridized probes, which could be distinguished by their color, showed that some sequences (e.g., A, B, and C), separated from each other by millions of base pairs, appear located near each other within nuclei. For some sets of sequences, the measured distances in nuclei between one probe (e.g., C) and sequences successively farther away initially appear to increase (e.g., D, E, and F) and then appear to decrease (e.g., G and H). The measured distances between probes are consistent with loops ranging in size from one to four million base pairs.*

cell metaphase chromosome. We need very likely a topological model that accounts for the several ordered transformations that are required for the dimensions of metaphase chromosomes, which are 10,000-fold shorter and 400 to 500-fold thicker than the double stranded DNA helices contained within them. Loops-like arrangement of chromatin and their stacking into a cylinder of 800 to 1000 nm in thickness, which is in good agreement with the diameter of the metaphase chromosome, and twisting the cylinder into a superhelix would further compact it, is a model that account well for the corkscrew appearance of metaphase chromosome.

Happily cells achieve this tight packing of DNA while still maintaining the chromosome in a form that allows regulatory proteins to gain access to the DNA to turn on (or off) specific genes or to duplicate the chromosomal DNA. This means that all epigenetic states and processes have to be established, inherited, controlled and modified in such a way as to permit that their integrity is maintained while preserving the possibility of accessibility and deformability. Thus, the topological flexibility of the many-levels structure of chromosomes, the chromatin dynamics and the gene's regulatory modifications are intimately interconnected processes and determinant factors of cellular organization.

Condensation of genetic material appears to be a very fundamental mechanism of life. Now, since condensation realize as a kind of topological embedding of one space, the restrained linear DNA, into another space, the 3-dimesional chromosome structure in the nucleus, it seems reasonable to think that topological embeddings and transformations are dynamic processes that are essential for the maintain and the integrity of life. One demonstration of that is the fact that the exotic supercoiled forms that double helix can assume are tertiary structures which have an important effect on the molecule's secondary structure and its function. DNA and chromosome organization must fulfill precise topological prerequisite in order to achieve certain functional processes. In particular, DNA transcription and replication can both be enhanced and regulated by topological supercoiling. It now appear clear, for example, that for replication to be completed, the linking number of the DNA, Lk, must be reduced from its vast (+) value to exactly zero. In bacteria, DNA gyrase introduces (–) supercoils and thereby removes parental Lk. Moreover, in certain cases, the severity of the phenotype can be controlled by changing the level of supercoiling in the cell.

We have thus three interrelated theoretical and experimental facts, which we would like to stress: 1) DNA condensation is a driving force for double helix unlinking and chromosome portioning, by folding, in topological domains. 2) Condensation is achieved by supercoiling, which is a topological state of macromolecules enhanced by three kinds of deformations (embeddings): twisting, writhing and knotting. If the DNA is modeled as a ribbon in three-space whose axis is not flat in the plane, we can define the *twist* of the ribbon abstractly as the integral of the incremental twist of the ribbon about the axis, integrated as we traverse the axis once; so it simply measures how much the ribbon twists about the axis from the frame of reference of the axis: it need not be an integer (for a rigorous mathematical definition of these notions, see Sternberg [5], and Boi [6]). The *writhe* measures how much the axis of the ribbon is contorted in space. Because (-) supercoiling in bacteria arises from a topological misalignment and not a protein corset, it has the flexibility to do work. 3) Supercoiling results from topological strain and the contortion of DNA by proteins, notably the nucleosomal histone octet and the structural maintenance of chromosomes (SMC) proteins.

To conclude, it must be emphasized that the complex topology of DNA is essential for the life of all organisms. In particular, it is needed for the process known as DNA replication, whereby a replica of the DNA is made and one copy is passed on to each daughter cell. The most direct evidence for the vital role played by DNA topology is provided by the results of attempts to change the topology of DNA inside cells. Two related questions arise immediately from the recognition that DNA topology is essential for life. How did the complex topology of DNA evolve, and why is it so important for cells and organisms? We foresee that a great deal of the future research will relate to these two fundamental issues.

References

[1] Widom, J., "Structure, dynamics, and function of chromatin in vitro", Annu. Rev. Biophys. Biomol. Struct., 27 (1998), 285-327.
[2] Manueledis, L., "A view of interphase chromosomes", Science, 250 (1990), 1533-1538.
[3] Sedat, J. and Manuelidis, L., "A direct approach to the structure of eukaryotic chromosomes", Cold Spring Harbor Symp. Quant. Biol., 42 (1978), 331-342.

[4] Belmont, A.S. and Bruce, K., "Visualisation of G1 chromosome: a folded, twisted, supercoiled chromonema model of interphase chromatid structure", J. Cell Biol., 127(1994), 287-299.
[5] Sternberg, S., "Lectures on Differential Geometry", Chelsea, New York, 1983.
[6] Boi, L., "Topological knots models in physics and biology", in Geometries of Nature, Living Systems and Human Cognition, L. Boi (ed.), World Scientific, Singapore, 2005, 203-278.

PART C

Methods and Techniques

OPTIMISATION STRATEGIES FOR MODELLING AND SIMULATION

JEAN LOUCHET

COMPLEX Team, Institut National de Recherche en Informatique et Automatique
Rocquencourt, 78153 Le Chesnay cedex, France
E-mail: Jean.Louchet@gmail.com

Progress in computation techniques has been dramatically reducing the gap between modeling and simulation. Simulation as the natural outcome of modeling is used both as a tool to predict the behavior of natural or artificial systems, a tool to validate modeling, and a tool to build and refine models - in particular identify model internal parameters. In this paper we will concentrate upon the latter, model building and identification, using modern optimization techniques, through application examples taken from the digital imaging field. The first example is given by Image Processing with retrieval of known patterns in an image. The second example is taken from synthetic image animation: we show how it is possible to learn the model's internal physical parameters from actual trajectory examples, using Darwin-inspired evolutionary algorithms. In the third example, we will demonstrate how it is possible, when the problem cannot easily be handled by a reasonably simple optimization technique, to split the problem into simpler elements which can be efficiently evolved by an evolutionary optimization algorithm - which is now called "Parisian Evolution". The "Fly algorithm" is a realtime stereovision algorithm which skips conventional preliminary stages of image processing, now applied into mobile robotics and medical imaging. The main question left is now, to which degree is it possible to delegate to a computer a part of the physicist's role, which is to collect examples and build general laws from these examples?

Keywords: Model building, optimization techniques, digital imaging

1. Exploring Parameter Spaces using Evolutionary Programming

Obviously, artificial evolution - also known as evolutionary programming - is not the only way to function optimization. However, this presentation will focus on artificial evolution as promising tools to resolve problems related to data analysis. Evolutionary programming is essentially based on Charles Darwin's theory on the evolution of biological species. Its philosophy is to consider that the evolution of species is driven by a combination of the following factors: heredity, selective pressure of the environment in favor of the fittest, random mutations, (optional) crossover. More thorough descriptions of how these principles have been transposed into a family of programming techniques can easily be found in textbooks - I will only summarize the main points here.

The evolutionary computation (EC) domain may be structured into two main schools, emerged at about the same time in the 60s. The American school developed genetic algorithms and genetic programming – the best known figures are probably David Goldberg [1] and John Koza [2]. In parallel, the German school developed the Evolution Strategies approach – the best known figures are Hans-Paul Schwefel [3] and Ingo Rechenberg. The scheme is basically the same; the main difference relies in the search space: Booleans or finite alphabets with genetic algorithms; source code with genetic programming; real numbers with evolution strategies.

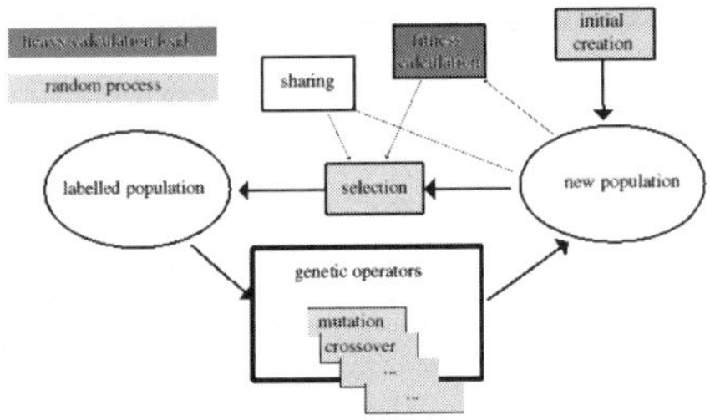

Fig. 1. One of the possible architectures for artificial evolution.

The main common features of EC are:

- a search space E (also called the *"parameter space"*) - elements of E are called *individuals;*
- a function f to be optimized (the *"fitness function"*):

$$f : (x_1, x_2, ..., x_n) \in E \rightarrow f\,(x_1, x_2, ..., x_n) \in R$$

- genetic operators, such as mutation and crossover.

In order to find *argmax(f)*, a population $P \subset E$ is created randomly. Fitness is calculated over this population in order to apply a *selection* operator (ranking, tournament, etc.) and select which individuals will survive to the next generation. In most implementations, eliminated individuals will be replaced by an equal number of new individuals created through the application of genetic operators to the former population. When properly adjusted, the population will converge and concentrate around the main maxima of the fitness function, and after a sufficient number of generations the best individual will give a good approximation of the global optimum.

Discussing in more detail the details, pros and cons of the different approaches to EC would be far beyond the scope of this presentation. However I will outline the fact that they do not deserve the reputation of clumsiness and inefficiency some people give them. My experience shows that most disappointments with EC come from applying one variant to a given real-world problem without a proper examination of which variant is most appropriate. One cannot stress too much the fact that careful parameter space choice and genetic parameters adjustment are the key to success. There is no magical "all-purpose, self-adjustable and fast" evolutionary algorithm.

Of course, not all optimization problems are relevant of EC. Some problems are better solved by classical methods, others may require some degree of hybridization in order to combine the robustness of EC with the accuracy of convex optimization methods, and in some cases EC may provide the fastest resolution method - or even the only one, as this will be illustrated by the following examples.

2. Retrieving Patterns: following the Hough Transform

The Hough transform [4] is classically used in image analysis in order to detect alignments. Straight lines being represented by the equation:

$$\rho = x \, \cos\theta \, + \, y \, \sin\theta$$

each point (x, y) in the image votes for the set of all the (θ, ρ) in the parameter space verifying the equation above.

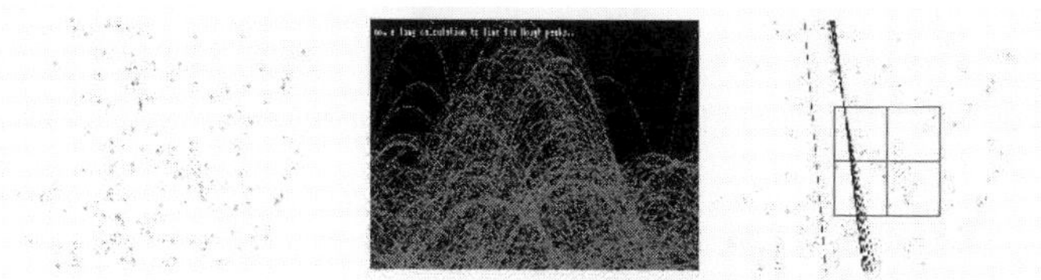

Fig. 2. Detection of alignments of Neolithic monuments in South-West England.

This process is fairly slow and memory consuming. It is virtually impossible to transpose it to cases where the dimension of search space is higher than 3, as it leads to calculate vote values over the whole parameter space then explore it exhaustively in order to detect the best solutions. An appealing alternative is to directly explore the search space [Roth 1992, Lutton 1994, Ser 1999] rather than the data space. Evolution Strategies [Rechenberg 1994, Bäck 1995] give an interesting opportunity to only calculate values in the part of the parameter space where population individuals actually are, rather than on the whole parameter space.

It is thus possible to build an "evolutionary version" of the Hough transform. In the classical context of the Hough transform, detecting straight lines in images, the evolutionary version only gives a slight advantage, as the classical method where each point (x, y) in the image, votes for the set of the (θ, ρ) in the parameter space such that $\rho = x \, \cos\theta \, + \, y \, \sin\theta$ is reasonably fast and memory consuming.

However, the real plus of evolutionary parameter space exploration comes to light when searching patterns in high dimensional spaces.

In the "balls" image sequence, the problem consists in detecting circles with unknown diameters (a moving ball) in an image sequence. The individuals are triplets (a, b, r) containing the parameters of the circles $(x-a)^2+(y-b)^2 = r^2$. An individual's fitness is defined as the average gradient norm taken on 40 points randomly chosen on the circle. The algorithm parameters are:

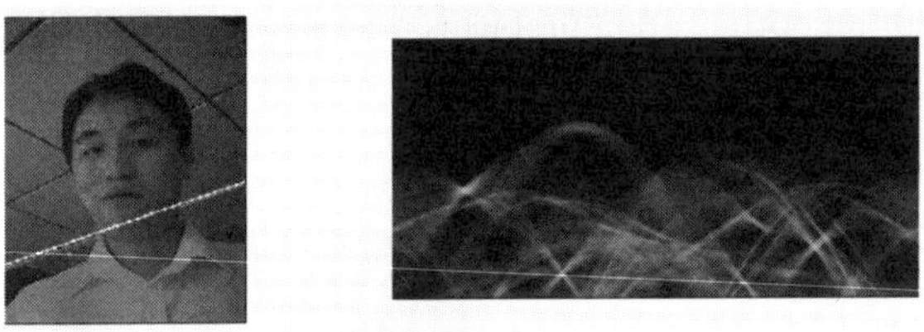

Fig. 3. Straight line detection in a natural image.

Fig. 4. Evolutionary Hough giving the two main lines on Cheng image.

Population size	100
Selection	2 – tournament
Mutation rate (%)	15
Mutation amplitude r	10
Mutation amplitude a, b	40
Crossover rate (barycentric) (%)	5
Generations per frame	800 to initialise. then 240 per frame

If the object's motion is small enough between two consecutive frames, the algorithm will track the object's movement, unlike the deterministic version which has to begin calculations from scratch at every new frame whatever the degree of consistency between consecutive images. In other terms, while the Hough transform has never been seen as fast and evolutionary algorithms are generally regarded as slow, the evolutionary Hough transform gets true realtime properties. This simple

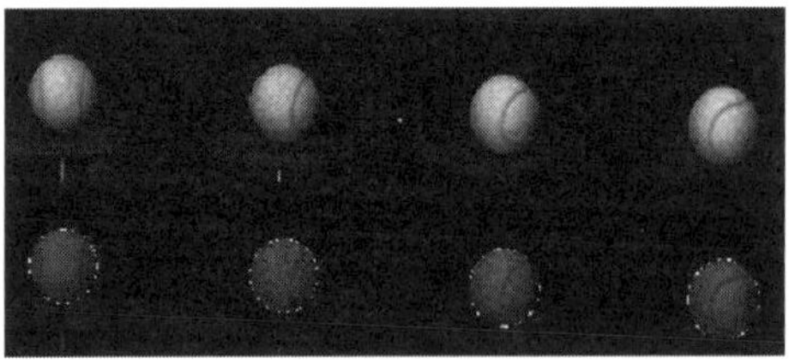

Fig. 5. Circle tracking on the ball image sequence.

version of the algorithm has been written in C without any specialized library.

3. From Phenomena to Processes: Identifying Internal Parameters

The following examples move one step further into data interpretation. After motion detection (*how* does it move?), we may get into interpretation (*why* does it move?) though the identification of the internal parameters of mechanical models able to explain and predict motion. In this application, we use physical models of elastic or fluid objects using pointwise masses and bonds [5, 6].

Particles are described by their masses, positions and speeds. Motion results from forces applied by internal bonds or the environment (gravitation...). The main bond types involved are; unary bonds, between a particle and the medium (e.g. viscosity, gravitation...); and binary bonds, between particle pairs. Each bond type may consist of two components: the first component is defined as the gradient of an energy potential which depends on the relative positions of the particles involved; the second one generates damping forces, which depend on the particles' mutual positions and speeds. Conditional bonds can be used, which allow to simulate fluids in addition to deformable solids.

The task of the identification algorithm is, from given kinematic data, to find out the bonds' parameters. To this end, a cost function evaluates the quality of candidate physical models (i.e. object description files). This cost function measures a generalized distance between the trajectory predicted by the model to be evaluated, and the real given trajectory. The problem consists in finding, among all possible models, the one with the lowest cost. Conventional numerical optimization techniques are unsuccessful on these cost functions, probably lacking the desirable mathematical regularity properties. Even simulated annealing cannot realistically cope with objects containing more than about half a dozen particles.

A conventional evolutionary scheme succeeds in identifying small particle systems (i.e. objects with simple behaviors), but loses efficiency and precision with increasing particle numbers. This is why we designed a cost function which calculates *short-term* (rather than long-term) trajectory differences: the cost function is the quadratic sum of differences between the real reference trajectory and the points extrapolated from the preceding time step. This has important consequences on the shape of the cost function [7].

Second, we exploited a topological property: thanks to the fact that in our "discrete-time physics" the position of a particle at a given time only depends on the recent history of its neighborhood, it is possible to split the cost function into the sum of each particle's contribution. These contributions are called *"local cost functions"* and defined as the temporal sum of the prediction errors concerning its coordinates. We integrated these local cost functions into the crossover and mutation processes, resulting in the fact that the algorithm converges independently on each region of the object: the local cost functions are not influenced by remote regions as the global cost function would be. The fundamental consequence is that *the number of generations (calculation steps) required for convergence does no more depend on the number of unknowns or the object's complexity.*

We tested this method on a variety of examples, among which: fabrics [8], fluid flows [9], and a human heart. In the first two cases, we successfully reverse engineered the parameters used by colleagues from Image Animation (fabrics) and fluid mechanics simulation, only using the image sequences generated.

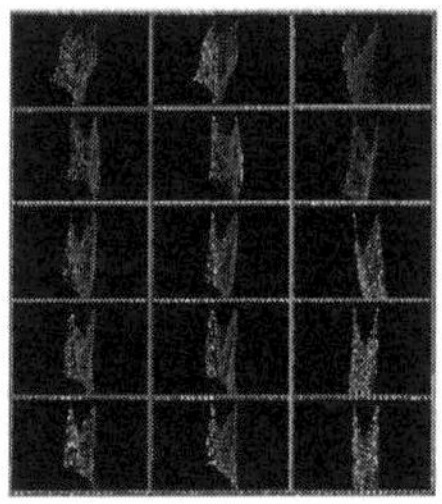

Fig. 6. The original animation (left column), then two reconstructions using the reverse engineered parameters with two qualities of identification (centre: cost = 0.014; right: cost = 0.642).

The third case is the localization of long-term heart scars following a heart attack. Data used are a set of points which have been matched manually from a sequence of MR images. Reverse engineering of the internal mechanical parameters of a purely elastic model shows higher bond stiffnesses in the necrosed regions.

Fig. 7. Two 2-dimensional images of a jet penetrating into a fluid (frames nos. 300 and 400 from a sequence).

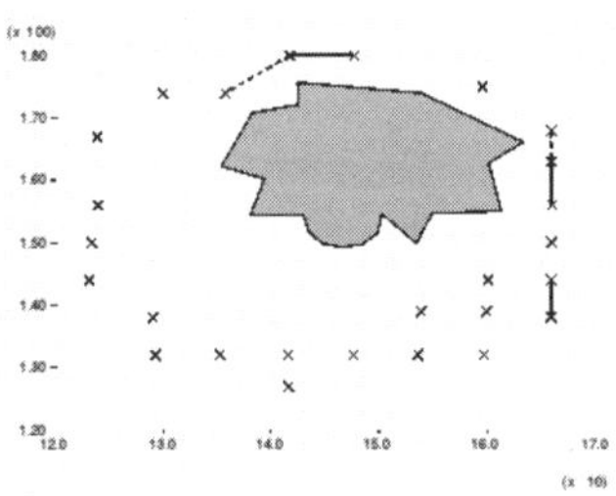

Fig. 8. Detection of necrosis on a human heart from MR image sequences using mass and spring model identification.

4. Parisian Evolution or How To Split Optimisation: The Fly Algorithm

In many applications it is not possible to interpret the image or the image sequence just as the sum of separable image entities, each one corresponding to a particular point to be discovered in the parameter space of a suitable model. It is often more convenient to describe the image as the combination of the outputs of a model which has been applied to a large subset of a certain parameter space. In the following, we will examine how it is possible to exploit the "Parisian" approach [10] which considers the solution will not be represented by a single individual in the population but by the whole population (or at least by an important fraction of it).

The Fly algorithm is the best known application of Parisian Evolution [11, 12]. It consists of evolving a population of 3-D points so that they will concentrate onto the visible surfaces of the objects in the scene. An individual (a fly) is defined as a point in space, whose chromosome is its set of co-ordinates (x, y, z). The co-ordinates of the fly's projections are (x_L, y_L) in the image given by the left camera and (x_R, y_R) in the right camera. The calibration parameters of the cameras are known, therefore x_L, y_L, x_R, y_R may be readily calculated from x, y, z using projective geometry.

The fitness function is designed in order to give higher fitness values to flies whose projections into the different cameras are consistent. The fitness function translates the degree of similarity of the fly's projections onto the different cameras used (usually two).

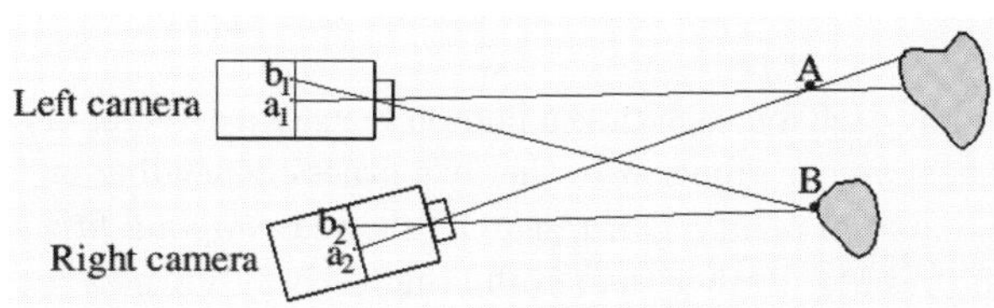

Fig. 9. Pixels b_1 and b_2, projections of fly B, have identical grey levels, while pixels a_1 and a_2, projections of fly A, which receive their illumination from two different physical points on the objects surface, have different grey levels.

The fitness function exploits this property and evaluates the degree of similarity of the pixel neighborhoods of the projections of the fly onto each image, giving highest fitness values for the flies lying on objects surfaces.

In parallel to these robotic applications, we are currently extending the scope of the Fly algorithm to 3-D reconstruction in medical imaging[a]. The goal is to reconstruct the 3-D shape of a radioactive organ from its projections, while taking in account Compton diffusion of high energy photons. Here, a fly is considered as being a photon source. The program simulates the random trajectories of the

[a]in cooperation with Dr Jean-Marie Rocchisani, Universit Paris-13.

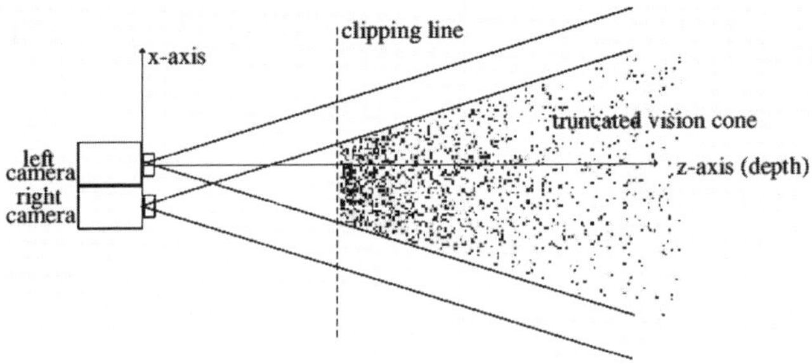

Fig. 10. The initial fly population inside the intersection of the camera fields of view.

Fig. 11. Running the Fly algorithm in real-time processing on a highway. The flies (black dots) concentrate on contrasted road edges and other cars, giving their distances.

photons emitted by each fly: each time a photon reaches a detector crystal, this gives a positive or negative contribution to the fly's fitness depending on how much the crystal is illuminated in reality. Our research[b] showed *marginal fitness* gives a huge improvement on results. Rather than evaluating each fly's fitness independently of the others, marginal evolution means that the fitness of a given fly is relative to the rest of the population. The global population fitness reflects the degree of similarity of the images given by the population of flies, with the given images. The fitness of an individual fly is now defined as the fitness of the population, minus the fitness of the same population without this particular fly.

5. Conclusion

Cheap, massive computing power is merging closer modeling and simulation. This opens the way to new tools for physicists: powerful computing and matching evolutionary algorithms may help the physicist identify model parameters, recover processes from noisy observable phenomena, detect and recognize objects which would

[b]with a decisive contribution from Aurelie Bousquet, Institut National de Sciences Appliques, Rouen (France).

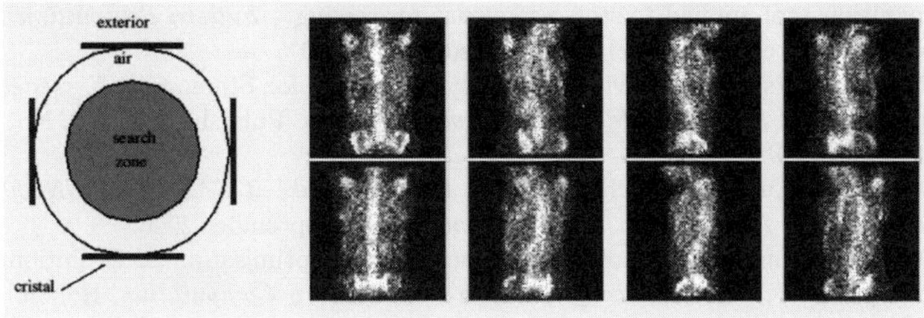

Fig. 12. 3-D reconstruction using Flies: reconstructed skeleton.

otherwise not be detectable. Evolutionary computation, which should not be reduced to elementary genetic algorithms, provides a set of easy to manipulate and often surprisingly efficient tools for optimization of complex functions, model fitting and model construction. Recent tools like the EASEA package [13] enable a wide choice of approaches and operators while freeing the user from the hassle of rewriting his own algorithm. The main keys to success are careful coding of data, and rather than forcing ready-made solutions into a given problem [14], use evolutionary concepts as a Pandora toolbox and custom build one's algorithm.

References

[1] D.A. Goldberg, *Genetic Algorithms in Search, Optimization and Machine Learning*, Addison-Wesley 1989

[2] J. R. Koza, *Genetic Programming*, MIT Press, 1992

[3] T. Bäck and H.-P. Schwefel, "Evolution Strategies I: Variants and their computational implementation", *Genetic Algorithms in Engineering and Computer Science*, John Wiley & Sons.

[4] P.V.C. Hough, *Method and Means of Recognising Complex Patterns*, U.S. Patent n ° 3 069 654, 18 December 1962

[5] A. Luciani, S. Jimenez, J.L. Florens, C. Cadoz, O. Raoult, "Computational Physics: a Modeller Simulator for Animated Physical Objects", *Proc. Eurographics Conference*, Vienna, Elsevier, September 1991

[6] N. Szilas and C. Cadoz, "Physical Models That Learn", International Computer Music Conference, Tokyo, 1993

[7] B. Stanciulescu and J. Louchet, "Evolving Physical Models to Understand Motion in Image Sequences", European Symposium on Intelligent Techniques ESIT'2000, Aachen, Germany, September 2000

[8] J.Louchet, X.Provot, D.Crochemore, "Evolutionary identification of cloth animation models", Eurographics Workshop on Animation and Simulation, Maastricht, September 1995

[9] J. Louchet and L. Jiang, "An Identification Tool to build Physical Models for Virtual Reality", IWSIP Manchester, UK, November1996.

[10] P. Collet, E. Lutton, F. Raynal, M. Schoenauer, "Individual GP: an Alternative Viewpoint for the Resolution of Complex Problems", *GECCO99*, Orlando, Florida, July 1999

[11] J. Louchet, M. Guyon, M.-J. Lesot, A. Boumaza, "Dynamic Flies: a new pattern

recognition tool applied to stereo sequence processing", *Pattern Recognition Letters*, Elsevier Science B.V., March 2001, revised June 2001

[12] J. Louchet, "Using an Individual Evolution Strategy for Stereovision", *Genetic Programming and Evolvable Machines*, Kluwer Academic Publishers, Vol. 2, No. 2, 101-109, March 2001

[13] P. Collet, E. Lutton, M. Schoenauer, J. Louchet, *Take it EASEA*, *PPSN2000 Conference on Parallel Problem Solving From Nature*, September 2000

[14] P. Collet, E. Lutton, J.n Louchet, "Issues on the Optimisation of Evolutionary Algorithm Code", *CEC2002 conference on Evolutionary Computation*, Honolulu, May 2002

MODELING COMPLEXITY USING HIERARCHICAL MULTI-AGENT SYSTEMS

JEAN-CLAUDE HEUDIN
International Institute of Multimedia, Pôle Universitaire Léonard de Vinci, Paris La Défense - France
Jean-CLaude.Heudin@devinci.fr

Existing approaches to modeling natural systems are inadequate to exhibit all the properties of complex phenomena. Current models in the field of complex systems, such as cellular automata, are straightforward to understand and give interesting theoretical results, but they are not very pertinent for a real case study. We show that hierarchical multi-agent modeling provides an interesting alternative for modeling complex systems through two case examples: a virtual ecosystem for studying species dynamics and the simulation of gravitational structures in cosmology.

Keywords: complexity, hierarchical multi-agents systems

1. Introduction

Modeling complex systems is crucial for many domains of science. No consensus exists on how to model complex natural phenomena. The dominating approach tries to reduce a complex system to its underlying fundamental parts and processes. A dynamical model is then formulated with variables related by these processes, and the model is used to predict the evolution of the system. This method decreases the high number of operative degrees of freedom by simplifying the underlying processes using approximations, averaging or smoothing parameters. However, when these simplifications are too strong, results become inadequate and do not reflect the real behaviors. In addition, some global properties, called emergent, simply disappear. This is due to the fact that collective patterns or behaviors emerge from the non-linear interactions between parts of the system. Because they are generally the result of a high number of interacting elements, such global properties cannot be captured by only considering the local properties of individual elements. Recently, the study of complex systems in a unified framework has become recognized as a new interdisciplinary scientific field [1]. Alternative approaches to reductionism have been proposed and many models have been successfully tested [2] such as cellular automata [3]. These models are adequate to study general laws and tendencies, but are inadequate to simulate and predict the behavior of a "real" system. In addition, these models are generally limited to two complexity levels: (1) at the lower level, the network of elementary automata and (2) at the higher level, the emerging global properties. In this paper, we present an alternative based on hierarchical multi-agent modeling. This approach will be illustrated by two examples. The first one is a virtual ecosystem for studying species dynamics, the second one is a simulation of gravitational structures in cosmology.

2. The Approach

Our approach to modeling complex natural systems is based on two phases. The first one is a top-down analysis that defines the different complexity levels and their related elementary parts. This hierarchical decomposition could be done on spatial and/or temporal axes. This takes advantage of the tendency of natural systems to self-organize in space-time hierarchies [4]. The second phase is a bottom-up simulation that attempts to capture the behavioral essence of the phenomena [5]. There must be no centralized structure or control in the simulation program. Global patterns and properties emerge from the interactions between the elementary parts. If defined and organized correctly, the resulting system should exhibit the same dynamic behavior as the natural system. In some cases, these two phases of top-down analysis and bottom-up synthesis must be iterated a few times to converge toward a satisfying model.

3. Hierarchical Multi-agent Systems

The ideal simulation tool for this approach is to implement each complexity level as a multi-agent system. An agent could be described as an autonomous thread or program. It is characterized by a set of properties and behaviors, generally implemented using an object-oriented programming language. We can describe the essential features of the hierarchical multi-agent system by the following rules:

- The complex system is modeled as a set of hierarchical levels.
- Each level is composed of a dynamical network of agents with similar behavioral repertoires.
- Each agent details the way in which it reacts to local situation and interactions with other agents of the same level.
- Each agent can reflect the behaviors of a sub-network of agents at a lower level.
- There is no agent that directs all the other agents. Any behavior or global pattern is therefore emergent.
- Global effects or constraints can be grouped in a specific environmental level.

Such a system can be implemented in a distributed programming environment, exploiting hierarchy and concurrency to perform large-scale simulations. To illustrate, we consider one example in biology and another one in cosmology.

4. Virtual Ecosystem Example

4.1. *Overview*

LifeDrop is a virtual world simulating an ecosystem consisting of small creatures evolving in a minimalist environment, a virtual drop of water [6]. One of the main objectives of this virtual ecosystem is to provide a framework for studying species dynamics in an artificial environment in comparison with biology. The morphology of the LifeDrop creatures is inspired by the original work done by Richard Dawkins with the Blind Watchmaker [7]. While the original Biomorph creatures in the Blind Watchmaker were limited to fixed 2D shapes, they have been extended to 3D animated autonomous agents showing a variety of life-like behaviors in LifeDrop.

4.2. *The hierarchical model*

LifeDrop is a hierarchical multi-agent system composed of three complexity levels. The lower level corresponds to the internal structure and local behaviors of biomorphs. The second level deals with biomorphs interactions. The third level manages all the physical and chemical properties of the environment.

4.2.1. *level 1*

Each creature includes a set of four agents organized as a layered model inspired by the subsumption architecture [8]. In this model, a given level relies on the existence of its sub-levels and all levels executes in parallel:

- *Genotype* : the digital DNA of the creature. It contains most of morphological and behavioral parameters of the creature.
- *Metabolism* : it manages the essentials cycles and states of the creature such as lifetime, energy, reproduction, stress, etc.
- *Reactive behaviors* : it manages all the basic reactive behaviors, like obstacle avoidance, fleeing, searching food, etc.
- *Cognitive behaviors* : it manages more elaborated behaviors such as the selection of a mating partner.

4.2.2. *level 2*

Each creature is also an individual autonomous agent. It is the creature's "incarnation" into the environment. Agents have an elementary perception system. They perceive all other creatures close enough, most perception parameters being encoded in their genotype. Depending on the perceived information and internal state, an agent chooses an action and perform it using the layered agent system described before. This level is also responsible for drawing the 3D animated shape of the creature.

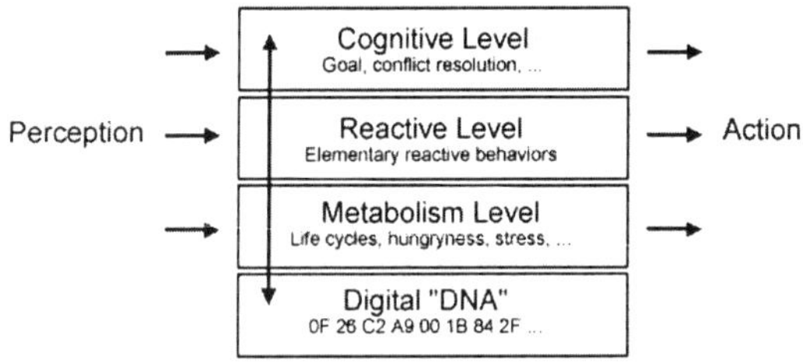

Fig. 1. Each biomorph is composed of four agents organized as a layered architecture.

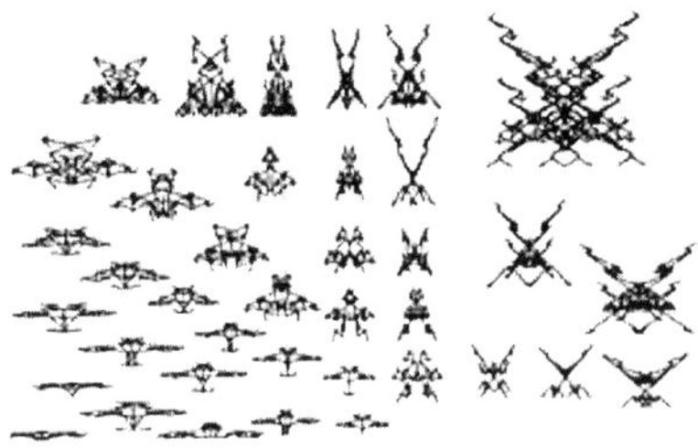

Fig. 2. Some examples of biomorph shapes (after R. Dawkins).

4.2.3. *level 3*

This level corresponds to the environment, which is a virtual drop of water. It is implemented as a closed world with boundaries acting as wall for the creatures. It integrates a physical engine, managing forces and a set of global states such as water viscosity, acidity, etc. It simulates the impact of physical and chemical laws on the creature and, in return, the impact of the creatures on the environment.

4.3. *Experimental results*

One of the key features of LifeDrop is its speciation mechanism. Unlike most of the earlier artificial ecosystem experiments, such as Polyworld [9] for example, species are not only defined as a set of observed emerging properties but also with a "species barrier". The widely accepted definition of a species is the one proposed by Ernst Mayr. He defined a species as a set of individuals that can mate and be fecund between them, and only between them [10]. This notion is implemented in LifeDrop using a stress-based genetic model inspired by the works done by Matic et al. on bacteria [11]. We have conducted many experiments for studying species dynamics with LifeDrop. The following figure (cf. Fig. 4) shows a typical result obtained with an initial randomized population and the simulation of two crisis periods. These crisis correspond to a brutal and global change in the quality of the environment.

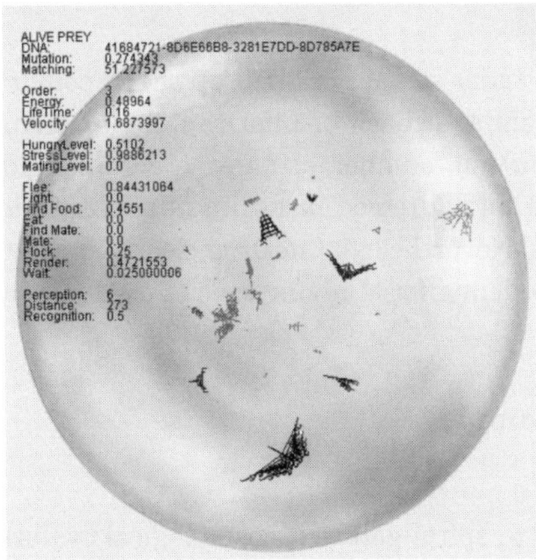

Fig. 3. A general view of the LifeDrop environment.

It is important to note, that there is no explicit nor implicit implementation of any evolution scheme within LifeDrop. All the species dynamics emerge from the interactions between agents in the environment. Results show that the LifeDrop virtual ecosystem exhibits all the classical types of natural species dynamics depending on the simulation parameters: gradual evolution, punctuated equilibrium, natural drift, etc. We have argued that all these variants of the Darwinian natural selection theory could be seen as cases of a undefined evolution theory [12].

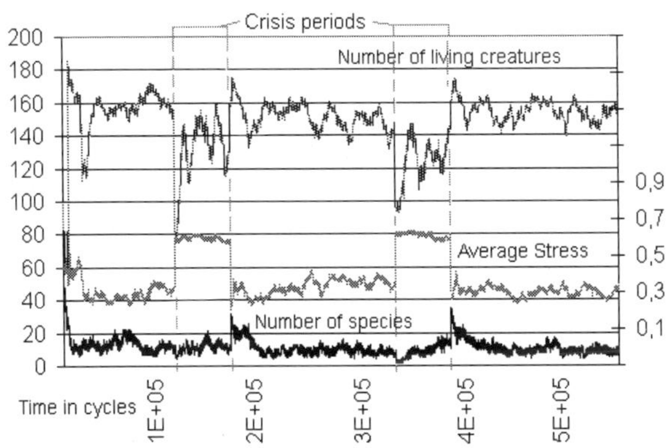

Fig. 4. A typical time series in an experiment with simulated crisis periods.

4.4. *Discussion*

The hierarchical multi-agent system in Lifedrop shows convincing life-like creatures and behaviors. The main drawback of LifeDrop is in the complexity of its model. This leads to an important number of parameters and difficulties for analyzing results. This point will be addressed in future implementations by suppressing all non-crucial parameters. Nevertheless, LifeDrop clearly shows the advantages of hierarchical multi-agent modeling for studying species dynamics and evolution theories.

5. Cosmological Example

5.1. *Overview*

From globular clusters to spiral galaxies, cosmological evolution shows a wide variety of patterns and complex behaviors. For years, numerical simulation in cosmology has tried to reproduce and explain these behaviors using a large number of point-mass particles subjected to gravitation. This approach has carried out to successes, but one important problem is that complex patterns do not appear until a very high number of particles [13]. As an example, a realistic spiral galaxy should theoretically require 10^{41} point-mass particles, which is not yet possible [14]. Since, computing time increases as the square of the particle number, most interesting structures cannot be successfully studied. Few algorithms, like the TreeCode [15] for example, significantly reduce the required computing time. However, even with these advanced algorithms, the average number of particles currently used remains between 128^3 and 512^3. We have designed a hierarchical multi-agent system that solves this problem by allowing to dramatically decrease the number of required agents while maintaining the emergence of complex patterns.

5.2. *The hierarchical model*

Our approach takes advantages of a hierarchical multi-agent system while keeping a strong bond with physics [16]. Each complexity level of our model uses physical laws and interactions on a number of agents lower than was necessary in earlier approaches. We have defined four complexity levels [17].

5.2.1. *Level 1*

Level one agents are composed of a set of point-mass particles subjected to gravitational interaction. Each agent is characterized by three evolving parameters:

- *mass* m_1
- *position* (x_1, y_1, z_1)
- *radius* r_1

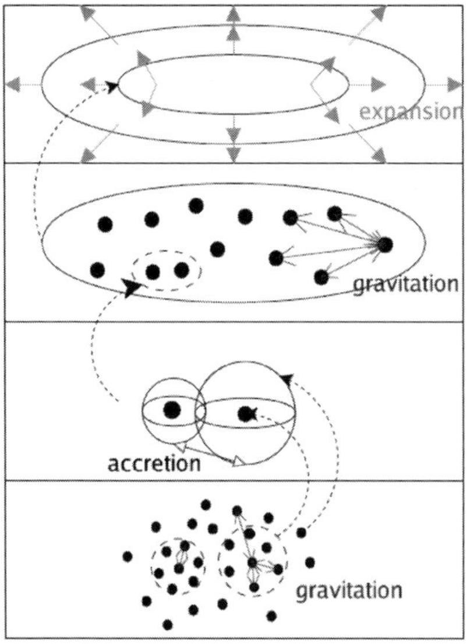

Fig. 5. The four complexity levels of the hierarchical multi-agent system.

The evolution of the agent is calculated using the following classical gravitational equation:

$$\frac{dx_i}{dt^2} = G \cdot \sum_{i,j=1 \ / \ i \neq j}^{M} \left(\frac{m_j(x_j - x_i)}{(d_{ij})^3} \right) \tag{1}$$

Where M is the total number of agents, G is the gravitational constant, mi is the mass of agent i, and d_{ij} the distance between agent i and agent j.

5.2.2. *Level 2*

This level is dedicated to local interactions between agents such as accretion [18]. An agent at this level is surrounded by a less dense matter halo. Each agent is then characterized by five evolving parameters:

- *mass* m_2
- *position* (x_2, y_2, z_2)
- *radius including vicinity* r_2
- *linear velocity* (Vx_2, Vy_2, Vz_2)
- *spin* (S_x, S_y, S_z)

The velocity modification of an agent is calculated using the following equation:

$$\delta V x_i = \sum_{j=1}^{M} \left[\mu_j^t \cdot \left(V_j^t(x) + s_j^t(x) \right) \right] \tag{2}$$

Where s_j^t is the spin of agent j at time t, $V_j^t(x)$ is the velocity of agent j at time t, and μ_j^t the accretion parameter of the agent j at time t.

5.2.3. *Level 3*

The third level is dedicated to long range interactions between agents such as gravity. Each agent is characterized by only two evolving parameters:

- *mass* m_3
- *position* (x_3, y_3, z_3)

The evolution of the agent is calculated using the same equation as for Level 1. Note that we don't use a brute force calculation but an implementation of the TreeCode algorithm [17] in both cases.

5.2.4. *Level 4*

This level is dedicated to environmental actions such as expansion. This is computed by applying a radial force to all existing agents. Each agent is characterized by the following parameters:

- *position* (x_4, y_4, z_4)

The evolution of the agent is the calculated using the classical Hubble equation:

$$v = H \cdot r \tag{3}$$

Where v is velocity of an unspecified point in the universe, H is the Hubble constant, and r is the radius of curvature of the universe.

5.3. *Experimental results*

We have conducted many experiments in order to validate the hierarchical model. We describe here one example concerning the collision of two galaxies that took place 160 Million years ago. This case study, called The Mice, has been often studied in cosmology as in [19] for example. We used 300.000 agents in our simulation with the following distribution: 50% of gas and 50% of dark matter, which is only affected by gravity. The other parameters are :

- First galaxy : $m_{G1} = 3.95\ 10^{41}$ kg, $r_{G1} = 9.40$ kpc
- Second galaxy: $m_{G2} = 4.05\ 10^{41}$ kg, $r_{G2} = 11.0$ kpc

The total simulation time was of 69 seconds corresponding to 220 Million years. We obtained the following result (Fig. 6c), which is compared to an earlier simulation [20] (Fig. 6b) and an image of the real phenomena from the Hubble spatial telescope [21] (Fig. 6a). These three images show clearly the same global emerging pattern. In order to show the properties of our hierarchical model we have conducted many

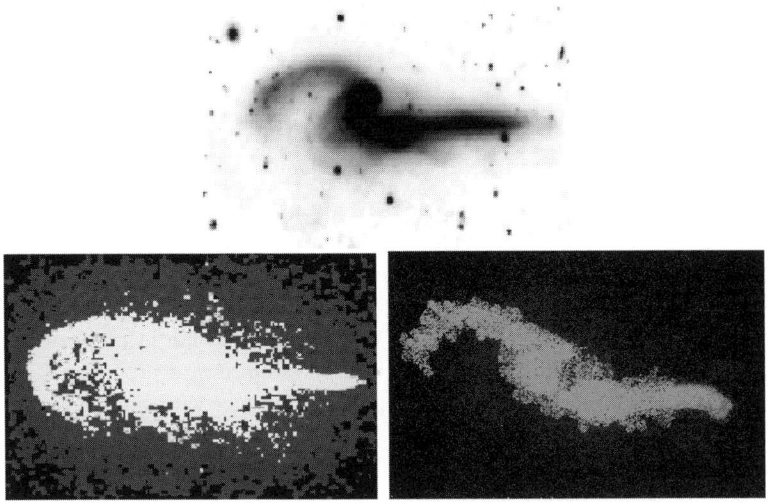

Fig. 6. On top, an image of the Mice from Hubble (a). The bottom left image is a result of a simulation by J. E. Hibbard (b). The bottom right image is a result of our hierarchical multi-agent system with 300.000 agents (c).

additional simulations with reduced numbers of agents. The following figures show a comparison between a classical TreeCode and our hierarchical model with only 3.000 agents, that is 100 times less than in the previous experiment. While the global pattern is unable to emerge with the TreeCode, our hierarchical multi-agent system continues to exhibit the same emerging behavior.

6. Discussion

These qualitative results with the Mice experiment show that, in contrast to classical simulation methods, the number of agents in our hierarchical model does not have a strong influence on the emergence of complex structures. However, it is clear that reducing the number of agents has an impact on the precision and quality of results. Also, there is a threshold, around 2.000 agents in this experiment, under which the hierarchical model fails to exhibit the expected pattern.

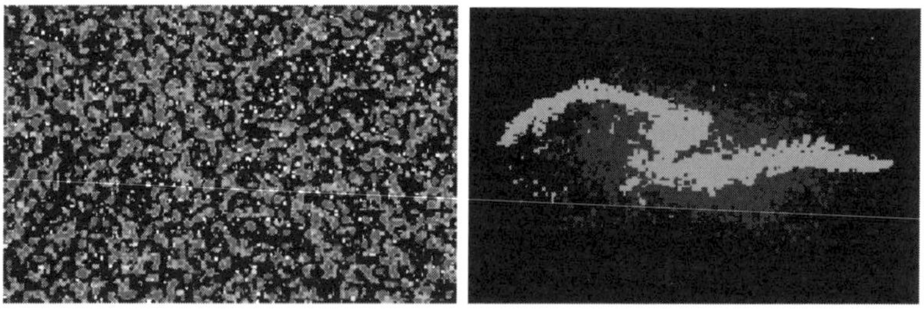

Fig. 7. The right image shows the simulation result with our hierarchical multi-agent system and the left image using a dedicated TreeCode algorithm. Both use 3.000 particles or agents.

7. Conclusion

In this paper we have presented some of the advantages of hierarchical multi-agent modeling through two case examples: a virtual ecosystem and the simulation of large structures in cosmology. Many models classically used in the field of complex systems are simple to understand and universalistic. However, their disadvantage is that they generally don't give useful and detailed information in real case studies. Hierarchical multi-agent modeling constitutes one step toward providing a complementary approach for simulating the complexity of nature with more accuracy and relevance.

References

[1] S. Wolfram, "A New Kind of Science", *Wolfram Media Campaign*, 2002

[2] G. Weisbuch, "Dynamique des systèmes complexes, une introduction aux réseaux d'automates", *InterEditions/CNRS*, Paris, 1989

[3] T. Toffoli and N. Margolus, "Cellular Automata Machines, a New Environnement for Modeling", *The MIT Press*, Cambridge, 1987

[4] B.T. Webner, "Complexity in Natural Landform Patterns", *Science*, 284, 102-104, 1999

[5] C.G. Langton, "Artificial life", *Artificial Life, SFI Studies in Sciences of Complexity*, 6, 1-47, 1989

[6] M. Métivier, C. Lattaud and J.C. Heudin", "A Stress based Spéciation Model in LifeDrop", *"Proceedings 8th, International Conference on Artificial Life*, MIT Press, Cambridge, 121-126, 2002

[7] R. Dawkins, *The Blind Watchmaker: why the Evidence of Evolution Reveals a Universe without Design*, Penguin Books, London, 1986

[8] R.A. Brooks, "How to build complete creatures rather than isolated cognitive simulators", *Architectures for Intelligence*, Lawrence Erlbaum Associates, Hillsdale, 225-239, 1991

[9] L. Yaeger, "Computation Genetics, Pysiology, Metabolism, Neural Systems, Learning, Vision and Behavior or Polyword Life in a Context", *Artificial Life III, SFI Studies in Sciences of Complexity*, 17, 263-298, 1994

[10] E. Mayr, *The Growth of Biological Thought: Diversity, Evolution and Inheritance*, Harvard University Press, Cambridge, 1982

[11] C. Matic, C. Rayssiguier and M. Radman , "Interspecies Gene Exchange in Bacteria:

the Role of SOS and Mismatch repair Systems in Evolution of Species", *Cell*, 80, 507-515, 1995

[12] J.C. Heudin, "L'évolution au bord du chaos", *Hermès Science*, Paris, 135-156, 1998

[13] E.A. Kuksheva et al, "Numerical Simulation of Self Organization in Gravitationally Unstable", *Media on Supercomputers*, LNCS-2763, Springer, 350-368, 2003

[14] J.M. Dawson, "Computer Simulation of Plasmas", *Gravitational N Bode Problem*, 315, 1972

[15] J. Barnes and Hut, "A Hierarchical O(n log n) Force Calculation Algorithm", *Nature*, 324, 446-449, 1986

[16] J.C. Heudin, "Complexity Classes in three dimensional gravitational agents", *Proceedings 8th International Conference on Artificial Life*, MIT Press, Cambridge, 9-13, 2002

[17] J.C. Torrel, C. Lattaud and J.C. Heudin, "Complex Stellar Dynamics Using a Hierarchical Multi Agent Model", *NECSI International Conference on Complex Systems*, 2006

[18] J. Frank, A. King and D. Raine, *Accretion Power in Astrophysics*, Cambridge University Press, 2002

[19] J.C. Mihos, G.D. Bothun and D.O. Richstone, "Modeling the Spatial Distribution os Star Formation in Interacting Disk Galaxies", *Astrophysical Journal*, 418, 82-89, 1993

[20] J. Barnes and J.E Hibbard, "Models of Interacting Galaxies", www.ifa.hawaii.edu, 2004

[21] www.hubblesite.org

TOPOLOGICAL APPROACHES TO SEARCH AND MATCHING IN MASSIVE DATA SETS

FIONN MURTAGH

Department of Computer Science,
Royal Holloway, University of London,
Egham, Surrey TW20 0EX, England.
E-mail: fmurtagh@acm.org

We begin with pervasive ultrametricity due to high dimensionality and/or spatial sparsity. How extent or degree of ultrametricity can be quantified leads us to the discussion of varied practical cases when ultrametricity can be partially or locally present in data. We show how the ultrametricity can be assessed in time series signals. An aspect of importance here is that to draw benefit from this perspective the data may need to be recoded. Such data recoding can also be powerful in proximity searching, as we will show, where the data is embedded globally and not locally in an ultrametric space.

Keywords: Multivariate data analysis; Cluster analysis; Hierarchy; Factor analysis; Correspondence analysis; Ultrametric; p-Adic; Phylogeny.

1. Introduction

The topology or inherent shape and form of an object is important. In data analysis, the inherent form and structure of data clouds are important. Quite a few models of data form and structure are used in data analysis. One of them is a hierarchically embedded set of clusters, – a hierarchy. It is traditional (since at least the 1960s) to impose such a form on data, and if useful to assess the goodness of fit. Rather than fitting a hierarchical structure to data, our recent work has taken a different orientation: we seek to find (partial or global) inherent hierarchical structure in data. As we will describe in this article, there are interesting findings that result from this, and some very interesting perspectives are opened up for data analysis.

A formal definition of hierarchical structure is provided by ultrametric topology (in turn, related closely to p-adic number theory). We will return to this in section 2 below. First, though, we will summarize some of our findings.

Ultrametricity is a pervasive property of observational data. It arises as a limit case when data dimensionality or sparsity grows. More strictly such a limit case is a regular lattice structure and ultrametricity is one possible representation for it. Notwithstanding alternative representations, ultrametricity offers computational efficiency (related to tree depth/height being logarithmic in number of terminal nodes), linkage with dynamical or related functional properties (phylogenetic interpretation), and processing tools based on well known p-adic or ultrametric theory (examples: deriving a partition, or applying an ultrametric wavelet transform).

Local ultrametricity is also of importance. Practical data sets (derived from, or observed in, databases and data spaces) present some but not exclusively ultrametric characteristics. This can be used for forensic data exploration (fingerprinting data sets, as we discuss below). Local ultrametricity has been used to expedite search and discovery in information spaces (in Chávez et al. [5] as discussed by us in Murtagh [12], which we will not discuss further here). Such proximity searching and matching has traditionally been addressed ultrametrically by fitting a hierarchy to data. Below, we show a different way to embed the data (in a computationally highly efficient way) in an ultrametric space, using a principle employed in our local ultrametric work: namely, data recoding.

Our ultimate aim in this work is to proceed a lot further, and gain new insights into data (and observed phenomena and events) through ultrametric (topology) or equivalently p-adic (algebra) representation theory.

2. Quantifying Degree of Ultrametricity

Summarizing a full description in Murtagh [12] we explored two measures quantifying how ultrametric a data set is, – Lerman's and a new approach based on triangle invariance (respectively, the second and third approaches described in this section).

The triangular inequality holds for a metric space: $d(x,z) \leq d(x,y) + d(y,z)$ for any triplet of points x, y, z. In addition the properties of symmetry and positive definiteness are respected. The "strong triangular inequality" or ultrametric inequality is: $d(x,z) \leq \max \{d(x,y), d(y,z)\}$ for any triplet x, y, z. An ultrametric space implies respect for a range of stringent properties. For example, the triangle formed by any triplet is necessarily isosceles, with the two large sides equal; or is equilateral.

- Firstly, Rammal et al. [20] used discrepancy between each pairwise distance and the corresponding subdominant ultrametric. Now, the subdominant ultrametric is also known as the ultrametric distance resulting from the single linkage agglomerative hierarchical clustering method. Closely related graph structures include the minimal spanning tree, and graph (connected) components. While the subdominant provides a good fit to the given distance (or indeed dissimilarity), it suffers from the "friends of friends" or chaining effect.
- Secondly, Lerman [10] developed a measure of ultrametricity, termed H-classifiability, using ranks of all pairwise given distances (or dissimilarities). The isosceles (with small base) or equilateral requirements of the ultrametric inequality impose constraints on the ranks. The interval between median and maximum rank of every set of triplets must be empty for ultrametricity. We have used extensively Lerman's measure of degree of ultrametricity in a data set. Taking ranks provides scale invariance. But the limitation of Lerman's approach, we find, is that it is not reasonable to study ranks of real-valued

(values in non-negative reals) distances defined on a large set of points.

- Thirdly, our own measure of extent of ultrametricity [12] can be described algorithmically. We examine triplets of points (exhaustively if possible, or otherwise through sampling), and determine the three angles formed by the associated triangle. We select the smallest angle formed by the triplet points. Then we check if the other two remaining angles are approximately equal. If they are equal then our triangle is isosceles with small base, or equilateral (when all triangles are equal). The approximation to equality is given by 2 degrees (0.0349 radians). Our motivation for the approximate ("fuzzy") equality is that it makes our approach robust and independent of measurement precision.

A supposition for use of our measure of ultrametricity is that we can define angles (and hence triangle properties). This in turn presupposes a scalar product. Thus we presuppose a normed vector space with a scalar product – a Hilbert space – to provide our needed environment. Quite a general way to embed data, to be analyzed, in a Euclidean space, is to use correspondence analysis [14]. This explains our interest in using correspondence analysis quite often in this work: it provides a convenient and versatile way to take input data in many varied formats (e.g., ranks or scores, presence/absence, frequency of occurrence, and many other forms of data) and map them into a Euclidean, factor space.

3. Ultrametricity and Dimensionality

3.1. *Distance properties in very sparse spaces*

Murtagh [12], and earlier work by Rammal et al. [19, 20], has demonstrated the pervasiveness of ultrametricity, by focusing on the fact that sparse high-dimensional data tend to be ultrametric. In Murtagh [12] it is shown how numbers of points in our clouds of data points are irrelevant; but what counts is the ambient spatial dimensionality. Among cases looked at are statistically uniformly (hence "unclustered", or without structure in a certain sense) distributed points, and statistically uniformly distributed hypercube vertices (so the latter are random 0/1 valued vectors). Using our ultrametricity measure, there is a clear tendency to ultrametricity as the spatial dimensionality (hence spatial sparseness) increases.

As Hall et al. [8] also show, Gaussian data behave in the same way and a demonstration of this is seen in Table 1. For uniform data (results not shown here), the data are generated on $[0,1]^m$; hypercube vertices are in $\{0,1\}^m$, and for the Gaussian clouds, the data are of mean 0, and variance 1, on each of m coordinates. "Dimen." is the ambient dimensionality, m. "Isosc." is the number of isosceles triangles with small base, as a proportion of all triangles sampled. "Equil." is the number of equilateral triangles as a proportion of triangles sampled. "UM" is the proportion of ultrametricity-respecting triangles (= 1 for all ultrametric). To provide an idea of consensus of these results, the 200,000-dimensional Gaussian was repeated and yielded on successive runs values of the ultrametricity measure of: 0.96, 0.98, 0.96.

Table 1. Typical results, based on 300 sampled triangles from triplets of points.

No. points	Dimen.	Isosc.	Equil.	UM
Hypercube				
100	20	0.14	0.02	0.16
100	200	0.16	0.21	0.36
100	2000	0.01	0.86	0.87
100	20000	0	0.96	0.96
100	200000	0	0.97	0.97
Gaussian				
100	20	0.12	0.01	0.13
100	200	0.23	0.14	0.36
100	2000	0.04	0.77	0.80
100	20000	0	0.98	0.98
100	200000	0	0.96	0.96

In the following, we explain why high dimensional and/or sparsely populated spaces are ultrametric.

As dimensionality grows, so too do distances (or indeed dissimilarities, if they do not satisfy the triangular inequality). The least change possible for dissimilarities to become distances has been formulated in terms of the smallest additive constant needed, to be added to all dissimilarities [3, 4, 18, 21]. Adding a sufficiently large constant to all dissimilarities transforms them into a set of distances. Through addition of a larger constant, it follows that distances become approximately equal, thus verifying a trivial case of the ultrametric or "strong triangular" inequality. Adding to dissimilarities or distances may be a direct consequence of increased dimensionality.

For a close fit or good approximation, the situation is not as simple for taking dissimilarities, or distances, into ultrametric distances. A best fit solution is given by De Soete [6] (and software is available in R [9]). If we want a close fit to the given dissimilarities then a good choice would avail either of the maximal inferior, or subdominant, ultrametric; or the minimal superior ultrametric. Stepwise algorithms for these are commonly known as, respectively, single linkage hierarchical clustering; and complete link hierarchical clustering. (See [2, 10, 11] and other texts on hierarchical clustering.)

3.2. *No "curse of dimensionality" in very high dimensions*

Bellman's [1] "curse of dimensionality" relates to exponential growth of hypervolume as a function of dimensionality. Problems become tougher as dimensionality increases. In particular problems related to proximity search in high-dimensional spaces tend to become intractable.

In a way, a "trivial limit" [22] case is reached as dimensionality increases. This makes high dimensional proximity search very different, and given an appropriate data structure – such as a binary hierarchical clustering tree – we can find nearest neighbors in worst case $O(1)$ or constant computational time [12]. The proof is simple: the tree data structure affords a constant number of edge traversals.

The fact that limit properties are "trivial" makes them no less interesting to study. Let us refer to such "trivial" properties as (structural or geometrical) regularity properties (e.g. all points lie on a regular lattice). First of all, the symmetries of regular structures in our data may be of importance. Secondly, "islands" or clusters in our data, where each "island" is of regular structure, may be exploitable. Thirdly, the mention of exploitability points to the application areas targeted: in this article, we focus on search and matching and show some ways in which ultrametric regularity can be exploited in practice. Fourthly, and finally, regularity by no means implies complete coverage (e.g., existence of all pairwise linkages) implying that interesting or revealing structure will be present in observed or recorded data sets.

Thus we see that in very high dimensions, and/or in very (spatially) sparse data clouds, there is a simplification of structure, which can be used to mitigate any "curse of dimensionality". The distances within and between clusters become tighter with increase in dimensionality [16].

4. Increasing Ultrametricity Through Data Recoding: Ultrametricity of Time Series

In Murtagh [13] we use the following coding to show that chaotic time series are less ultrametric than, say, financial (futures, FTSE – Financial Times Stock Exchange index, stock price index), biomedical (EEG for normal and epileptic subjects, eyegaze trace), telecoms (web traffic) or meteorological (Mississippi water level, sunspots) time series; random generated (uniformly distributed) time series data are remarkably similar in their ultrametric properties; and ultrametricity can be used to distinguish various types of biomedical (EEG) signals.

A time series can be easily embedded in a space of dimensionality m, by taking successive intervals of length m, or a delay embedding of order m. Thus we define points

$$\mathbf{x}_r = (x_{r-m+1}, x_{r-m+2}, \ldots, x_{r-1}, x_r)^t \in \mathbb{R}^m$$

where t denotes vector transpose.

Given any $\mathbf{x}_r = (x_{r-m+1}, x_{r-m+2}, \ldots, x_{r-1}, x_r)^t \in \mathbb{R}^m$, let us consider the set of s such contiguous intervals determined from the time series of overall size n. For convenience we will take $s = \lfloor n/m \rfloor$ where $\lfloor . \rfloor$ is integer truncation. The contiguous intervals could be overlapping but for exhaustive or near-exhaustive coverage it is acceptable that they be non-overlapping. In our work, the intervals were non-overlapping. The quantification of the ultrametricity of the overall time series is provided by the aggregate over s time intervals of the ultrametricity of each $\mathbf{x}_r$, $1 \leq r \leq s$.

We seek to directly quantify the extent of ultrametricity in time series data. Earlier in this article we have seen how increase in ambient spatial dimensionality leads to greater ultrametricity. However it is not satisfactory from a practical point of view to simply increase the embedding dimensionality m insofar as short memory relationships are of greater practical relevance (especially for prediction). The greatest possible value of $m > 1$ is the total length of the time series, n. Instead we will look for an ultrametricity measurement approach for given and limited sized dimensionalities m. Our experimental results for real and for random data sets are for "window" lengths $m = 5, 10, \ldots, 105, 110$.

We seek local ultrametricity, i.e. hierarchical structure, by studying the following: Euclidean distance squared, $d_{jj'} = (x_{rj} - x_{rj'})^2$ for all $1 \leq j, j' \leq m$ in each time window, $\mathbf{x}_r$. It will be noted below in this section how this assumption of Euclidean distance squared has worked well but is not in itself important: in principle any dissimilarity can be used.

We enforce sparseness [12, 19, 20] on our given distance values, $\{d_{jj'}\}$. We do this by thresholding each unique value $d_{jj'}$, in the range $\max_{jj'} d_{jj'} - \min_{jj'} d_{jj'}$, by an integer in $\{1, 2\}$. Note that the range is chosen with reference to the currently considered time series window, $1 \leq j, j' \leq m$. Thus far, the recoded value, $d'_{jj'}$ is not necessarily a distance. With the extra requirement that $d'_{jj'} \longrightarrow 0$ whenever $j = j'$ it can be shown that $d'_{jj'}$ is a metric [13].

To summarize, in our coding, a small pairwise transition is mapped onto a value of 1; and a large pairwise transition is mapped onto a value of 2. A pairwise transition is defined not just for data values that are successive in time but for any pair of data values in the window considered.

This coding can be considered as (i) taking a local region, defined by the sliding window, and (ii) coding pairwise "change" = 2, versus "no change" = 1, relationships. Then, based on these new distances, we use the ultrametric triangle properties to assess conformity to ultrametricity. The average overall ultrametricity in the time series, quantified in this way, allows us to fingerprint our time series.

A wide range of window sizes (i.e., lengths), m, was investigated. Window size is not important: in relative terms the results found remain the same. Taking part of a time series and comparing the results to the full time series gave similar outcomes, thus indicating that the fingerprinting was an integral property of the data.

Our "change/no change" metric is crucial here, and not the input dissimilarity which is mapped onto it. Note too that generalization to multivariate time series is straightforward.

Eyegaze trace signals were found to be remarkably high in ultrametricity, which may be due to extreme values (truncated off-scale readings resulting from the subject's blinking) that were not subject to preprocessing. Web traffic was also very high in ultrametricity, due to to extreme values. All EEG data sets were close together, with clear separation between the normal sleep subject, and the epilepsy cases. The lowest ultrametricity was found for chaotic time series.

5. Fast Clustering through Baire Space Embedding

The clustering of chemical compounds, based on chemical descriptors or representations, is important in the pharmaceutical and chemical sectors. It is used for screening and knowledge discovery in large databases of chemical compounds. A chemical compound is encoded (through various schemes that are not of relevance to us here) as a fixed length bit string (i.e. a set of boolean or 0/1 values). We have started to look at a set of 1.2 million chemical compounds, each characterized (in a given descriptor or coding system, the Digital Chemistry bci1052 dictionary of fragments) by 1052 variables.

While attributes per chemical compound are roughly Gaussian in distribution, chemicals per attribute follow a power law. We found the probability of having more than p chemicals per attribute to be approximately $c/p^{1.23}$ for large p and for constant, c. This warrants normalization, which we do by dividing attribute/chemical presence values by the attribute marginal (i.e., attribute column sum). Any presence value is now a floating point value.

Consider now the very simplified example of two chemicals, x and y, with just one attribute, whose maximum precision of measurement is K. So let us consider $x_K = 0.478$; and $y_K = 0.472$. In these cases, maximum precision, $|K| = 3$. For first decimal place $k = 1$, we find $x_k = y_k = 4$. For $k = 2, x_k = y_k$. But for $k = 3, x_k \neq y_k$. We now introduce the following distance:

$$d_B(x_K, y_K) = \begin{cases} 1 & \text{if } x_1 \neq y_1 \\ \inf 2^{-n} & x_n = y_n \quad 1 \leq n \leq |K| \end{cases}$$

So here $d_B(x_K, y_K) = 2^{-3}$. This distance is a greatest common prefix metric, and indeed ultrametric. Its maximum value is 1, i.e. it is a 1-bounded ultrametric. Our reason for use of d_B to denote this distance is due to it endowing a metric on the Baire space, the space of countably infinite sequences.

The case of multiple attributes is handled as follows. We have the set J of attributes. Hence we have $|J|$ values for each chemical structure. So the ith chemical structure, for each $j \in J$ value with precision $|K|$, is x_{iJK} . Collectively, all our data are expressed by x_{IJK}. As before, we normalize by column sums to work therefore on x^J_{IJK}. To find the Baire distance properties we work simultaneously on all J values, corresponding to a given chemical structure. Therefore the partition at level $k = 1$ has clusters defined as all those numbers indexed by i that share the same $k = 1$, or 1st, digit *in all J values.*

Table 2 demonstrates how this works. In Table 2, each of the three data sets consists of 7500 chemicals, and they are shown in immediate succession. The number of significant decimal digits is 4 (more precise, and hence more different clusters found), 3, 2, and 1 (lowest precision in terms of significant digits).

In Table 3 we look at k-means, using as input the cluster centers provided by the 1-significant digit Baire approach. In Table 3, results are shown for k-means, based on the same three data sets used heretofore, each relating to 7500 chemi-

Table 2. Results for the three different data sets.

Sig. dig. k	No. clusters
4	6591
4	6507
4	5735
3	6481
3	6402
3	5360
2	2519
2	2576
2	2135
1	138
1	148
1	167

cal structures, with 1052 descriptors. "Sig. dig.": number of significant digits used. "No. clusters": number of clusters in the data set of 7500 chemical structures, associated with the number of significant digits used in the Baire scheme. "Largest cluster": cardinality. "No. discrep.": number of discrepancies found in k-means clustering outcome. "No. discrep. cl.": number of clusters containing these discrepant assignments.

Relatively very few changes were found. We note that the partitions in each case are dominated by a very large cluster. Further details on this work can be found in Murtagh et al. [17].

Table 3. Results of k-means for the same three data sets.

Sig. dig.	No. clusters	Largest cluster	No. discrep.	No. discrep. cl.
1	138	7037	3	3
1	148	7034	1	1
1	167	6923	9	7

6. Conclusions

It has been our aim in this work to link observed data with an ultrametric topology. The traditional approach in data analysis, of course, is to impose structure on the data. This is done, for example, by using some agglomerative hierarchical clustering algorithm. We can always do this (modulo distance or other ties in the data). Then we can assess the degree of fit of such a (tree or other) structure to our data.

For our purposes, here, this is unsatisfactory.

Firstly, our aim was to show that ultrametricity can be naturally present in our data, globally or locally. We did not want any "measuring tool" such as an agglomerative hierarchical clustering algorithm to overly influence this finding. (Unfortunately Rammal et al. [20] suffers from precisely this unhelpful influence of the "measuring tool" of the subdominant ultrametric. In other respects, Rammal et al. [20] is a seminal paper.)

Secondly, let us assume that we did use hierarchical clustering, and then based our discussion around the goodness of fit. This again is a traditional approach used in data analysis, and in statistical data modeling. But such a discussion would have been unnecessary and futile. For, after all, if we have ultrametric properties in our data then many of the widely used hierarchical clustering algorithms will give precisely the same outcome, and furthermore the fit is by definition exact.

In linking data with an ultrametric embedding, whether local only, or global, we have, in this article, proceeded also in the direction of exploiting this achievement. While some applications, like discrimination between time series signals, or texts, have been covered here, other applications like bioinformatics database search and discovery, and analysis of large scale cosmological structures [15], have just been opened up. In Ezhov and Khrennikov [7] this methodology is applied to quantum statistics, and their application to multi-agent systems. There is a great deal of work to be accomplished.

References

[1] R. Bellman, Adaptive Control Processes: A Guided Tour, Princeton University Press, 1961.

[2] J.P. Benzécri, L'Analyse des Données, Tome I Taxinomie, Tome II Correspondances, 2nd ed., Dunod, Paris, 1979.

[3] F. Cailliez and J.P. Pagès, Introduction à l'Analyse de Données, SMASH (Société de Mathématiques Appliquées et de Sciences Humaines), Paris, 1976.

[4] F. Cailliez, The analytical solution of the additive constant problem, Psychometrika, 48, 305–308, 1983.

[5] E. Chávez, G. Navarro, R. Baeza-Yates and J.L. Marroquín, Proximity searching in metric spaces, ACM Computing Surveys, 33, 273–321, 2001.

[6] G. de Soete, A least squares algorithm for fitting an ultrametric tree to a dissimilarity matrix, Pattern Recognition Letters, 2, 133–137, 1986.

[7] A.A. Ezhov and A.Yu. Khrennikov, On ultrametricity and a symmetry between Bose-Einstein and Fermi-Dirac systems, in A.Yu. Khrennikov, Z. Rakić and I.V. Volovich, Eds., p-Adic Mathematical Physics, American Institute of Physics Conf. Proc. Vol. 826, 55–64, 2006.

[8] P. Hall, J.S. Marron and A. Neeman, Geometric representation of high dimension low sample size data, *Journal of the Royal Statistical Society B*, **67**, 427–444, 2005.

[9] K. Hornik, A CLUE for CLUster Ensembles, Journal of Statistical Software, 14 (12), 2005.

[10] I.C. Lerman, Classification et Analyse Ordinale des Données, Paris, Dunod, 1981.

[11] F. Murtagh, Multidimensional Clustering Algorithms, Physica-Verlag, 1985.

[12] F. Murtagh, On ultrametricity, data coding, and computation, Journal of Classification, 21, 167–184, 2004.

[13] F. Murtagh, Identifying the ultrametricity of time series, European Physical Journal

B, 43, 573–579, 2005.

[14] F. Murtagh, Correspondence Analysis and Data Coding with R and Java, Chapman & Hall/CRC, 2005.

[15] F. Murtagh, From data to the physics using ultrametrics: new results in high dimensional data analysis, in A.Yu. Khrennikov, Z. Rakić and I.V. Volovich, Eds., p-Adic Mathematical Physics, American Institute of Physics Conf. Proc. Vol. 826, 151–161, 2006.

[16] F. Murtagh, Hilbert space becomes ultrametric in the high dimensional limit: application to very high frequency data analysis, arXiv:physics/0702064v1, 2007.

[17] F. Murtagh, G. Downs and P. Contreras, Hierarchical clustering of massive, high dimensional data sets by exploiting ultrametric embedding, 2007, submitted.

[18] E. Neuwirth and L. Reisinger, Dissimilarity and distance coefficients in automation-supported thesauri, Information Systems, 7, 47–52, 1982.

[19] R. Rammal, J.C. Angles d'Auriac and B. Doucot, On the degree of ultrametricity, Le Journal de Physique – Lettres, 46, L-945–L-952, 1985.

[20] R. Rammal, G. Toulouse and M.A. Virasoro, Ultrametricity for physicists, Reviews of Modern Physics, 58, 765–788, 1986.

[21] W.S. Torgerson, Theory and Methods of Scaling, Wiley, 1958.

[22] A. Treves, On the perceptual structure of face space, BioSystems, 40, 189–196, 1997.

DATA MINING: COMPUTATIONAL THEORY OF PERCEPTIONS AND ROUGH-FUZZY GRANULAR COMPUTING

SANKAR K. PAL
Indian Statistical Institute, Kolkata, India
E-mail: sankar@isical.ac.in

Data mining and knowledge discovery is described from pattern recognition point of view along with the relevance of soft computing. Key features of the computational theory of perceptions (CTP) and its significance in pattern recognition and knowledge discovery problems are explained. Role of fuzzy-granulation (f-granulation) in machine and human intelligence, and its modeling through rough-fuzzy integration are discussed. New concept of rough-fuzzy clustering is introduced with algorithm. Merits of fuzzy granular computation, in terms of performance and computation time, for the task of case generation in large scale case based reasoning systems are illustrated through examples. Rough-fuzzy clustering is applied for segmenting brain MR images. Results show the superior performance in terms of β index.

Keywords: soft computing, fuzzy granulation, Guidelines

1. Introduction

In recent years, the rapid advances being made in computer technology have ensured that large sections of the world population have been able to gain easy access to computers on account of falling costs worldwide, and their use is now commonplace in all walks of life. Government agencies, scientific, business and commercial organizations are routinely using computers not just for computational purposes but also for storage, in massive databases, of the immense volumes of data that they routinely generate, or require from other sources. Large-scale computer networking has ensured that such data has become accessible to more and more people. In other words, we are in the midst of an information explosion, and there is urgent need for methodologies that will help us bring some semblance of order into the phenomenal volumes of data that can readily be accessed by us with a few clicks of the keys of our computer keyboard. Traditional statistical data summarization and database management techniques are just not adequate for handling data on this scale, and for extracting intelligently, information or, rather, knowledge that may be useful for exploring the domain in question or the phenomena responsible for the data, and providing support to decision-making processes. This quest had thrown up some new phrases, for example, *data mining* [1, 2] and *knowledge discovery in databases (KDD)* which are perhaps self-explanatory, but will be briefly discussed in the following few paragraphs. Their relationship with the discipline of pattern recognition (PR), certain challenging issues, and the role of soft computing will also be mentioned.

The massive databases that we are talking about are generally characterized by the presence of not just numeric, but also textual, symbolic, pictorial and aural data. They may contain redundancy, errors, imprecision, and so on. KDD is aimed at discovering natural structures within such massive and often heterogeneous data. Therefore PR plays a significant role in KDD process. However, KDD is being visualized as not just being capable of knowledge discovery using generalizations and magnifications of existing and new pattern recognition algorithms, but also the adaptation of these algorithms to enable them to process such data, the storage and accessing of the data, its preprocessing and cleaning, interpretation, visualization and application of the results, and the modeling and support of the overall human-machine interaction. What really makes KDD feasible today and in the future is the rapidly falling cost of computation, and the simultaneous increase in computational power, which together make possible the routine implementation of sophisticated, robust and efficient methodologies hitherto thought to be too computation-intensive to be useful. A block diagram of KDD is given in Figure 1 [3].

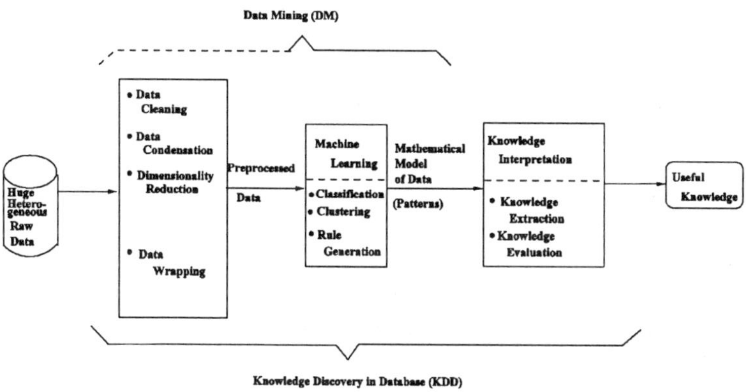

Fig. 1. Block diagram for knowledge discovery in databases [3]

Data mining is that part of knowledge discovery which deals with the process of identifying valid, novel, potentially useful, and ultimately understandable patterns in data, and excludes the knowledge interpretation part of KDD. Therefore, as it stands now, data mining can be viewed as applying PR and machine learning principles in the context of voluminous, possibly heterogeneous data sets. Furthermore, soft computing-based (involving fuzzy sets, neural networks, genetic algorithms and rough sets) PR methodologies and machine learning techniques hold great promise for data mining. The motivation for this is provided by their ability to handle imprecision, vagueness, uncertainty, approximate reasoning and partial truth and lead to tractability, robustness and low-cost solutions [4]. An excellent survey demonstrating the significance of soft computing tools in data mining problem is recently provided by Mitra et al. [5]. Some of the challenges arising out of those posed by massive data and high dimensionality, nonstandard and incomplete data, and over-fitting problems deal mostly with issues like user interaction, use of prior knowledge, assessment of statistical significance, learning from mixed media data, management of changing (dynamic) data and knowledge, integration of different classical and modern soft computing tools, and making knowledge discovery more understandable to humans by using linguistic rules, visualization, etc.

Web mining can be broadly defined as the discovery and analysis of useful information from the web or WWW which is a vast collection of completely uncontrolled heterogeneous documents. Since the web is huge, diverse and dynamic, it raises the issues of scalability, heterogeneity and dynamism, among others. Recently, a detailed review explaining the state of the art and the future directions for web mining research in soft computing framework is provided by Pal et al. [6]. One may note that web mining, although considered to be an application area of data mining on the WWW, demands a separate discipline of research. The reason is that web mining has its own characteristic problems (e.g., page ranking, personalization), because of the typical nature of the data, components involved and tasks to be performed, which can not be usually handled within the conventional framework of data mining and analysis. Moreover, being an interactive medium, human interface is a key component of most web applications. Some of the issues which have come to light, as a result, concern with – (a) need for handling context sensitive and imprecise queries, (b) need for summarization and deduction, and (c) need for personalization and learning. Accordingly, *web intelligence* became an important and urgent research field that deals with a new direction for scientific research and development by exploring the fundamental roles and practical impacts of machine intelligence and information technology (IT) on the next generation of web-empowered products, systems, services and activities. It plays a key role in today's IT in the era of WWW and agent intelligence.

Bioinformatics which can be viewed as a discipline of using *computational methods to make biological discoveries* [7] has recently been considered as another important candidate for data mining applications. It is an interdisciplinary field mainly involving biology, computer science, mathematics and statistics to analyze biological sequence data, genome content and arrangement, and to predict the function and structure of macromolecules. The ultimate goal is to enable the discovery of new biological insights as well as to create a global perspective from which unifying principles in biology can be derived. There are three major sub-disciplines dealing with the following three tasks in bioinformatics:

a) Development of new algorithms and models to assess different relationships among the members of a large biological data set;

b) Analysis and interpretation of various types of data including nucleotide and amino acid sequences, protein domains, and protein structures; and
c) Development and implementation of tools that enable efficient access and management of different types of information.

First one concerns with the mathematical and computational aspects, while the other two are related to the biological and data base aspects respectively. Data analysis tools used earlier in bioinformatics were mainly based on statistical techniques like regression and estimation. With the need of handling large heterogeneous data sets in biology in a robust and computationally efficient manner, soft computing, which provides machinery for handling uncertainty, learning and adaptation with massive parallelism, and powerful search and imprecise reasoning, has recently gained the attention of researchers for their efficient mining.

While talking about pattern recognition and data mining in the 21st century, it will remain incomplete without the mention of the *Computational Theory of Perceptions (CTP),* recently explained by Zadeh [8, 9], which has a significant role in the said tasks. In the following section we discuss its basic concepts and features, and relation with soft computing.

The organization of the paper is as follows. Section 2 introduces the basic notions of computational theory of perceptions and f-granulation, while Section 3 presents rough-fuzzy approach to granular computation. Section 4 reports the application of rough-fuzzy granulation in case based reasoning. New concept of rough-fuzzy clustering is introduced in Section 5 along with its application for segmenting brain MR images. Concluding remarks are given in Section 6.

2. Computational Theory of Perceptions and F-granulation

Computational theory of perceptions (CTP) [8, 9] is inspired by the remarkable human capability to perform a wide variety of physical and mental tasks, including recognition tasks, without any measurements and any computations. Typical everyday examples of such tasks are parking a car, driving in city traffic, cooking meal, understanding speech, and recognizing similarities. This capability is due to the crucial ability of human brain to manipulate perceptions of time, distance, force, direction, shape, color, taste, number, intent, likelihood, and truth, among others.

Recognition and perception are closely related. In a fundamental way, a recognition process may be viewed as a sequence of decisions. Decisions are based on information. In most realistic settings, decision-relevant information is a mixture of measurements and perceptions; e.g., the car is six year old but looks almost new. An essential difference between measurement and perception is that in general, measurements are crisp, while perceptions are fuzzy. In existing theories, perceptions are converted into measurements, but such conversions in many cases, are infeasible, unrealistic or counterproductive. An alternative, suggested by the CTP, is to convert perceptions into propositions expressed in a natural language, e.g., it is a warm day, he is very honest, it is very unlikely that there will be a significant increase in the price of oil in the near future.

Perceptions are intrinsically imprecise. More specifically, perceptions are f-granular, that is, both fuzzy and granular, with a granule being a clump of elements of a class that are drawn together by indistinguishability, similarity, proximity or functionality. For example, a perception of height can be described as very tall, tall, middle, short, with very tall, tall, and so on constituting the granules of the variable 'height'. F-granularity of perceptions reflects the finite ability of sensory organs and, ultimately, the brain, to resolve detail and store information. In effect, f-granulation is a human way of achieving data compression. It may be mentioned here that although information granulation in which the granules are crisp, i.e., c-granular, plays key roles in both human and machine intelligence, it fails to reflect the fact that, in much, perhaps most, of human reasoning and concept formation the granules are fuzzy (f-granular) rather than crisp. In this respect, generality increases as the information ranges from singular (age: 22 yrs), c-granular (age: 20-30 yrs) to f-granular (age: "young"). It means CTP has, in principle, higher degree of generality than qualitative reasoning and qualitative process theory in AI [10, 11]. The types of problems that fall under the scope of CTP typically include: perception based function modeling, perception based system modeling, perception based time series analysis, solution of perception based equations, and computation with perception based probabilities where perceptions are described as a collection of different linguistic *if-then* rules.

F-granularity of perceptions puts them well beyond the meaning representation capabilities of predicate logic and other available meaning representation methods. In CTP, meaning representation is based on the use of so called constraint-centered semantics, and reasoning with perceptions is carried out by goal-directed propagation of generalized constraints. In this way, the CTP adds to existing theories the capability to operate on and reason with perception-based information.

This capability is already provided, to an extent, by fuzzy logic and, in particular, by the concept of a linguistic variable and the calculus of fuzzy if-then rules. The CTP extends this capability much further and in new directions. In application to pattern recognition and data mining, the CTP opens the door to a much wider and more systematic use of natural languages in the description of patterns, classes, perceptions and methods of recognition, organization, and knowledge discovery. Upgrading a search engine to a question- answering system is another prospective candidate in web mining for CTP application. However, one may note that dealing with perception-based information is more complex and more effort-intensive than dealing with measurement-based information, and this complexity is the price that has to be paid to achieve superiority.

3. Granular Computation and Rough-Fuzzy Approach

Rough set theory [12] provides an effective means for analysis of data by synthesizing or constructing approximations (upper and lower) of set concepts from the acquired data. The key notions here are those of "information granule" and "reducts". Information granule formalizes the concept of finite precision representation of objects in real life situation, and reducts represent the core of an information system (both in terms of objects and features) in a granular universe. *Granular computing* refers to that where computation and operations are performed on information granules (clump of similar objects or points). Therefore, it leads to have both data compression and gain in computation time, and finds wide applications. An important use of rough set theory and granular computing in data mining has been in generating logical rules for classification and association. These logical rules correspond to different important regions of the feature space, which represent data clusters.

For the past few years, rough set theory and granular computation has proven to be another soft computing tool which, in various synergistic combinations with fuzzy logic, artificial neural networks and genetic algorithms, provides a stronger framework to achieve tractability, robustness, low cost solution and close resembles with human like decision making. For example, rough-fuzzy integration can be considered as a way of emulating the basis of f-granulation in CTP, where perceptions have fuzzy boundaries and granular attribute values. Similarly, rough neural synergistic integration helps in extracting crude domain knowledge in the form of rules for describing different concepts/classes, and then encoding them as network parameters; thereby constituting the initial knowledge base network for efficient learning. Since in granular computing computations/operations are performed on granules (clump of similar objects or points), rather than on the individual data points, the computation time is greatly reduced. The results on these investigations, both theory and real life applications, are being available in different journals and conference proceedings. Some special issues and edited volumes have also come out [13–15].

4. Rough-Fuzzy Granulation and Case Based Reasoning

Case based reasoning (CBR) [16], which is a novel Artificial Intelligence (AI) problem-solving paradigm, involves adaptation of old solutions to meet new demands, explanation of new situations using old instances (called cases), and performance of reasoning from precedence to interpret new problems. It has a significant role to play in today's pattern recognition and data mining applications involving CTP, particularly when the evidence is sparse. The significance of soft computing to CBR problems has been adequately explained in a recent book by Pal, Dillon and Yeung [17] and Pal and Shiu [18]. In this section we demonstrate an example [19] of using the concept of f-granulation, through rough-fuzzy computing, for performing an important task, namely, *case generation,* in large scale CBR systems.

A case may be defined as a contextualized piece of knowledge representing an evidence that teaches a lesson fundamental to achieving goals of the system. While case selection deals with selecting informative prototypes from the data, case generation concerns with construction of 'cases' that need not necessarily include any of the given data points. For generating cases, linguistic representation of patterns is used to obtain a fuzzy granulation of the feature space. Rough set theory is used to generate dependency rules corresponding to informative regions in the granulated feature space. The fuzzy membership functions corresponding to the informative regions are stored as cases. Figure 2 shows an example of such case generation for a two dimensional data having two classes. The granulated feature space has 3^2=9 granules. These granules of different sizes are characterized by three membership functions along each axis, and have ill-defined (overlapping) boundaries. Two dependency rules: $class_1 \leftarrow L_1 \wedge H_2$ and $class_2 \leftarrow H_1 \wedge L_2$ are obtained using rough set theory. The fuzzy membership functions, marked bold, corresponding to the attributes appearing in the rules for a class are stored as its case.

Unlike the conventional case selection methods, the cases here are cluster granules and not sample points. Also, since all the original features may not be required to express the dependency rules, each case involves a reduced number of relevant features. The methodology is therefore suitable for mining data sets, large both in dimension and size, due to its low time requirement in case generation as well as retrieval.

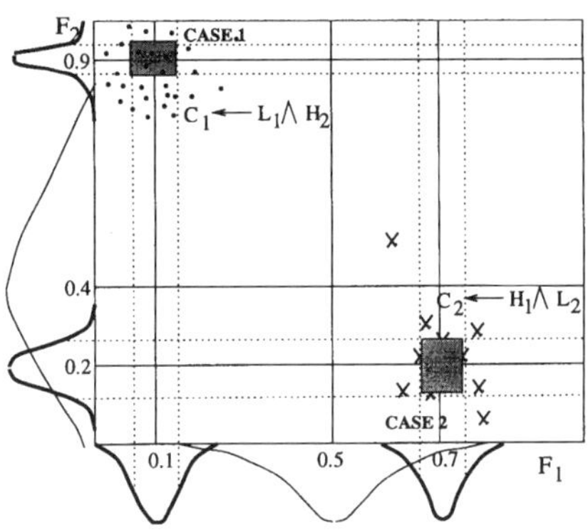

Fig. 2. Rough-fuzzy case generation for a two dimensional data [15]

The aforesaid characteristics are demonstrated in Figures 3 and 4 [19] for two real life data sets with features 10 and 649 and number of samples 586012 and 2000 respectively. Their superiority over IB3, IB4 [16] and random case selection algorithms, in terms of classification accuracy (with one nearest neighbor rule), case generation (t_{gen}) and retrieval (t_{ret}) times, and average storage requirement (average feature) per case, are evident. The numbers of cases considered for comparison are 545 and 50 respectively.

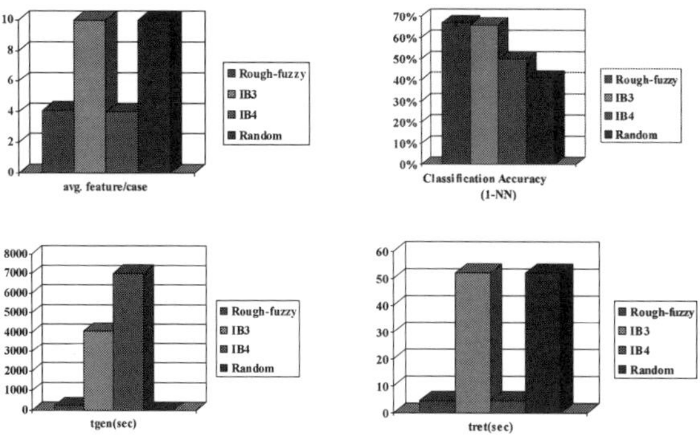

Fig. 3. Performance of different case generation schemes for the forest cover-type GIS data set with 7 classes, 10 features and 586012 samples.

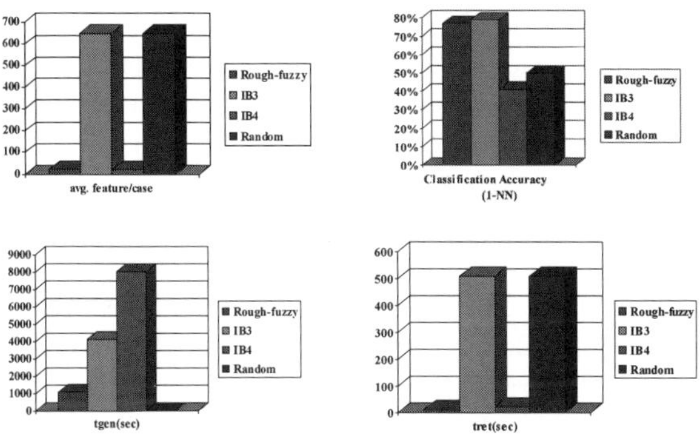

Fig. 4. Performance of different case generation schemes for the handwritten numeral recognition data set with 10 classes, 649 features and 2000 samples.

5. Rough-Fuzzy Clustering and Segmentation of Brain MR Images

Incorporating both fuzzy and rough sets, a new clustering algorithm is described, termed as rough-fuzzy c-means (RFCM). The proposed c-means adds the concept of fuzzy membership of fuzzy sets, and lower and upper approximations of rough sets into c-means algorithm. While the membership of fuzzy sets enables efficient handling of overlapping partitions, the rough sets deal with uncertainty, vagueness, and incompleteness in class definition [20].

In proposed RFCM, each cluster is represented by a centroid, a crisp lower approximation, and a fuzzy boundary. The lower approximation influences the fuzziness of final partition. According to the definitions of lower approximations and boundary of rough sets, if an object belongs to lower approximations of a cluster, then the object does not belong to any other clusters. That is, the object is contained

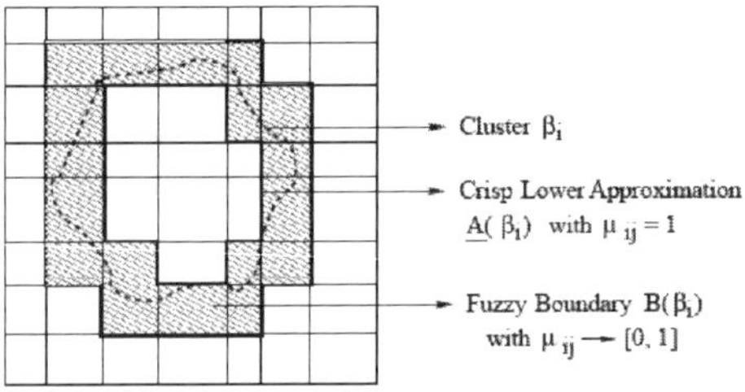

Fig. 5. Rough-fuzzy c-means: each cluster is represented by crisp lower approximations and fuzzy boundary [20].

in that cluster definitely. Thus, the weights of the objects in lower approximation of a cluster should be independent of other centroids and clusters, and should not be coupled with their similarity with respect to other centroids. Also, the objects in lower approximation of a cluster should have similar influence on the corresponding centroids and cluster. Whereas, if the object belongs to the boundary of a cluster, then the object possibly belongs to that cluster and potentially belongs to another cluster. Hence, the objects in boundary regions should have different influence on the centroids and clusters. So, in RFCM, the membership values of objects in lower approximation are 1, while those in boundary region are the same as fuzzy c-means. In other word, the proposed c-means first partitions the data into two classes - lower approximation and boundary. Only the objects in boundary are fuzzified. The new centroid is calculated based on the weighting average of the crisp lower approximation and fuzzy boundary. Computation of the centroid is modified to include the effects of both fuzzy memberships and lower and upper bounds.

The performance of RFCM algorithm is presented using Iris data set and segmentation of brain MR images. The Iris data set is a four-dimensional data set containing 50 samples each of three types of Iris flowers. One of the three clusters (class 1) is well separated from the other two, while classes 2 and 3 have some overlap. The data set can be downloaded from http://www.ics.uci.edu/~mlearn.

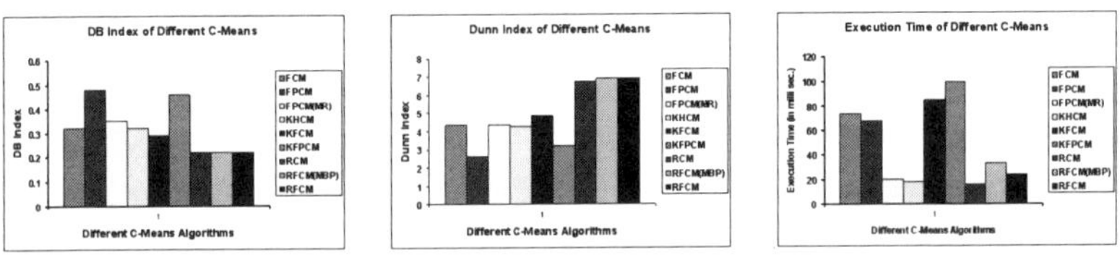

Fig. 6. Comparison of DB and Dunn Index [25], and execution time of HCM, FCM [21], RCM [22], RFCM^{MBP} [23], and RFCM.

The performance of different c-means algorithms is reported with respect to DB and Dunn index [25]. The results reported above three figures establish the fact that RFCM provides best result having lowest DB index and highest Dunn index with lower execution time.

Next, the results on segmentation of brain MR images are presented for RFCM algorithm. Above 100 MR images with different sizes and 16 bit gray levels are tested with different c-means algorithms. All the brain MR images are collected from Advanced Medicare and Research Institute (AMRI), India. The comparative performance of different c-means is reported with respect to β index.

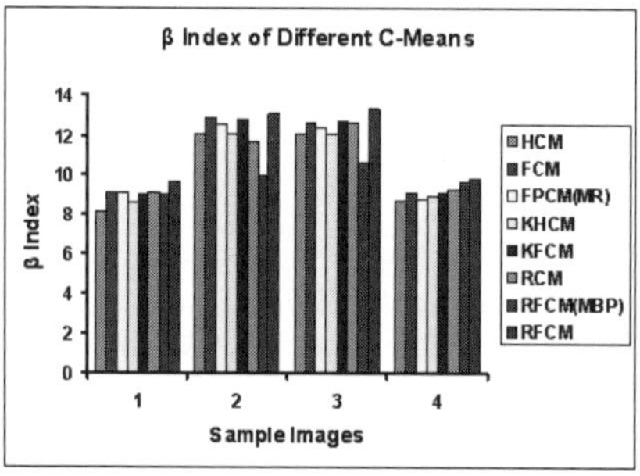

Fig. 7. Comparison of β index [24]] of HCM, FCM [21], RCM [22], RFCM^{MBP} [23], and RFCM.

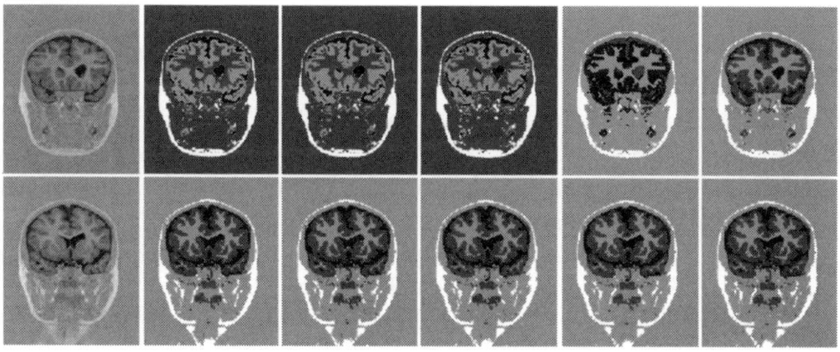

Fig. 8. Some original and segmented images of HCM, FCM [21], RCM [22], RFCM^{MBP} [23], and RFCM.

The performance of different c-means algorithms on some brain MR images with respect to β index [24] is presented here. The original images along with the segmented versions of different c-means are shown above. All the results reported here confirm that the proposed algorithm produces segmented images more promising than do the conventional methods. Also, the values of β index of RFCM are better compared to other c-means algorithms.

6. Conclusions

Data mining and knowledge discovery in databases, which has recently drawn the attention of researchers significantly, have been explained from the view-point of pattern recognition. As it appears, soft computing methodologies, coupled with computational theory of perception (CTP), have great promise for efficient mining of large, heterogeneous data and solution of real-life recognition problems. Fuzzy granulation, through rough-fuzzy computing, and performing operations on fuzzy granules provide both information compression and gain in computation time; thereby making it suitable for data mining applications. We believe the next decade will bear testimony to this in several fields including web intelligence/mining which is considered to be a forefront research area in today's era of IT.

References

[1] J.G. Shanahan, *Soft Computing for Knowledge Discovery: Introducing Cartesian Granule Feature*, Kluwer Academic, Boston, 2000.

[2] S.K. Pal and A. Pal, Eds., *Pattern Recognition: From Classical to Modern Approaches*, World Scientific, Singapore, 2002.

[3] A. Pal and S.K. Pal, Pattern recognition: Evolution of methodologies and data mining, In *Pattern Recognition: From Classical to Modern Approaches*, Eds., S.K. Pal and A. Pal, World Scientific, Singapore, 2002, pp. 1-23.

[4] L.A. Zadeh, Fuzzy logic, neural networks and soft computing, *Communications of the ACM*, vol. 37, pp. 77-84, 1994.

[5] S. Mitra, S.K. Pal and P. Mitra, Data Mining in Soft Computing Framework: A Survey, *IEEE Trans. Neural Networks*, vol. 13, no. 1, pp.3-14, 2002.

[6] S.K. Pal, V. Talwar and P. Mitra, Web Mining in Soft Computing Framework: Relevance, State of the Art and Future Directions, *IEEE Trans. Neural Networks*, vol.13, no.5, pp.1163-1177, 2002.

[7] P. Baldi and S. Brunak, Bioinformatics: The Machine Learning Approach, MIT Press, Cambridge, MA, 1998.

[8] L.A. Zadeh, A new direction in AI: Toward a computational theory of perceptions, *AI Magazine*, vol. 22, pp. 73-84, 2001.

[9] L.A. Zadeh, Foreword, *Neuro-Fuzzy Pattern Recognition: Methods in Soft Computing*, (Authors: S.K. Pal and S. Mitra), Wiley, New York, 1999.

[10] B.J. Kuipers, *Qualitative Reasoning*, MIT Press, Cambridge, 1984.

[11] R. Sun, *Integrating Rules and Connectionism for Robust Commonsense Reasoning*, Wiley, N.Y., 1994.

[12] Z. Pawlak, *Rough Sets: Theoretical Aspects of Reasoning about Data*, Kluwer Academic, Dordrecht, 1991.

[13] S.K. Pal and A. Skowron, Eds., *Rough-Fuzzy Hybridization: A New Trend in Decision Making*, Springer-Verlag, Singapore, 1999.

[14] S.K. Pal, L. Polkowski and A. Skowron, Eds., Rough-neuro Computing: A Way to Computing with Words, Springer, Berlin, 2003.

[15] S.K. Pal and A. Skowron, Eds., Special issue on Rough Sets, Pattern Recognition and Data Mining, *Pattern Recognition Letters*, vol. 24, no. 6, 2003.

[16] J.L. Kolodner, *Case-Based Reasoning*, Morgan Kaufmann, San Mateo, CA, 1993.

[17] S.K. Pal, T.S. Dillon, and D.S. Yeung, Eds., *Soft Computing in Case Based Reasoning*, Springer, London, 2001.

[18] S.K. Pal and S.C.K. Shiu, *Foundations of Soft Case Based Reasoning*, John Wiley, NY, 2003.

[19] S.K. Pal and P. Mitra, Case generation using rough sets with fuzzy discretization, *IEEE Trans. Knowledge and Data Engineering*, 16(3), p.p: 292-300, 2004.

[20] P. Maji and S. K. Pal, Rough-Fuzzy C-Medoids Algorithm and Selection of Bio-Basis for Amino Acid Sequence Analysis, *IEEE Trans. Knowledge and Data Engineering* (to appear).

[21] J. C. Bezdek, *Pattern Recognition with Fuzzy Objective Function Algorithm*, New York: Plenum, 1981.

[22] P. Lingras and C. West, Interval Set Clustering of Web Users with Rough K-Means, *Journal of Intelligent Information Systems*, 23(1), p.p: 5-16, 2004.

[23] S. Mitra, H. Banka, and W. Pedrycz, Rough-Fuzzy Collaborative Clustering, *IEEE Transactions on Systems, Man, and Cybernetics - Part B: Cybernetics*, vol. 36, p.p: 795-805, 2006.

[24] S. K. Pal, A. Ghosh, and B. U. Sankar, Segmentation of Remotely Sensed Images with Fuzzy Thresholding, and Quantitative Evaluation, *International Journal of Remote Sensing*, 21(11), p.p: 2269-2300, 2000.

[25] J. C. Bezdek and N. R. Pal, Some New Indexes for Cluster Validity, *IEEE Transactions on System, Man, and Cybernetics, Part B*, vol. 28, p.p: 301-315, 1988.

BICLUSTERING BIOINFORMATICS DATA SETS: A POSSIBILISTIC APPROACH

FRANCESCO MASULLI

Dipartimento di Informatica e Scienze dell'Informazione,
Università di Genova, and CNISM,
Via Dodecaneso 35, I-16146 Genova, Italy
E-mail: masulli@disi.unige.it
http://www.disi.unige.it/person/MasulliF/

The analysis of genomic data from DNA microarray can produce a valuable information on the biological relevance of genes and on correlations among them. In the last few years some biclustering techniques have been proposed and applied to this analysis. Biclustering is a learning task for finding clusters of samples possessing similar characteristics together with features creating these similarities. When applied to genomic data it can allow us to identify genes with similar behavior with respect to different conditions. In this paper a new approach to the biclustering problem will be introduced extending the Possibilistic Clustering paradigm. The proposed Possibilistic Biclustering algorithm finds one bicluster at a time, assigning a membership to the bicluster for each gene and for each condition. Some results on oligonucleotide microarray data sets will be presented and compared with those obtained using other biclustering methods.

Keywords: Bioinformatics data sets; Data analysis; Biclustering; Possibilistic Biclustering algorithm.

1. Introduction

Nowadays, in the Post-Genomic era, many Bioinformatics data sets are available (most of them released in public domain on the Internet), but often the information embedded in them has no yet completely exploited, due to the lack of accurate machine learning tools and/or of their diffusion in the Bioinformatics community.

Most of Bioinformatics data set come from DNA microarray experiments and are normally given as a rectangular matrix, where each column represents a feature (e.g., a gene) and each row represents a data sample or condition (e.g., a patient). The analysis of microarray data sets can give a valuable information on the biological relevance of genes and correlations between them [11].

The analysis of microarray data sets involves many Machine Learning tasks [15], including:

- Clustering (or unsupervised classification): *Given a set of sample partition them into groups of similar data samples according to some similarity criteria.*
- Classification (or supervised classification): *Find classes of the test data set*

using known classification of training data set.

- Feature selection (or dimensionality reduction): *For each of the classes, select a subset of features responsible for creating the condition corresponding to the class .*
- Outlier detection: *Some of the data samples are not good representative of any of the classes. Therefore, it is better to disregard them while performing data analysis.*

Clustering algorithms are applied for solving *class discovering* problems, while classification methods are suited for *class prediction* problems, and feature selection procedures can perform *biomarker selection.*

Bioinformatics data set analysis pose some major challenges in Machine Learning, due, e.g. to (a) the typical *noisiness* of those data that complicates the solution of Machine Learning tasks (requesting methods with hight *robustness to noise*); (b) the *high-dimensionality* of data making complete search in most of data mining problems computationally infeasible (*curse of dimensionality*); (c) *inaccurate* or *missing* data values; (d) *not sufficient* available data to obtain statistically significant conclusions.

In this paper we shall focus the following problem: *How to identify in a DNA microarray genes with similar behavior with respect to different conditions ?.* This problem is an instance of the *problem of biclustering* (also known as *co-clustering* or *two-way clustering*) [4, 7, 17] that is a methodology allowing for feature set and data points clustering simultaneously, i.e., searching for clusters of samples possessing similar characteristics together with features creating these similarities.

Starting from the seminal paper by Cheng and Church [4], in the last few years many biclustering algorithms have been proposed for the analysis of Bioinformatics data sets (see, e.g., [11]). This paper presents a new development of a biclustering algorithm problem based on the possibilistic clustering paradigm [8] named the Possibilistic BiClustering (PBC) algorithm [6]. PBC algorithm finds one bicluster at a time and assigns to each data matrix element a possibilistic membership to the bicluster.

In the next section we introduce the problem of Biclustering. In Sect. 3 we present the Possibilistic BiClustering algorithm in two variants, using respectively the average and the product fuzzy aggregators. The experimental validation is presented in Sect.4. The section of conclusions ends the paper.

2. Biclustering Problem

Let x_{ij} be the expression level of the i-th gene in the j-th condition. A *bicluster* is defined as a subset of the $m \times n$ data matrix X. A bicluster [4, 7, 17] is a pair $(\mathbf{g}, \mathbf{c})$, where $\mathbf{g} \subset \{1, \ldots, m\}$ is a subset of genes and $\mathbf{c} \subset \{1, \ldots, n\}$ is a subset of conditions. We are interested in largest biclusters from DNA microarray data that do not exceed an assigned homogeneity constraint [4] as they can supply relevant biological information.

Let n be the size or volume of a bicluster defined as the number of cells in the gene expression matrix X belonging to it, and let

$$d_{ij}^2 = \frac{(x_{ij} + x_{IJ} - x_{iJ} - x_{Ij})^2}{n} \tag{1}$$

where the elements x_{IJ}, x_{iJ} and x_{Ij} are respectively the bicluster mean, the row mean and the column mean of X for the selected genes and conditions:

$$x_{IJ} = \frac{1}{n}\sum_{i\in\mathbf{g}}\sum_{j\in\mathbf{c}} x_{ij} \qquad x_{iJ} = \frac{1}{n_c}\sum_{j\in\mathbf{c}} x_{ij} \qquad x_{Ij} = \frac{1}{n_g}\sum_{i\in\mathbf{g}} x_{ij} \tag{2}$$

We can define now G as the mean square residual, a quantity that measures the bicluster homogeneity [4]:

$$G = \sum_{i\in\mathbf{g}}\sum_{j\in\mathbf{c}} d_{ij}^2 \tag{3}$$

The residual quantifies the difference between the actual value of an element x_{ij} and its expected value as predicted from the corresponding row mean, column mean, and bicluster mean.

To the aim of finding large biclusters we must perform an optimization maximizing the bicluster cardinality n and at the same time minimizing the residual G, that is reported to be an NP-complete task [14]. The high complexity of this problem has motivated researchers to apply various approximation techniques to generate near optimal solutions. In the present work we take the approach to combine the criteria in a single objective function.

3. Possibilistic Approach to Biclustering

In order to generalize the concept of biclustering in a fuzzy set theoretical approach [19], for each bicluster we assign two vectors of membership, one for the rows and one other for the columns, denoting them respectively $\mathbf{a}$ and $\mathbf{b}$. In a crisp set framework row i and column j can either belong to the bicluster ($a_i = 1$ and $b_j = 1$) or not ($a_i = 0$ or $b_j = 0$). An element x_{ij} of X belongs to the bicluster if both $a_i = 1$ and $b_j = 1$, i.e., its membership u_{ij} to the bicluster is:

$$u_{ij} = \text{and}(a_i, b_j) \tag{4}$$

The cardinality of the bicluster is defined as:

$$n = \sum_i \sum_j u_{ij} \tag{5}$$

A fuzzy formulation of the problem can help to better model the bicluster and also to improve the optimization process. In a fuzzy setting we allow membership u_{ij}, a_i and b_j to belong in the interval $[0, 1]$. The membership u_{ij} of a point to the bicluster can be obtained by aggregating row and column memberships, for example by:

$$u_{ij} = a_i b_j \qquad \text{(product)} \tag{6}$$

or

$$u_{ij} = \frac{a_i + b_j}{2} \qquad \text{(average)} \tag{7}$$

Many other aggregation operators can be applied, see, e.g., [5]. The fuzzy cardinality of the bicluster is defined as the sum of the memberships u_{ij} for all i and j as in Eq. 5. In the same manner, Eq. 1 remains unchanged, while Eq.s 2 and 3 are generalized as follows:

$$d_{ij}^2 = \frac{(x_{ij} + x_{IJ} - x_{iJ} - x_{Ij})^2}{n} \tag{8}$$

where:

$$x_{IJ} = \frac{\sum_i \sum_j u_{ij} x_{ij}}{\sum_i \sum_j u_{ij}} \qquad x_{iJ} = \frac{\sum_j u_{ij} x_{ij}}{\sum_j u_{ij}} \qquad x_{Ij} = \frac{\sum_i u_{ij} x_{ij}}{\sum_i u_{ij}} \tag{9}$$

$$G = \sum_i \sum_j u_{ij} d_{ij}^2 \tag{10}$$

Then we can tackle the problem of maximizing the bicluster cardinality n and minimizing the residual G using the fuzzy possibilistic paradigm [8, 9, 13]. To this aim we make the following assumptions:

- we treat one bicluster at a time;
- the fuzzy memberships a_i and b_j are interpreted as typicality degrees of gene i and condition j with respect to the bicluster.

Moreover, depending on the aggregator operator we chose, we will obtain a different implementation of the Possibilistic Biclustering algorithm. In the following, we will denote the variables with av when we will use Eq. 7 for computing the membership u_{ij} (PBC-av algorithm [6]), and with pr when we will use Eq. 6 (PBC-pr algorithm).

For PBC-av all the requirements are fulfilled by minimizing the following functional J_{B}^{av} with respect to $\mathbf{a}$ and $\mathbf{b}$:

$$J_{\mathrm{B}}^{av} = \sum_{ij} \left(\frac{a_i + b_j}{2} \right) d_{ij}^2 + \lambda \sum_i (a_i \ln(a_i) - a_i) + \mu \sum_j (b_j \ln(b_j) - b_j), \tag{11}$$

while for PBC-pr we must minimize:

$$J_{\mathrm{B}}^{pr} = \sum_{ij} a_i b_j d_{ij}^2 + \lambda \sum_i (a_i \ln(a_i) - a_i) + \mu \sum_j (b_j \ln(b_j) - b_j) \tag{12}$$

In both cases, the parameters λ and μ control the size of the bicluster by penalizing to small values of the memberships. Their value can be estimated by simple statistics over the training set, and then hand-tuned to incorporate possible a-priori knowledge and to obtain the desired results.

Setting the derivatives of J_{B}^{av} with respect to the memberships a_i and b_j to zero we obtain these necessary conditions:

$$a_i^{av} = \exp\left(-\frac{\sum_j d_{ij}^2}{2\lambda}\right) \tag{13}$$

$$b_j^{av} = \exp\left(-\frac{\sum_i d_{ij}^2}{2\mu}\right), \tag{14}$$

while the necessary conditions for the minimization of J_{B}^{pr} are:

$$a_i^{pr} = \exp\left(-\frac{\sum_j b_j d_{ij}^2}{\lambda}\right) \tag{15}$$

$$b_j^{pr} = \exp\left(-\frac{\sum_i a_i d_{ij}^2}{\mu}\right) \tag{16}$$

Those necessary conditions for the minimization of J_{B} together with the definition of d_{ij}^2 (Eq. 8) can be used by an algorithm able to find a numerical solution for the optimization problem (Picard iteration). In Fig. 1 we show the Possibilistic Biclustering using Eq. 6 (PBC-pr). In PBC-av algorithm, instead, Eq.s 13, 14 must be used in place of Eq.s 15, 16.

(1) Initialize the memberships **a** and **b**
(2) Compute d_{ij}^2 $\forall i, j$ using Eq. 8
(3) Update a_i $\forall i$ using Eq. 15
(4) Update b_j $\forall j$ using Eq. 16
(5) **if** $\|\mathbf{a}' - \mathbf{a}\| < \varepsilon$ and $\|\mathbf{b}' - \mathbf{b}\| < \varepsilon$ then **stop**
(6) **else jump** to step 2

Fig. 1. PBC-pr algorithm.

The parameter ε is a threshold controlling the convergence of the algorithm. The initialization of memberships can be made randomly or using some a priori information about relevant genes and conditions. Moreover, the PBC algorithm can be used as a refinement step for other algorithms using as initialization the results already obtained from them.

After convergence of the algorithm the memberships **a** and **b** can be defuzzified using an α-cut (e.g. $\alpha = 0.5$). In this way the results obtained with PBC can be compared with those of other techniques.

4. Results

4.1. *Experimental validation*

We applied the PBC algorithm to the *Yeast* database [a] which is a genomic database composed by 2884 genes and 17 conditions [1, 2, 16]. We removed from the database all genes having missing expression levels for all the conditions, obtaining a set of 2879 genes.

We performed many runs varying the parameters λ and μ and considering a thresholding for the memberships **a** and **b** of 0.5 for the defuzzification. In figure 2 the effect of the choice of these two parameters on the size of the bicluster can be observed. Increasing them results in a larger bicluster.

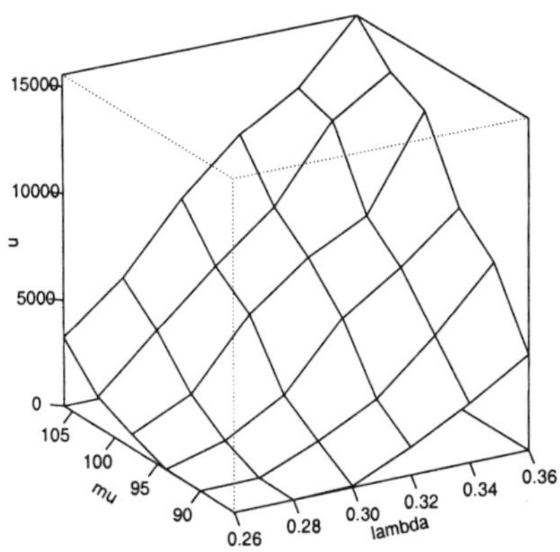

Fig. 2. Size of the biclusters vs. parameters λ and μ (PBC-av).

In Fig. 2 each result corresponds to the average on 20 runs of PBC-av algorithm. Note that, even if the memberships are initialized randomly, starting from the same set of parameters, it is possible to achieve almost the same results. Thus PBC is slightly sensitive to initialization of memberships while strongly sensitive to parameters λ and μ. The parameter ε can be set considering the desired precision on the final memberships. Here it has been set to 10^{-2}.

In Tab. 1 a set of obtained biclusters is shown with the achieved values of G. In particular it is very interesting the ability of PBC to find biclusters of a desired size just tuning the parameters λ and μ. A plot of a small and a large biclusters can be

[a] http://arep.med.harvard.edu/biclustering/yeast.matrix

found in Fig. 3.

Table 1. Comparison of the biclusters obtained by our algorithms on yeast data (PBC-av). The G value, the number of genes n_g, the number of conditions n_c, the cardinality of the bicluster n are shown with respect to the parameters λ and μ.

λ	μ	n_g	n_c	n	G
0.25	115	448	10	4480	56.07
0.19	200	457	16	7312	67.80
0.30	100	654	8	5232	82.20
0.32	100	840	9	7560	111.63
0.26	150	806	15	12090	130.79
0.31	120	989	13	12857	146.89
0.34	120	1177	13	15301	181.57
0.37	110	1309	13	17017	207.20
0.39	110	1422	13	18486	230.28
0.42	100	1500	13	19500	245.50
0.45	95	1622	12	19464	260.25
0.45	95	1629	13	21177	272.43
0.46	95	1681	13	21853	285.00
0.47	95	1737	13	22581	297.40
0.48	95	1797	13	23361	310.72

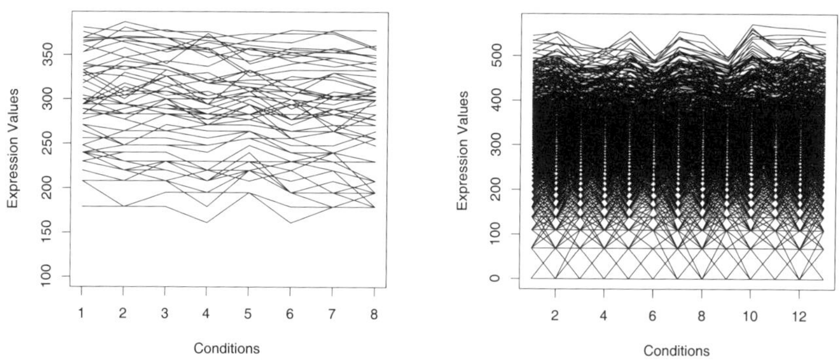

Fig. 3. Plot of a small and a large bicluster (PBC-av).

4.2. *Comparative study*

Table 2 lists a comparison of results on *Yeast* data, involving performance of other, related biclustering algorithms with a $\delta = 300$ (δ is the maximum allowable residual for G). The deterministic DBF [20] discovers 100 biclusters, with half of these lying in the size range 2000 to 3000, and a maximum size of 4000. FLOC [18] uses a probabilistic approach to find biclusters of limited size, that is again dependent on the initial choice of random seeds. FLOC is able to locate large biclusters. However DBF generates a lower mean squared residue, which is indicative of increased similarity between genes in the biclusters. Both these methods report an improvement over the pioneering algorithm by Cheng et al. [4], considering mean squared residue as well as bicluster size.

Single-objective GA with local search has also been used [3], to generate considerably overlapped biclusters.

Table 2. Comparative study on *Yeast* data $\lambda = 0.69$ and $\mu = 119$.

Method	avg. G	avg. n	avg. n_g	avg. n_c	Largest n
DBF [20]	115	1627	188	11	4000
FLOC [18]	188	1826	195	12.8	2000
Cheng-Church [4]	204	1577	167	12	4485
Single-objective GA [12]	52.9	571	191	5.13	1408
Multi-objective GA [12]	235	10302	1095	9.29	14828
PBC-pr	289.6	13823	1228	11.38	14755
PBC-av	297	22571	1736	13	22607

The average results reported in Tab. 2 concerning the Possibilistic Biclustering algorithm have been obtained involving 20 runs over the same set of parameters λ and μ. The biclusters obtained where very similar, obtaining G close to $\delta = 300$ for all of them and the achieved bicluster size is on average very high. From Tab. 2, we see that the Possibilistic Approach has better performances in finding large biclusters in comparison with others methods.

5. Conclusions

In the last few years, some biclustering techniques have been proposed and applied to the analysis of DNA microarray data sets. Biclustering algorithms allow us to identify genes with similar behavior with respect to different conditions.

In this paper we have extended a new approach to the biclustering problem extending the Possibilistic Clustering paradigm already presented in [6]. The proposed Possibilistic Biclustering algorithm finds one bicluster at a time, assigning a membership to the bicluster for each gene and for each condition. The membership of an element of the data matrix X to the bicluster is obtained by *aggregation* of memberships of his row (gene) and column (condition) with respect to bicluster. The results show the ability of the PBC algorithm to find biclusters with low residuals. The quality of the large biclusters obtained is better in comparison with other biclustering methods.

Acknowledgments

Work funded by a grant of the University of Genova.

References

[1] Aach, J., Rindone, W., Church, G.: Systematic management and analysis of yeast gene expression data (2000)

[2] Ball, C.A., et al.: Integrating functional genomic information into the saccharomyces genome database. Nucleic Acids Research **28**(1) (2000) 77–80

[3] Bleuler, S., Prelić, A., Zitzler, E.: An EA framework for biclustering of gene expression data. In: Congress on Evolutionary Computation (CEC-2004), Piscataway, NJ, IEEE (2004) 166–173.

[4] Cheng, Y., Church, G.M.: Biclustering of expression data. In: Proceedings of the Eighth International Conference on Intelligent Systems for Molecular Biology, AAAI Press (2000) 93–103.

[5] Klir, G., Yuan, B.: Fuzzy Sets and Fuzzy Logic: Theory and Applications. Prentice Hall (1995).

[6] Filippone, M., Masulli, F., Rovetta, S., Mitra, S., Banka, H.: "Possibilistic Approach to Biclustering: An Application to Oligonucleotide Microarray Data Analysis", Computational Methods in Systems Biology, LNCS/LNBI4210 (2006) 312–322, Springer-Verlag, Heidelberg (Germany)

[7] Hartigan, J.A.: Direct clustering of a data matrix. Journal of American Statistical Association **67(337)** (1972) 123–129

[8] Krishnapuram, R., Keller, J.M.: A possibilistic approach to clustering. Fuzzy Systems, IEEE Transactions on **1**(2) (1993) 98–110.

[9] Krishnapuram, R., Keller, J.M.: The possibilistic c-means algorithm: insights and recommendations. Fuzzy Systems, IEEE Transactions on **4**(3) (1996) 385–393.

[10] Liu,J., W. Wang,W.: Op-Cluster: Clustering by Tendency in High Dimensional Space, Proc. Third IEEE Int'l Conf. Data Mining, (2003) 187–194.

[11] Madeira, S.C., Oliveira, A.L.: Biclustering algorithms for biological data analysis: A survey. IEEE Trans. Computational Biology and Bioinformatics **1** (2004) 24–45.

[12] Mitra, S., Banka, H.: Multi-objective evolutionary biclustering of gene expression data. Pattern Recognition **39**(12) (2006)2464–2477.

[13] Nasraoui, O., Krishnapuram, R.: Crisp interpretations of fuzzy and possibilistic clustering algorithms. Volume 3., Aachen, Germany (1995) 1312–1318.

[14] Peeters, R.: The maximum edge biclique problem is NP-Complete. Discrete Applied Mathematics **131** (2003) 651–654

[15] Simon,R., Radmacher,M.D., Dobbin,K., McShane, L.M., Pitfalls in the use of DNA microarray data for diagnostic and prognostic classification. J. Natl Cancer Inst., 95,(2003) 14-18.

[16] Tavazoie, S., Hughes, J.D., Campbell, M.J., Cho, R.J., Church, G.M.: Systematic determination of genetic network architecture. Nature Genetics **22**(3) (1999)

[17] Turner, H., Bailey, T., Krzanowski, W.: Improved biclustering of microarray data demonstrated through systematic performance tests. Computational Statistics and Data Analysis **48**(2) (2005) 235–254

[18] Yang, J., Wang, H., Wang, W., Yu, P.: Enhanced biclustering on expression data. In: Proceedings of the Third IEEE Symposium on BioInformatics and Bioengineering (BIBE'03). (2003) 1–7

[19] Zadeh, L.: Fuzzy sets. Information and Control,**48**(3) (1965) 338-353.

[20] Zhang, Z., Teo, A., Ooi, B.C.a.: Mining deterministic biclusters in gene expression data. In: Proceedings of the Fourth IEEE Symposium on Bioinformatics and Bioengineering (BIBE'04). (2004) 283–292

SUPERVISED AUTOMATIC LEARNING MODELS: A NEW PERSPECTIVE

EUGENIO F. SÁNCHEZ-ÚBEDA

Instituto de Investigación Tecnológica, Universidad Pontificia Comillas,
c/ Alberto Aguilera 23, 28015 Madrid, Spain
E-mail: Eugenio.Sanchez@iit.upcomillas.es
www.iit.upcomillas.es

Huge amounts of data are available in many disciplines of Science and Industry. In order to extract useful information from these data, a large number of apparently very different learning approaches have been created during the last decades. Each domain uses its own terminology (often incomprehensible to outsiders), even though all approaches basically attempt to solve the same generic learning tasks. The aim of this paper is to present a new perspective on the main existing automatic learning strategies, by providing a general framework to handle and unify many of the existing supervised learning models. The proposed taxonomy allows highlighting the similarity of some models whose original motivation comes from different fields, like engineering, statistics or mathematics. Common supervised models are classified into two main different groups: *structured* and *unstructured* models. Multidimensional models are shown as a composition of one-dimensional models, using the latter as elementary building blocks. In order to clarify ideas, examples are provided.

Keywords: Automatic learning; Supervised learning; Machine learning.

1. Introduction

A large number of apparently very different learning approaches have been created during the last decades to extract useful information from the available data. There are learning methods developed in the context of machine learning, neural networks, statistics or neurofuzzy systems, to name a few. Each domain uses its own terminology (often incomprehensible to outsiders), even though all approaches basically attempt to solve the same generic learning tasks.

Nowadays hundreds of papers related to the approach of "learning from data" are published each year. Although they are essential for the progress of Science, this also produces an increasing fragmentation of knowledge in automatic learning methods.

The aim of this paper is to present a new perspective on the main existing automatic learning strategies, by providing a general framework to handle and unify many of the existing supervised learning models. Furthermore, the paper focuses on the underlying ideas and their practical implications, rather than on purely theoretical aspects.

First the supervised learning problem is stated using an adequate level of generality. Next common supervised models are discussed. Although this review is incomplete and necessarily brief, the main purpose is to provide a simple framework to fit the ideas underlying most of the existing supervised models.

2. The supervised learning problem from the statistical point of view

The problem is simple to state but difficult to solve in general. Basically, the learning problem consists in finding a model of a desired dependence using a limited number of observations. Although this setting of the problem is really simple, there are learning paradigms providing different viewpoints of the problem. However, nowadays most of the authors agree that the probabilistic treatment provides the adequate level of generality required for a correct interpretation of the learning problems (see e.g. Refs. 1–6).

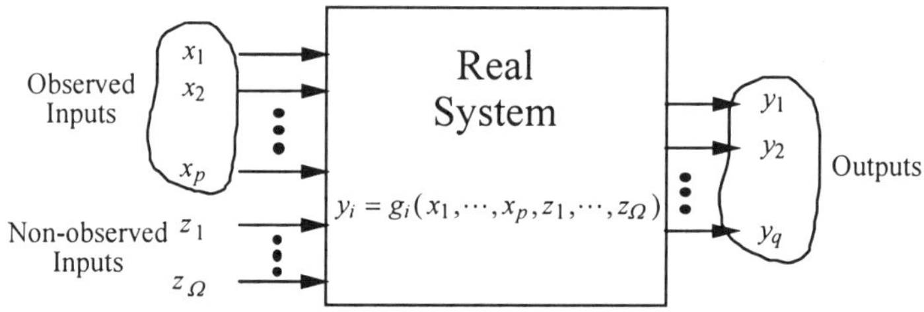

Fig. 1. System under study.

The problem to be solved using automatic learning states as follows [7]: we have a system (possibly a very complicated real one) and we want to automatically extract information about it in order to predict its behavior and, if possible, to understand this behavior. In the system under study (see Fig. 1) one can only measure or observe some quantities: the observed input variables $(x_1, \cdots, x_p)$ and the output variables $(y_1, \cdots, y_q)$. Thus, the goal consists in getting the hidden relationship between the inputs and the outputs, (i.e. determining the output values given only the values of the observed input variables). It is possible that there exist other (non-observed) input variables $(z_1, ..., z_\Omega)$, also related to the output, but we cannot access to their values because we do not know their existence or their possible influence. In practice, this means that the output values can not necessarily be uniquely specified by the observed input values due to the effect of non-observed inputs. Furthermore, it is also possible that the changes in some non-observed variables influence the value of both the observed inputs and outputs.

On the other hand, the arrows in Fig. 1 suggest that there exists a causal relationship between inputs and outputs: changing the input values causes a change in the value of the outputs. In general this is not always true, e.g. in diagnosis the existence of problems (diseases) is the cause of the symptoms. In this paper we consider the first case, where the goal consists in building a model of the relationship between the inputs and the outputs that allows estimating values for the outputs given only the values of the observed inputs.

The system under study (Fig. 1) could be represented by

$$y_i = g_i(x_1, \cdots, x_p, z_1, \cdots, z_\Omega) = g_i(\boldsymbol{x}, \boldsymbol{z}), \quad i = 1, q \tag{1}$$

where $g_i(\bullet)$ represents a simple valued function that relates the inputs (observed and non-observed) to the output y_i. This means that given values for all the inputs, the outputs are completely determined. On the other hand, the knowledge of the joint probability density function (pdf, for sort) $p(\boldsymbol{x}, \boldsymbol{z})$ allow us to know perfectly not only the distribution of input values but the distribution of output values (through a simple transformation using $g_i(\bullet)$). However, the presence of $\boldsymbol{z}$ implies that, in the best case, we will be able to know not the

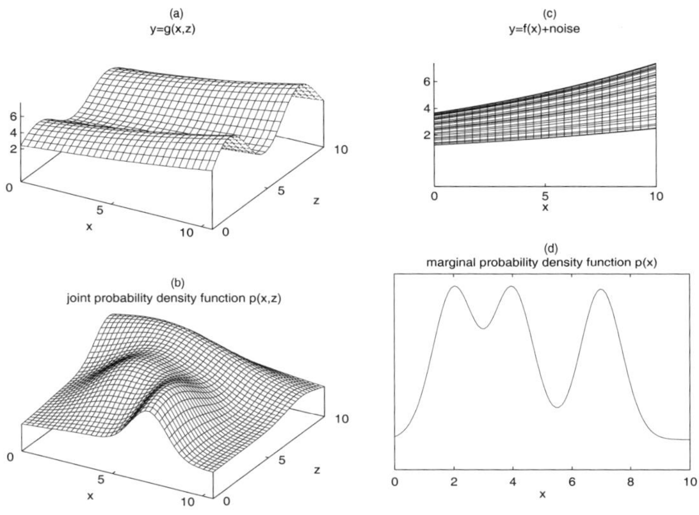

Fig. 2. An illustrative example of the learning problem. (a) The deterministic function $y = g(x, z)$. (b) The joint pdf $p(x, z)$ shows which are the more probable input values. (c) In practice, instead of $g(x, z)$ we have a family of functions $\{y = g_j(x)\}$. Here we have shown some of them. They have been directly obtained by projecting the surface of graph (a) along the plane (x, y). (d) Furthermore, instead of $p(x, z)$ we have the marginal pdf $p(x)$.

joint pdf $p(\boldsymbol{x}, \boldsymbol{z})$ but the marginal pdf $p(\boldsymbol{x})$, which is related to $p(\boldsymbol{x}, \boldsymbol{z})$ through

$$p(\boldsymbol{x}) = \int p(\boldsymbol{x}, \boldsymbol{z}) d\boldsymbol{z} \tag{2}$$

In practice, this distribution $p(\boldsymbol{x})$ will be obtained, when required, from a set of observations of the real system. Furthermore, because non-observed inputs are not accessible, instead of each deterministic function $g_i(\boldsymbol{x}, \boldsymbol{z})$ we will have a family of functions $\{g_{i,j}(\boldsymbol{x}); j = 1, \cdots, P\}$ connecting the output i to the observed inputs. The index j indicates one particular function $g_{i,j}(\boldsymbol{x})$ and it is directly related to the particular values of $\boldsymbol{z}$. Fig. 2 illustrates these ideas for a simple case where the system has one observed input variable, one non-observed input and one output. According to the plot (c) of this figure, even for a given value of $\boldsymbol{x}$ there is a clear uncertainty in the index j that we should use to get the correct value of the output $g_{i,j}(\boldsymbol{x})$. Moreover, depending on the value of $\boldsymbol{x}$ this uncertainty can be more or less large. Thus, it seems natural to express all this variability in probabilistic terms by building a statistical model.

In practice, to obtain a model of the real system we will have a set of observations (pairs of output/input values). This means that, if enough data are available, we could estimate not only $p(\boldsymbol{x})$ but the conditional pdf $p(y|\boldsymbol{x})$ which provides, for a given point $\boldsymbol{x}$, a reasonable criterion to handle the family of functions $\{g_{i,j}(\boldsymbol{x})\}$ at that point. Thus, for a given point $\boldsymbol{x}$ we will have a set of values $\{g_{i,j}(\boldsymbol{x})\}$ plus an estimation of its probability, given by $p(y|\boldsymbol{x})$. Notice that the expression

$$p(y|\boldsymbol{x}) = \int_{\forall \boldsymbol{z}/g(\boldsymbol{x},\boldsymbol{z})=y} p(\boldsymbol{z}|\boldsymbol{x}) d\boldsymbol{z}, \tag{3}$$

relates the conditional pdf $p(y|\boldsymbol{x})$ to the underlying function $g(\boldsymbol{x}, \boldsymbol{z})$ and the joint pdf $p(\boldsymbol{x}, \boldsymbol{z})$. Hence, for a given $\boldsymbol{x}$ point we can assume that the outputs are random variables defined by $p(y|\boldsymbol{x})$.

In order to handle these $p(y|\boldsymbol{x})$ the usual approach consists in summarizing them in terms of some descriptive statistics providing some idea of the distributions, like their average values and dispersions. This is carried out through a statistical model which breaks each output y_i down into two factors: the part showing the average value of y_i as a function of $\boldsymbol{x}$ (the so-called systematic component) and the part reflecting the variability of the output y_i due to the non-observed inputs (namely random component). In particular, in this (statistical) model the effect of the non-observed inputs on the outputs is taken into account through the additive model

$$y_i = \phi_i(x_1, \cdots, x_p) + \epsilon_i, \quad i = 1, q, \tag{4}$$

where the systematic component $\phi_i(\bullet)$ is a single valued deterministic function of the observed inputs and ϵ_i is the random stochastic component reflecting the variability of y_i. Thus, this model is an approximation of the reality given by Eq. (1). It takes explicitly into account the observed inputs, putting the effects of the rest into the random component. Notice that, by definition, $\phi_i(\boldsymbol{x}) = E_\epsilon(y_i|\boldsymbol{x})$ and $E_\epsilon(\epsilon_i|\boldsymbol{x}) = 0$, where the symbol $E_\epsilon(\bullet)$ denotes the average (the expected value) with respect to $p(y|\boldsymbol{x})$, that is

$$E_\epsilon(f) = \int f p(y|\boldsymbol{x}) d\boldsymbol{y}. \tag{5}$$

Thus, according to Eq. (4) the main goal of the automatic learning methods is to obtain a good model, (in terms of accuracy), of the true underlying functions $\phi_i(\bullet)$, able to provide a good prediction for the outputs y_i given only the values of the observed inputs $\boldsymbol{x}$. Interpretability of this model is highly valuable, but not provided by all automatic learning methods.

2.1. *Learning difficulties*

Although the learning problem can be easily stated, no completely accepted method exists to solve it. In this section we address reasons for this difficulty.

Note that, in general, we are trying to estimate an unknown function from one set of collected data where the number of examples is not infinite and with the additional difficulty of being possibly corrupted by noise.

2.1.1. *Finite size of the set of examples*

This is the main difficulty that any method should deal with. Here we consider the case of no noise, (i.e. $\epsilon = 0$ everywhere). In the hypothetical case where the set of examples were infinite in size, the model of the target function will consist in an infinite lookup table where each different input vector value has the corresponding correct value of the target function.

In practice, a close to infinite set of examples is not available, even more, we have a finite set of examples, but the model should be valid at any point. To extend the given information to the set of all possible input values some interpolation scheme is required (e.g. polynomial of low degree), that is decided in advance when the user selects the learning method to be applied. For example, 'classical' regression trees [8] use constant values to interpolate, one-dimensional cubic splines interpolate using polynomials of order three, whereas multi-layer perceptrons exploit combinations of sigmoids. Note that is generally not possible to decide in advance the best interpolation scheme for one particular problem, unless some assumptions (restrictions) about $\phi(\bullet)$ are made a priori.

2.1.2. *Random noise in the output values*

Additionally to the previous difficulty, it is possible that the examples are corrupted by noise, according to Eq. (4). If a large (infinite) set of examples is available, the noise can be 'removed' by computing the target function at any point (input) as the average value over all the examples with this input value. However, in practice the size of the set of examples is finite. This means that we cannot remove with complete guarantee the noise at the given points, making the problem even more difficult.

Concerning automatic learning methods, some of them try to deal with the noise and the lack of examples by extending the straightforward solution for the infinite size case: instead of removing the noise at one point x by computing the mean using an infinitesimal volume centered at x, use a finite volume centered at x, containing a reasonable number of examples. This is the case of regression trees [8], local averaging smoothers, or the well-known 'moving average' techniques. Other learning methods (most) try to protect against the noise by imposing some type of constraint in the set of possible solutions. Regularization based methods [9] are a good example of them. For example, in the one dimensional case, low order polynomials constrain the set of possible solutions to a set with some smoothness properties.

3. Supervised Learning Models

All existing supervised learning models have the common characteristic of dealing with labeled outputs and, in most of them, it is possible to identify two main parts: the internal structure of the model and its parameters. The (internal) structure of the model fixes or constraints the function space, i.e. the set of possible functions that it can implement varying the parameters of the model. This structure is the main responsible of the generalization capability of the model. The parameter space of the model allows selecting, among the set of possible functions constrained by its structure, the one which best agrees with the real system, i.e. with the learning set. In general, selecting the structure of the model is a difficult (arguably the most difficult) part, usually carried out by experts. A wrong selection of the structure can make the model useless. The selection of the best parameters consists in one (more or less) complex optimization process.

Figure 3 shows the taxonomy proposed in Ref. 7 in order to deal with the vast array of existing supervised learning models. This taxonomy discriminates between one-dimensional and multidimensional models. Moreover, for each class we distinguish between structured and unstructured models. This can be further enhanced by considering basic and sophisticated models. The later are more complex models, based on the first ones.

Multidimensional models are shown as a composition of one-dimensional models, using the latter as elementary building blocks. Neural networks such as RBFN or MLP are examples of sophisticated multidimensional models (based on gaussian-like and sigmoid-like basic models, respectively), whereas the standard K-NN model is a basic unstructured model.

Structured models are characterized by assuming some rigid form for the dependence of the outputs on the inputs. This means that they have some internal parameters that allow changing the outputs of the model, but constrained by its fixed structure. On the other hand, a fully unstructured model merely consists in one lookup table where each entry corresponds to a different input and it has associated one output value. This type of model is characterized by the lack of parameters, i.e., it is not possible to change sensibly the model by tuning some internal parameter. Note that these models are described not only by the tabulated function, but also by the (simple) interpolation scheme used to extend this model to the possible range of infinite input values. These models have the

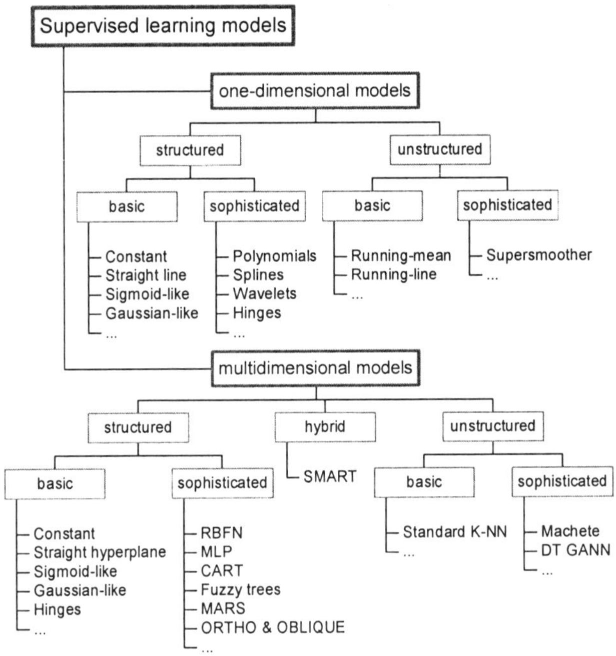

Fig. 3. Taxonomy proposed in Ref. 7 for classifying supervised learning models.

main disadvantage of requiring large amounts of storage resources, as well as their poor interpretability. Another drawback is that they cannot be re-tuned when required.

3.1. *From one to higher dimensions*

The difficulty of having a finite number of examples is strongly magnified by the high dimension of the input space, augmenting the sparsity of this space (the so-called "curse of dimensionality" [10]). The noise makes the problem even more complex. Basically, this curse replaces the geometrical intuition gained from low dimensional spaces with surprising and unexpected properties of the high dimensional ones (see Ref. 7 for illustrative experiments).

Three main techniques can be used to deal with the high dimensionality of the input space: partitioning the input space, projecting, and using norms. Most of the existing learning algorithms are based on some of them.

3.1.1. *Partitioning the input space*

In the automatic learning context, the divide-and-conquer strategy consists in splitting the full input space (containing the complete set of learning examples) into smaller subregions (containing only a subset of examples). Each subregion is specified using a membership function. Usually these subregions are disjoint and they have associated a single output value.

Methods based on this strategy assign an output value to a particular subregion using a process that, most of the times, is straightforward. For example, in classification the output is the majority class in the learning examples within that region. Thus, in the partitioning approach, the most relevant part is dividing the large space into smaller regions. In practice this process is usually implemented using a recursive approach. In fact, the nested nature of the (recursive) partitioning is usually represented graphically as a tree. This tree contains all of the information associated with the model in a single easily interpreted graphic. For this reason methods based on the recursive partitioning approach are sometimes called 'tree-structured' methods (see. e.g. Ref. 6). Examples of this approach are the well-known classification and regression trees, MARS or the non-parametric 'machete' method proposed by Friedman.

Usually the regions generated by partitioning the input space are constrained to be axis oriented hyper-rectangles. This allows not only the practical implementation of the partitioning approach but an automatic feature selection. The main feature of using this strategy is the interpretability of the results.

3.1.2. *Projecting*

This is the most commonly used dimension reducing transformation due to its simplicity and interpretability. Loosely speaking, a projection can be viewed as a shadow of an actual (usually sharper) structure in the full high dimensional space.

The basic assumption of projecting is that one can easily recover most of the actual structure in the high dimensional space from an acceptable (small) set of 'convenient' projections, i.e. a direction in the input space such that by fitting a one-dimensional model to the scatterplot obtained by projecting the data along that direction one obtains a relevant part of the structure in the full high dimensional space. This means that we should select among all the possible directions one such that the one-dimensional model fits better the projected scatterplot.

Concerning existing methods, the most popular and highly promoted multidimensional models deal with the high dimensionality by projecting. Examples are the well-known multilayer perceptrons (and most of its variants) as well as projection pursuit methods such as SMART [11] or the OBLIQUE model [7]. In the context of classification, linear discriminant functions like the Fisher's linear discriminant are also based on projecting.

The inner product between two vectors $\boldsymbol{u} = (u_1, \cdots, u_p)$ and $\boldsymbol{v} = (v_1, \cdots, v_p)$ is a single value defined as the linear combination

$$\boldsymbol{u}^T \boldsymbol{v} = \boldsymbol{v}^T \boldsymbol{u} = \sum_{k=1,p} u_k v_k. \tag{6}$$

To obtain the standard projection of $\boldsymbol{v}$ along $\boldsymbol{u}$ the previous quantity must be normalized (dividing by the module $\|\boldsymbol{u}\|$). Thus, if the normalized vector $\boldsymbol{\alpha}$ represents one direction of projection in the input space, then the quantity $\boldsymbol{\alpha}^T \boldsymbol{x}$ is the projection of the point $\boldsymbol{x}$ along that direction.

Sometimes, the transformation given by the inner product is slightly modified in order to collect in a compact form all the parameters describing the one-dimensional model as well as the direction of projection. This is the case, for example, of the typical multilayer perceptrons, where the activation functions of the hidden units are fixed sigmoids (i.e. without parameters to adjust). The weights of the network represent not only the directions selected for projecting but the parameters (centers and slopes) associated to the sigmoids, (see Ref. 7 for a complete explanation).

3.1.3. *Using norms*

Norms are usual mathematical concepts, they allow deciding if one vector (or matrix) is 'larger' or 'smaller' than another, according to some rule. In practice, these norms are slightly modified in order to be included inside the particular definition of the selected metric, (we will refer to them as norms). Thus, when we choose a particular norm, we are selecting implicity a particular type of 'radial' symmetry (i.e. the rule for associating a non-negative scalar with a vector representing the distance between two points).

Typically, in automatic learning these norms are used in combination with (decreasing) single argument functions to obtain local functions of multiple arguments. This is the case of the usual RBFN. It is also used in the context of non-parametric models, such as the k-Nearest Neighbor.

The Euclidean norm is the most popular choice in practice, due to its simplicity. The (squared) Euclidean distance (also called two-norm) from vector $\boldsymbol{u} = (u_1, \cdots, u_p)$ to vector $\boldsymbol{v} = (v_1, \cdots, v_p)$ is given by

$$\|\boldsymbol{u} - \boldsymbol{v}\|^2 = \sum_{k=1,p} (u_k - v_k)^2 = (\boldsymbol{u} - \boldsymbol{v})^T (\boldsymbol{u} - \boldsymbol{v}) \tag{7}$$

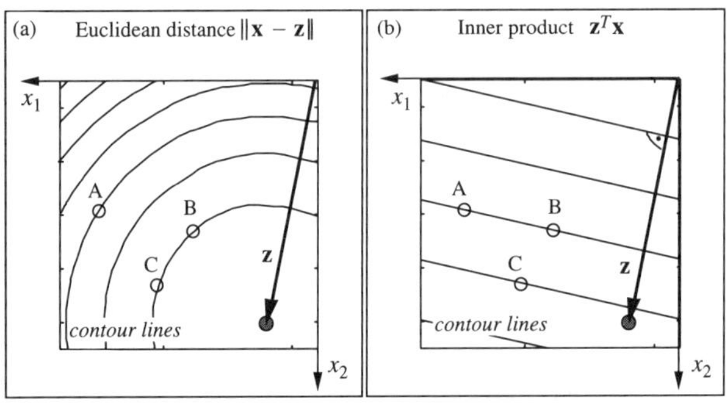

Fig. 4. Contour lines for both (a) the Euclidean distance and (b) the inner product. Points 'B' and 'C' ('A' and 'B') are equivalent according to the Euclidean distance (the inner product, respectively).

Thus, if the vector $\boldsymbol{\zeta} = (\zeta_1, \cdots, \zeta_p)$ represents one point (the center) in the input space, then the quantity $\|\boldsymbol{x} - \boldsymbol{\zeta}\|$ is the Euclidean distance between the point $\boldsymbol{x}$ and $\boldsymbol{\zeta}$. The contour lines of any one-dimensional function $f(\|\boldsymbol{x} - \boldsymbol{\zeta}\|)$ that uses the Euclidean norm as its single argument, will be radial in the input space, i.e. the contour lines will be always concentric hyperspheres.

The Euclidean norm is mainly used to obtain multidimensional Gaussian-like functions with variable spread. This is usually accomplished by slightly modifying the previous norm definition $\|\boldsymbol{u} - \boldsymbol{v}\|^2_{\boldsymbol{W}} = (\boldsymbol{u} - \boldsymbol{v})^T \boldsymbol{W} (\boldsymbol{u} - \boldsymbol{v})$, where the matrix $\boldsymbol{W}$ consists of additional adjustable parameters related to the spread of the contour lines.

3.1.4. *Underlying idea to norms and projections*

There exist mathematical relationships between the inner product and the Euclidean distance. Moreover, according to Ref. 7, there is an intuitive and clear point of view to understand the main difference between using the inner product or the Euclidean distance. The Euclidean distance $\|x - z\|$ imposes a (perfect) radial symmetry in the input space, where the vector z represents the center of that symmetry. On the other hand, it is not difficult to see that the inner product $z^T x$ produces contours that are parallel hyperplanes and that they are perpendicular to the direction given by the vector z. Figure 4 shows an example of these contour lines for the two-dimensional case. Thus, the inner product and the Euclidean distance can be viewed as two different ways of looking at the input space.

The inner product and the Euclidean distance deal with the high dimensionality of the input space by imposing (different) symmetries. According to this approach, norms and projections can be viewed as mathematical rules which provide not only a particular grouping of the data in the input space (clusters), but a ranking of these clusters [7]. Thus, all points within a particular cluster have the same index. Furthermore, these indexes provide an order relationship between clusters. For example, the Euclidean distance groups the data in hyperspheric clusters. The index of a cluster increases when the radius of the hypersphere increases. Note that in order to produce clusters with a complex geometry, sophisticated models combine several rules (usually of the same type).

4. Conclusions

A new perspective on the main existing automatic learning strategies has been presented, by providing a general framework to handle and unify supervised learning models. The proposed taxonomy allows highlighting the similarity of some models whose original motivation comes from different fields. Furthermore, we have identified three main strategies to deal with the high dimensionality of the input space: partitioning, projecting and use of weighted norms.

References

[1] T. Hastie, R. Tibshirani, and J.H. Friedman, *The Elements of Statistical Learning*, Springer, Augusr 2001

[2] V.S. Cherkassky and F. Mulier, *Learning from Data: Concepts, Theory, and Methods*, John Wiley & Sons, Inc., New York, NY, USA, 1998

[3] C.M. Bishop, *Neural Networks for Pattern Recognition*, Oxford University Press, New York, 1995

[4] V.N. Vapnik, *The Nature of Statistical Learning Theory*, Springer Verlag, New York, 1995

[5] L. Wehenkel, *Automatic learning techniques in power systems*, Kluwer Academic, Boston, 1997

[6] J.H. Friedman, *An overview of predictive learning and function approximation*, NATO ASI Series 136, 1994

[7] E.F. Sánchez-Úbeda, Models for data analysis: contributions to automatic learning, Comillas University, Madrid, Spain, October 1999

[8] L. Breiman, J.H. Friedman, R.A. Olshen, and C.J. Stone, *Classification and Regression Trees*, Belmont Wadsworth International, 1984

[9] F. Girosi, M. Jones, and T. Poggio, Regularization Theory and Neural Networks Architectures, *Neural Computation*, 7, 219, 1995

[10] R.E. Bellman, *Adaptive Control Processes*, Princeton Univ. Press, 1961

[11] J.H. Friedman, *Classification and multiple regression through projection pursuit*, Tech.Rep.12, Dept. of Statistics, Stanford University, January 1985

INTERACTIVE MACHINE LEARNING TOOLS FOR DATA ANALYSIS

ROBERTO TAGLIAFERRI*, FRANCESCO IORIO, FRANCESCO NAPOLITANO, GIANCARLO RAICONI

DMI, University of Salerno,
I-84084, via Ponte don Melillo, Fisciano (SA), Italy
**E-mail: robtag@unisa.it*
www.unisa.it

GENNARO MIELE

DSF, University of Naples "Federico II",
I-80136, via Cintia 6, Naples, Italy
E-mail: miele@na.infn.it

In this work we propose a scientific data exploration methodology and software environment that permits to obtain both data clustering/labeling and visualization. The approach is based on an elaboration pipeline, going from data importing to cluster analysis and assessment with each stage supported by dedicated visualization and interaction tools. Supported techniques include a stability based procedure for the algorithmic parameters estimation (i.e. number of centers in K-means or map dimension in Self-Organizing Maps); a set of models realizing the fuzzy membership analysis and allowing users to select sub-clusters according to a given stability threshold; a tool for studying cluster reliability based on resampling; a tool for the interactive hierarchical agglomeration of clusters. Finally, a novel technique allowing the exploration and visualization of the space of clustering solutions is introduced. All implemented techniques are supported through appealing visualizations in a highly interactive environment.

Keywords: Clusters analysis, cluster assessment, interactive visualizations, fuzzy clustering, clustering maps

1. Introduction

In these last years the field of Knowledge Discovery in Databases (KDD) is becoming of great importance for several fields of research. In genetics, for example, several data mining approaches are proposed to analyze catalogues obtained from genome sequencing projects, or in astronomy to many classification and regression aims. In Genetics and Bioinformatics, scientists have been successful in cataloguing genes through genome sequencing projects, and they can now generate vast quantities of gene expression data using microarrays. Many of these applications can suffer from poor data visualization techniques. On the other hand, data visualization is an important mean of extracting useful information from large quantities of raw data. The human eye and brain together make a formidable pattern detection tool, but for them to work the data must be represented in a low-dimensional space, usually two or three dimensions. Naive 2D or 3D visualizations alone are inadequate for exploring bioinformatics data. Biologists need a visual environment that facilitates exploring high-dimensional data dependent on many parameters. In this context, research needs further work into bioinformatics visualization to develop tools that will meet the upcoming genomic and proteomic challenges. Many algorithms for data visualization have been proposed by both neural computing and statistics communities, most of which are based on a projection of data into a two or three dimensional visualization space. We briefly review some of these advanced visualization techniques, and propose an integrated environment for clustering and 2D or 3D visualization of high dimensional biomedical data.

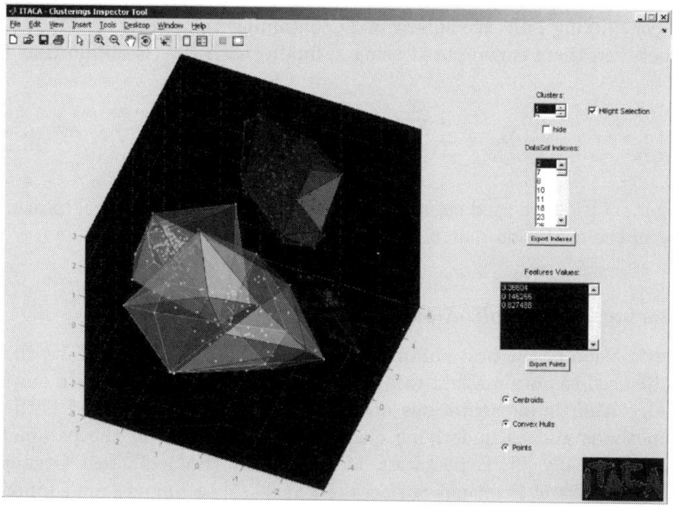

Fig. 1. The clusters inspector tool

2. Preliminary Analysis and Assessment Tools

Our approach enables the user to answer the following questions before trying blind-clustering on data: Is a dataset really clusterizable? In other words, does it contain localized uniformed groups and are these groups well separable ? How many clusters should we produce or, more generally, how should we set the parameters of the clustering procedures? Moreover, our tool enables the user to answer the following questions when the clustering is done: How much is the clusterization reliable? How much can we trust in the membership of a specific data-point to a cluster? If we manually reassign some point to a different cluster how much does the total reliability change? How much does each cluster reliability change? Can we extract, from the original dataset, a subset of points over which the clusterization has a fixed reliability? In which clusters do the points of this subset lie? And how do the points outlying this subset lead to belong to a given cluster?

2.1. *Parameter estimation*

The first step of the pipeline of the processes implemented consists of a procedure for testing the stability for any value on a prescribed range of parameters of the algorithm. This procedure is based on the Model Explorer algorithm [1] in which, given a dataset D, a parameterized clustering procedure $P(K, D)$ (where K is the parameter, i.e. the number of clusters to produce in K-means approach), a range of parameter values $[K_1, K_n]$, a similarity measure between clusterizations $s : \mathcal{P}(A) \times \mathcal{P}(A) \rightarrow [0, 1]$ (where $\mathcal{P}(A)$ is the set containing all the possible partitions of the set A), a number of trials T and a sampling ratio f, an estimation of the *average stability* granted for each parameter value in the testing range $\{Z_{K_1}, \ldots, Z_{K_n}\}$ is produced in output. Each Z_{K_i} is computed as follow: for each $t = 1, \ldots, T$, two sub samples $S_{t,1}$ and $S_{t,2}$, containing $|D|/f$ data points, are randomly chosen from D; two clusterizations are obtained applying the clustering procedure to each sub sample, $C_{t,1} = P(K_i, S_{t,1})$ and $C_{t,2} = P(K_i, S_{t,2})$; each clusterization is reduced considering only the subset of D containing the points of $S_{t,1} \cap S_{t,2}$ and the similarity between them is computed using s; finally, each Z_{K_i} is obtained as follows:

$$Z_{K_i} = \frac{1}{T} \sum_{t=1}^{T} s\left(C_{t,1}^{S_{t.1} \cap S_{t,2}}, C_{t,2}^{S_{t.1} \cap S_{t,2}}\right)$$

In our implementation we used classical similarity measures and a novel measure based on the entropy of the confusion matrix between partitions of the same set.

2.2. *Clustering and visualization*

After the estimation of the best parameter value of a clustering procedure, the same procedure can be used to obtain a first raw clusterization of the dataset D. In our tool several clustering algorithm implementations are provided, which can be used both directly as clustering methods and as underlying clustering procedure P in the parameter estimation process. K-Means [2], Expectation Maximization (EM) [2], Self Organizing Maps (SOM) [3], Probabilistic Principal Surfaces (PPS) [4] [5] techniques are supported.

The obtained clusterization can be inspected by a dedicated module offering the visualization of clusters convex hulls as they appear in projected 2D or 3D spaces obtained through PCA [2] and MDS [6] techniques. As an alternative, projections into arbitrary feature subspaces can be represented. An example of this kind of visualization is shown in Figure 1. The user can select each cluster and each point in a cluster checking all its features. Finally, the whole clusterization or specific clusters can be exported in the MATLAB workspace or saved on the disk.

3. Cluster Reliability

Given a clusterization $\mathcal{C} = \{C_1, \ldots, C_K\}$ on a dataset $D = \{\mathbf{x}_1, \ldots, \mathbf{x}_N\}$, the similarity matrix for $\mathcal{C}$ is the $N \times N$ matrix M with binary entries defined as follows:

$$M_{i,j} = \begin{cases} 1 & \text{if } \exists l \in [1, \ldots, K] | \, \mathbf{x}_i, \mathbf{x}_j \in C_l \\ 0 & \text{otherwise} \end{cases}, \quad \forall i, j = 1, \ldots, N.$$

For computing the reliability of each cluster in a clusterization we first produce several perturbed versions of the dataset, $D'_1, D'_2, \ldots, D'_P$, using a random projection obeying to Johnson-Lindenstrauss lemma [7], [8]. Then a clusterization is obtained over each of these new datasets and their similarity matrices $M_1, M_2, \ldots, M_P$ are built. Then, we build the *Fuzzy Similarity Matrix* (FSM) over these different clusterizations as follow:

$$\widetilde{M} = \frac{1}{P}(M_1 + M_2 + \ldots + M_P).$$

Considering, in the FSM, only the entries with the same indexes of the points lying in a cluster, observing that, for the cluster C_i^* always holds

$$|C_i^*| \leq \sum_{m,l} \widetilde{M}_{m,l}^{C_i^*} \leq |C_i^*|^2,$$

we can define a reliability measure for that cluster as:

$$r(C_i^*) = \frac{\sum_{l,m} \widetilde{M}_{l,m}^{C_i^*} - |C_i^*|}{|C_i^*|^2 - |C_i^*|}.$$

This measure assumes value in $[0, 1]$; its value is 0 if, considering all the clusterizations over the perturbed versions of the dataset, the points of C_i^* do never appear together. Conversely, its value is 1 if the points C_i^* ever appear in the same cluster.

4. Fuzzy Membership Analysis

In this model the clusters become fuzzy sets. Starting from the fuzzy similarity matrix, we can compute the value of each cluster membership function for each pattern of the dataset. Given a clusterization $\mathcal{C} = \{C_1, \ldots, C_K\}$ on the dataset $D = \{\mathbf{x}_1, \ldots, \mathbf{x}_N\}$, where $C_j = \{\mathbf{x}_{m_1}, \ldots, \mathbf{x}_{m_T}\}$ is a generic cluster, with $\mathbf{x}_{m_l} \in D$ and $m_l \in [1, \ldots, |D|], \forall l = 1, \ldots, T$, we define the value of the membership function of the cluster C_j for the pattern $\mathbf{x}_i$ as follows:

$$F_j(\mathbf{x}_i) = \begin{cases} \sum_{l=1}^{T} \widetilde{M}_{m_l,i} \Big/ |C_j| & \text{if } \mathbf{x}_i \notin C_j \\ \left(\sum_{l=1}^{T} \widetilde{M}_{m_l,i} - 1 \right) \Big/ \left(|C_j| - 1 \right) & \text{if } \mathbf{x}_i \in C_j \end{cases}$$

These functions quantify *how much* single points belong to a fixed group of other points over different trials of clustering on the perturbed versions of the dataset. It can be easily shown that always holds $0 \leq F_j(\mathbf{x}_i) \leq 1$, $\forall j = 1, \ldots, K$, $\forall i = 1, \ldots, N$.

The GUI of the tool for the fuzzy membership analysis is shown in Figure 2. In the left side sub-figure the convex hulls of the clusters composed by points with membership function greater than a given threshold are shown. The remaining points lie out of the convex hulls and are painted according to the values of the membership functions of each cluster, mixing the RGB composition of the cluster colors. The user is allowed to change the threshold and export the corresponding clusters. Basing on the values of the membership function for each outlier, the system suggests to manually reassign a point to a different cluster. The user can choose to accept such suggestions and explore alternative clusterizations.

5. Interactive Agglomerative Clustering

Since rarely the number of clusters is known in advance, we adopt an interactive agglomerative technique, performed on the clusters obtained in the assessment phase. The result is a dendrogram in which each leaf is associated to a cluster. The related tool supports the Negentropy distance, which can be computed directly between the clusters of the pre-clustering and that has several interesting properties [4]. Nonetheless, this step could employ any hierarchical clustering technique since the cluster linkage can be performed using any metric.

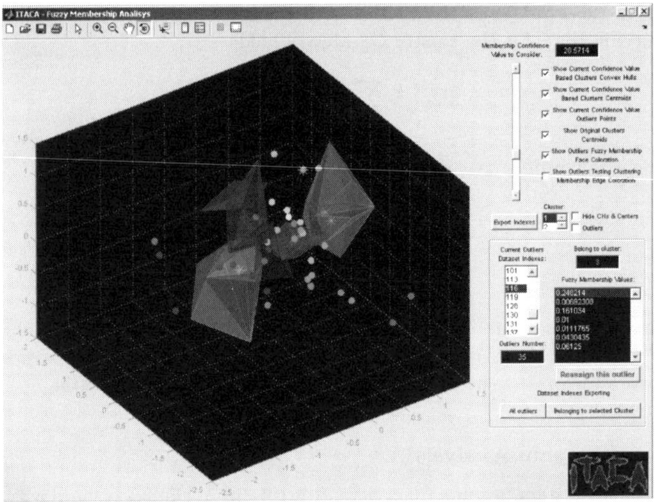

Fig. 2. The fuzzy membership analysis tool

The dendrogram plotted by our tool allows an interactive analysis (see fig. 3). In fact, a dashed line, representing the current threshold, is shown on it and the corresponding partitioning is made evident by the use of different colours for the leaves of the dendrogram belonging to different clusters (see Figure 2, right). Each color, in turn, is shaded from dark to bright according to its clustering stability: the darker is the leaf, the sooner it will become an outlier when lowering the threshold. In fact, the threshold can be directly dragged on the dendrogram while the colors corresponding to the new partitioning are updated in real time. Also, all currently open visualizations (see next section) are updated accordingly. We stress that the key point of this clustering approach is that the user can arrange together great amounts of data operating on a human-tractable amount of information.

In particular, we have considered visualizations based on three projection techniques: Multidimensional Scaling, Principal Component Analysis, Spherical Probabilistic Principal Surfaces. Often, a far more significant visualization can be obtained by merging the information given by such visualizations and the cluster information provided by the clustering phase. In particular, each of the points in the original data space can be associated with the color of the cluster it belongs to, which is provided by the dendrogram shaded coloring. After the association is obtained, each visualization method can be improved adding the clustering color to it. Here we stress that our tool does this in real time, giving immediate feedback to the user about the clustering he is taking into account. This technique can be applied to projected dataset in 2D or 3D. All visualizations can be shown contemporarily together with the dendrogram and updated in real time while acting on its threshold. Our tool also allows to select sub-trees of the dendrogram and to trace back the involved data points simultaneously on the other visualizations. Also subsets of points can be free-hand selected on the 2D visualizations to save portions of the dataset identified as interesting by visual inspection.

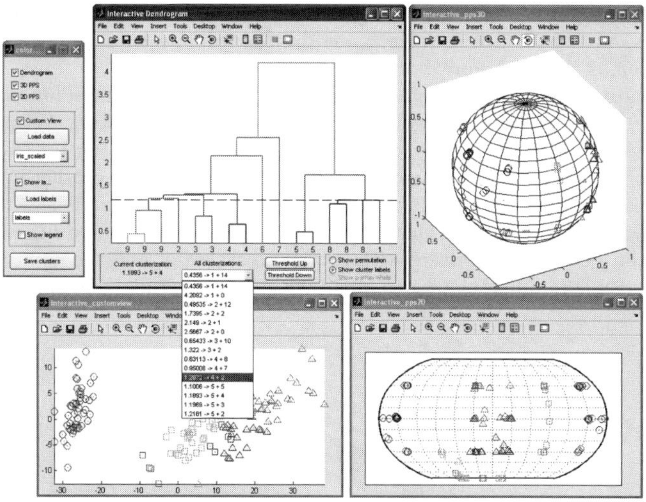

Fig. 3. Overview of the interactive hierarchical clustering tool.

5.1. *Visualization of prior knowledge*

It often happens that some or all the objects of a dataset are known to belong to some classes. This information can be used in at least two ways, to validate a clustering result and to infer new knowledge. The validation of a clustering can be obtained comparing the prior knowledge and the knowledge obtained from the clustering itself. We note that this can give a confidence degree about the use of the same partitioning with data on which no prior knowledge is available. The prior knowledge can also be used to produce new knowledge inferred by the presence or absence of objects of a certain class in a certain cluster. Different hypotheses can be made depending on the relations between the prior knowledge and the features used to cluster the objects. Prior knowledge can be visually shown using different forms of points in the dataset belonging to different classes. This permits to visualize, in all the visualization methods described, distance and cluster information together with the prior knowledge. Our tool allows the use of different sets of prior knowledge information and it permits to easily switch from one set to another one.

6. Clustering Maps

Many clustering algorithms are able to find local optima with respect to their objective function, depending on the initialization phase. Unfortunately there is no assurance that such local optima are good approximations of the global optimum and in practice they are often far from it. A simple way to deal with this problem is to compute a number of different clusterizations using different initializations and choose the one that better approximates a fitness function. This approach has at least one important limitation: whenever a dataset is reasonably clusterizable in more than one way, the process would blindly choose the one giving the best fitness and discard the others.

With the aim of further investigating the space of clusterizations for a given dataset, we produce a significant visualization for it. Given a measure assessing the difference between clusterizations (like the entropy of the corresponding confusion matrix), the MDS technique can be exploited to represent each clusterization as a point in the plane, with the

distance between points representing the distance between the corresponding clusterizations. Finally, adding the fitness value as the third coordinate for each point, we obtain a *clustering map* as that in Figure 4.
Clustering maps allow to intuitively represent the spaces of clustering solutions for a given dataset, independently from the algorithm producing the clusterization. Whenever a clustering map shows more than one region of good fitness value, it can be concluded that there are more than one reasonable clusterization for the data. Also other information is encoded in the clustering maps, like the distribution of the clusterizations with respect to the fitness function (how much a clusterization is likely to achieve good fitness value), the stability of the clustering algorithm with respect to the data, and the ability of different algorithms to explore different regions of the space of clustering solutions (see Figures 4 and 5).

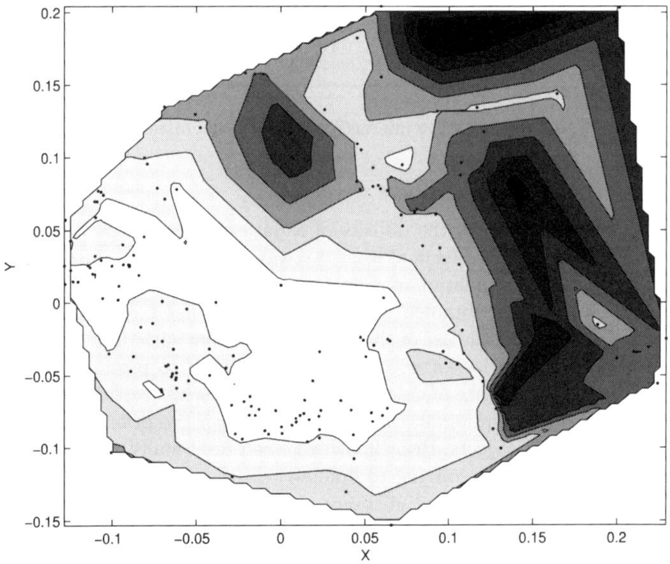

Fig. 4. Clustering Map for the IRIS dataset. Each point represent a clusterization obtained through a 2x3 SOM with random initialization. Brighter colors show regions of good fitness value (low distorsion).

7. Conclusions

We presented a set of visual and interactive tools for data clustering, including parameter estimation and clustering assessment and validation. The 2D and 3D visualizations, together with data interaction mechanisms, permit to easily produce reliable clusterizations and explore feasible alternatives. This last issue is also addressed through clustering maps, a novel visualization technique which is able to visualize the space of clustering solutions for a dataset in an information reach and easy to read representation.

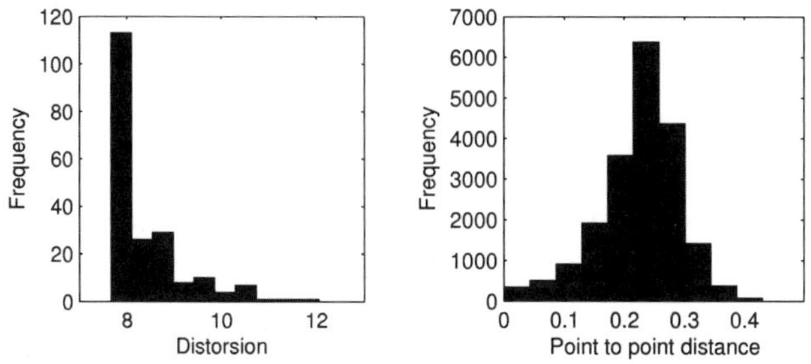

(a) Distorsion histogram

(b) Distance histogram

Fig. 5. Histograms for from the Clustering Map in 4. (a) Frequencies of the distorsion values. (b) Frequencies of the point-to-point distances.

References

[1] A. Ben-Hur, A. Elisseeff, I. Guyon: A stability based method for discovering structure in clustered data. Pacific Symposium on Biocomputing (2002)

[2] Duda, R.O., Hart, P.E., Stork, D.G. (2001) Pattern Classification, John Wiley and Sons Inc., Second Edition.

[3] Kohonen, T. (1995) Self-Organizing Maps, Springer-Verlag, Berlin.

[4] R. Amato, A. Ciaramella, N. Deniskina et al. , A Multi-Step Approach to Time Series Analysis and Gene Expression Clustering, Bioinformatics, Volume 22, N. 5, pp. 589-596, 2006.

[5] Chang, K., Ghosh, J. (2001) A unified Model for Probabilistic Principal Transactions on Pattern Analysis and Machine Intelligence, 23, n.1.

[6] J.B.Kruskal, M.Wish. (1978) Multidimensional Scaling, Sage.

[7] W.B. Johnson, J. Lindenstrauss: Extensions of Lipshitz mapping into Hilbert space. In Conference in modern analysis and probability, volume 26 of Contemporary Mathematics (1984)

[8] A. Bertoni, G. Valentini: Random projections for assessing gene expression cluster stability. IJCNN 2005, The IEEE-INNS International Joint Conference on Neural Networks, Montreal (2005)

DATA VISUALIZATION AND CLUSTERING: AN APPLICATION TO GENE EXPRESSION DATA

ANGELO CIARAMELLA, FRANCESCO IORIO, FRANCESCO NAPOLITANO, GIANCARLO RAICONI, ROBERTO TAGLIAFERRI*

DMI, University of Salerno, I-84084, via Ponte don Melillo, Fisciano (SA), Italy
**E-mail: rtagliaferri@unisa.it*

GENNARO MIELE*, ANTONINO STAIANO

DSF, University of Naples, I-80136, via Cintia 6, Napoli, Italy
**E-mail: gmiele@na.infn.it*

In this work we present a multi-step process that, starting from raw datasets, brings them through preprocessing, preclustering and agglomerative clustering stages, exploiting a visual and interactive environment for data analysis and exploration. At the core of the process lies the idea of subdividing the process of data clusterization into two main steps: the first one aimed to reduce the size of data and the second one to present the reduced and human-understandable dataset to the user. This last step allows him to participate in the process of clustering and helps him figuring out the data underlying structures. The approach is actually implemented in a group of user-friendly tools under the MATLAB environment, featuring a number of classical and novel data processing, visualization, assessment and interaction methods.

Keywords: Data Analysis, Data Visualization, Data Interaction, Clustering Assessment.

1. Introduction

In this work we propose a scientific data exploration software environment that permits to obtain both data clustering/labelling and visualization. The approach is based on a pipeline of elaborations, going from data importing to cluster analysis with each stage supported by a dedicated visualization and interaction tool. The human eye and brain together make a formidable pattern detection tool, but, for them to work, the data must be represented in a low-dimensional space, usually two or three-dimensional. Substantially the proposed approach enables the user to:

- Import data and visualize it with simple sub-space projections in order to make them human-understandable.
- Project and visualize data with more refined techniques in order to preserve its underlying structure. Supported visualizations are: 2D and 3D Multidimensional Scaling (MDS) [1], Spherical Probabilistic Principal Surfaces (PPS) [2], Robinson map projection of PPS [3] Surfaces plus custom 2D or 3D projections provided by the user.
- Perform clustering and assessment in order to obtain stable clusterizations. Visualize the clustering and perform clusters agglomeration interactively. Visually analize clusters.

- Perform deeper studies on the data by localizing regions of interests and interacting with the data itself.

The software implementing the process is developed under the MATLAB environment and is part of the Astroneural project [5]. As a real-world test for our tools, we used an excerpt from the HeLa database, found in Whitfield et al. [6]. We considered the third of the five experiments composing the dataset, basing on its high number of samples and low number of missing data (34 genes out of the total of 1099). The experiment is composed of expression values for 1099 genes of cancer cells monitored for 2 days and sampled once per hour, making up a 1099 × 48 initial dataset.

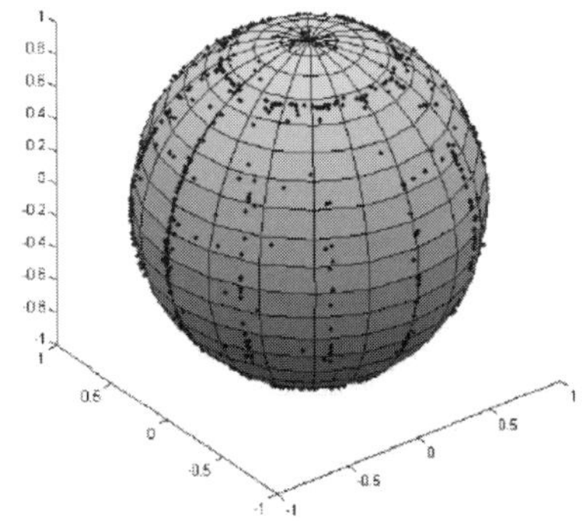

Fig. 1. 3D PPS projection of the HeLa Cell dataset.

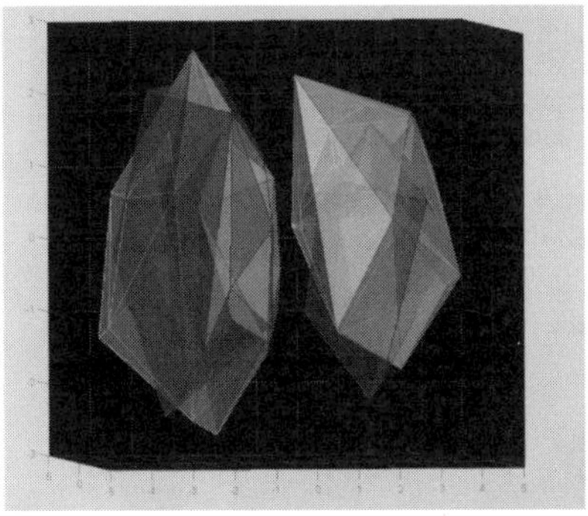

Fig. 2. Comparison of two clusterizations (yellow, red).

2. Preprocessing and Data Acquisition

Before the application of clustering or data mining algorithms, it is often useful and sometimes necessary to preprocess data. Most real-world datasets suffer from two major problems: noise and high dimensionality. These phenomena can derive from the method (or device) used to gain the data or from the nature of the data itself. In both cases a dataset may result to be intractable from the point of view of the accuracy of the results (compromised by noise) or the one of computational cost (exponentially growing with dimensionality). Astroneural includes a number of preprocessing methods ranging from data normalization to feature extraction with non-linear robust PCA (NL-RPCA) [7–9], which enable the user to prepare the data for the next processing steps by using a user-friendly interface. We used the NL-RPCA with the HeLa dataset to obtain a 20-dimensional reduction.

3. Clustering and Assessment

The clustering phase can be viewed both as the process of building the final clusterization for a dataset and the process of reducing the number of objects to present to the user for subsequent steps of the analysis. Most known classical methods are provided:

- K-Means;
- EM;
- Self Organizing Maps.

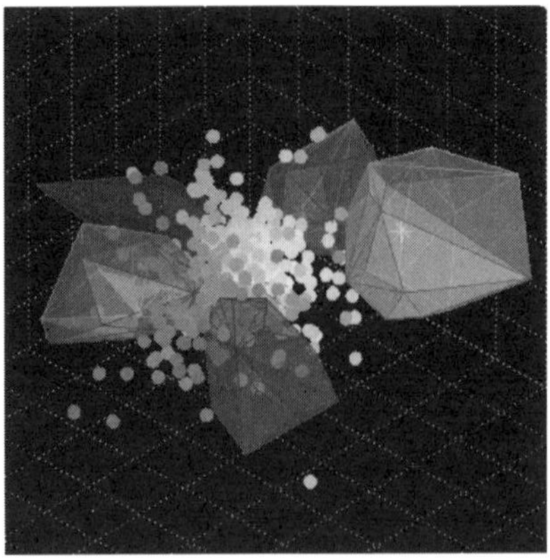

Fig. 3. Clusters and *fuzzy outliers* on HeLa Cell dataset.

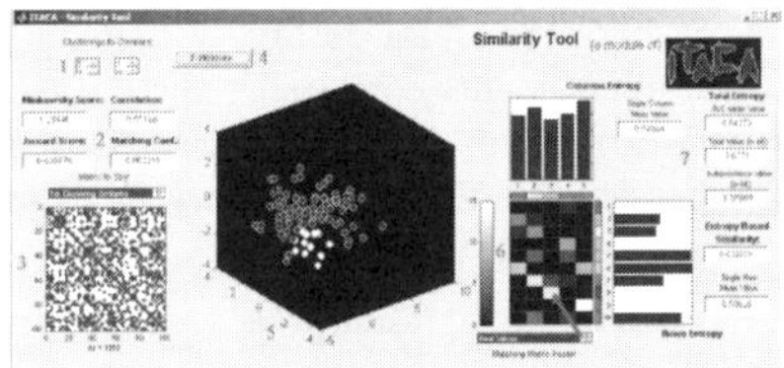

Fig. 4. Clusterization Similarity Module.

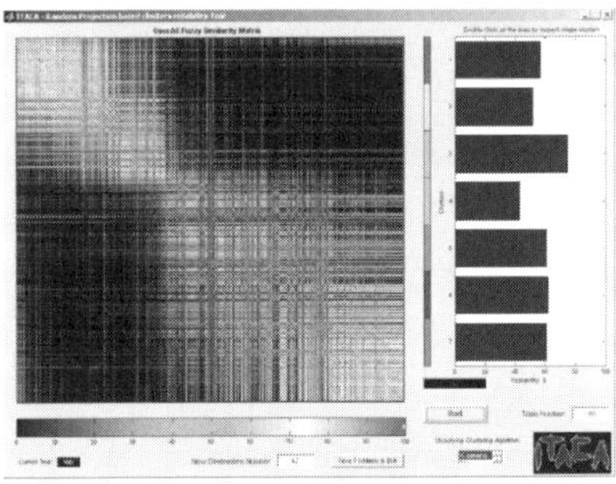

Fig. 5. Clusters reliability, on HeLa Cell Dataset, via the *fuzzy similarity matrix*.

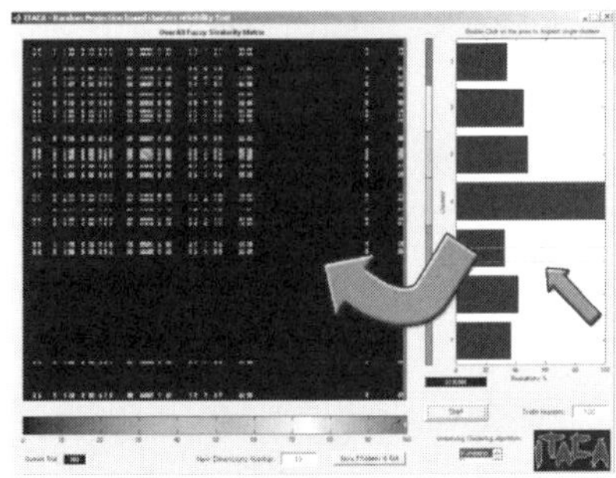

Fig. 6. Interactive selection of clusters elements matrix entries.

Special support is given for Probabilistic Principal Surfaces, which, besides being known to perform efficiently for clustering purposes, can be also exploited to build some appealing visualizations (figure 1). Also visualizations based on tridimensional convex hulls of the clusters points are provided (figures 2,3). Unfortunately, the clustering process is not straightforward and the problem of finding the best clusterization for a dataset is not easy to deal with. Usually different algorithms provide different clusterizations and even the same algorithm (like K-means) may provide different results depending on the initialization phase. The clustering assessment phase aims to help solving this problem searching for stable clusterizations between the ones generated on random projections of the dataset (figures 5,6) or searching for the best values of the clustering model parameters (in example, then number of clusters in K-Means). The best parameter value is the one that provides the maximum average similarity between clusterings on different sub samples of the original dataset over a number of iterations (maximum stability [10]). The similarity between different clusterings is measured using both classical measures, such as *Minkowski Index*, *Jaccard Coefficient*, *correlation* and *matching coefficient* (all found in Ben-Hur et al. [10]) and a novel measure based on the confusion matrix entropy [11] (the devoted module is shown in figure 4). Stability values, on HeLa Cell dataset, in function

of the number of clusters, using K-Means for sub samples clusterization, is shown in figure 7.
In addition, direct interaction with points in the clusters is also supported.

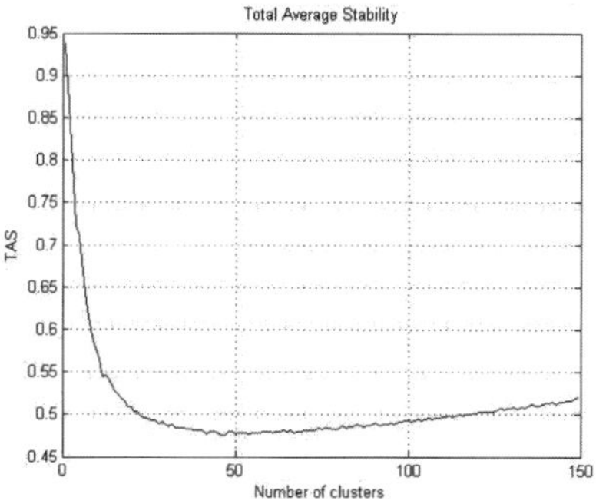

Fig. 7. Total Average Stability (mean of stability based on several classical similarity measures), on HeLa Cell dataset, in function of the clusters number.

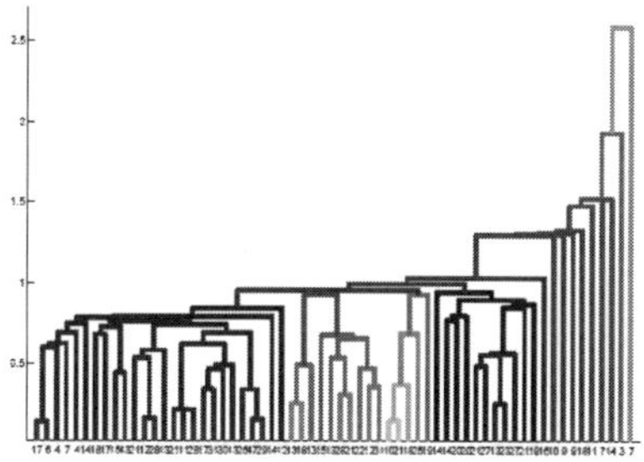

Fig. 8. Colored interactive dendrogram for hierarchical clustering.

4. Interactive Agglomerative Clustering

Once the dataset has gone through a first level of clustering, an interactive agglomerative approach is used to let the user choose the final clustering and explore relationships between clusters. Towards this aim a hierarchical clustering is performed on the clusters obtained in the previous step and the correspondent dendrogram visualization is shown (figure 8). Each leaf of the dendrogram is associated with a cluster from the preclustering phase. User is allowed to choose a distance threshold directly on the dendrogram visualization while the new colors obtained are mapped in real time to other visualization such as MDS and PPS projections (figures 9, 10). As before, this step is not bounded to any particular hierarchical clustering technique, since the clusters' linkage can be performed using any measure of distance between the preclusters' centroids. Anyway, besides classical agglomeration methods (such as single, complete and data average linkage, Ward's method, centroid, etc.) our tool supports Negentropy Clustering (NEC) [4, 5].

5. Results Analysis

Once a clusterization is automatically or interactively obtained, a more analytic approach is needed to quantify its significance. Typical measures used towards this objective are the mean and variance of the feature vectors of all objects in a cluster (figures 11,12), where a well characterized cluster is of course one with low variance, while clearly separated clusters must have also clearly different behaviors of the respective means. Our tool supports also *fuzzy membership analysis* that provides estimations of each cluster reliability and single points average tendency to belong to a cluster [11]. This estimates allow users to interactively reassign points to a different cluster. They are based on the *fuzzy similarity matrix* obtained performing clustering on several perturbed version of the original dataset. Used perturbations are random projections obeying Johnson-Lindenstrauss Lemma [12].

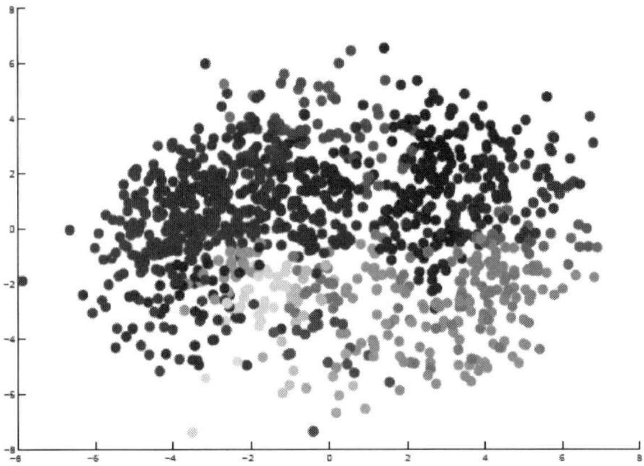

Fig. 9. The HeLa Cell dataset MDS projection with obtained by the interactive dendrogram.

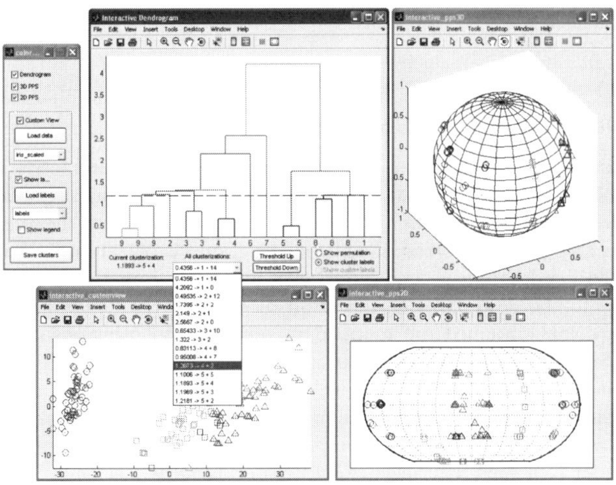

Fig. 10. Overview of the tools for interactive hierarchical clustering.

6. Conclusions

In this work we presented a multi-step approach to data exploration allowing the reduction of information contained in big datasets to a human-tractable size. We showed how a user is supported during the interactive part of the process with different kinds of data visualizations for the original data space which are updated as he interacts with the reduced data space. We also showed how the results of the process are analytically checked to assure the significance of the clusterization. For all the ideas presented we referred to an actual software implementation in the MATLAB environment which realizes the process and provides the interactive environment for the exploration of data and results.

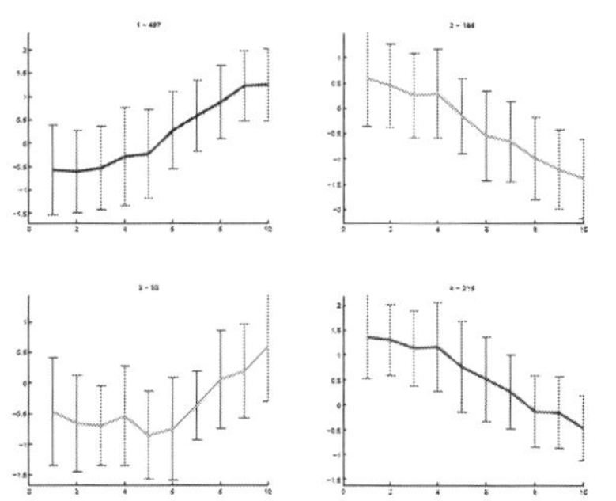

Fig. 11. Mean-Variance plot for the four principal clusters as shown in figure 9.

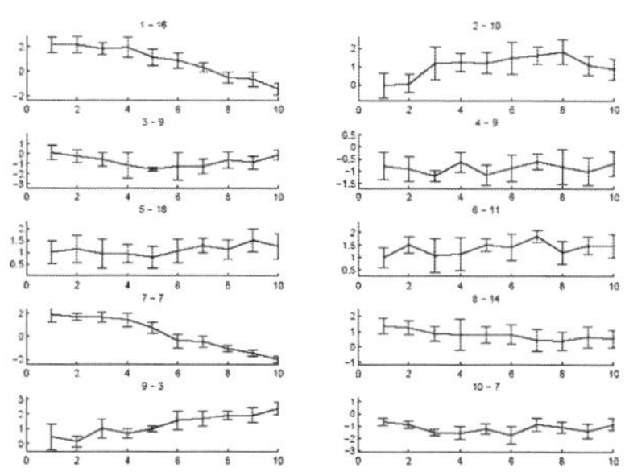

Fig. 12. Mean-Variance plot for the smaller clusters.

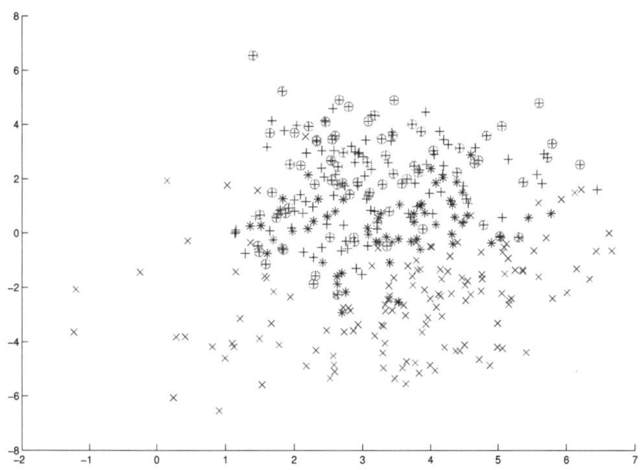

Fig. 13. Comparison of the clusters associated with DNA replication in Whitfield et al. [6](genes marked with 'x') and one of our clusters (genes marked with '+'). Genes not clustered at all in Whitfield [6] are marked with a circle.

References

[1] J.B.Kruskal, M.Wish. (1978) *Multidimensional Scaling*, Sage.

[2] K. Chang, J. Ghosh. (2001) *A unified Model for Probabilistic Principal Surfaces*, IEEE Transactions on Pattern Analysis and Machine Intelligence, 23, n.1.

[3] A. Robinson. (1974) *A New Map Projection: Its Development and Characteristics*, International Yearbook of Cartography 14, pp. 145-155.

[4] R. Amato, A. Ciaramella, N. Deniskina et al. (2006) *A Multi-Step Approach to Time Series Analysis and Gene Expression Clustering*, Bioinformatics, Volume 22, N. 5, pp. 589-596. dettagli, 1995.

[5] A.Ciaramella, G. Longo, A. Staiano, R. Tagliaferri. (2006) *NEC: A Hierarchical Agglomerative Clustering Based on Fisher and Negentropy Information*, Lecture Notes in Computer Science, vol. 3931, 49-56.

[6] M. L. Whitfield, G. Sherlock, A. J. Saldanha, J. I. Murray, C. A. Ball, K. E. Alexander, J. C. Matese, C. M. Perou, M. M. Hurt, P. O. Brown, D. Botstein. (2002) *Identification of Genes Periodically Expressed in the Human Cell Cycle and Their Expression in Tumors*, Molecular Biology of the Cell, vol. 13, 1977-2000.

[7] A. Ciaramella, C. Bongardo, H.D. Aller, M.F. Aller, G. De Zotti, A. Lahteenmaki, G. Longo, L. Milano, R. Tagliaferri, H. Terasranta, M. Tornikoski, S. Urpo (2004) *A Multifrequency Analysis of Radio Variability of Blazars*, Astronomy & Astrophysics Journal, 419, 485-500.

[8] R. Tagliaferri, A. Ciaramella, L. Milano, F. Barone, G. Longo. (1999) *Spectral Analysis of Stellar Light Curves by Means of Neural Networks, Astronomy & Astrophysics* Suppl.Series, 137, 391-405.

[9] R. Tagliaferri, N. Pelosi, A. Ciaramella, G. Longo, L. Milano, F. Barone. (2001) *Soft Computing Methodologies for Spectral Analysis in Cyclostratigraphy*, Computer and Geosciences, 27, 535-548.

[10] A. Ben-Hur, A. Elisseeff, I. Guyon (2002) *A stability based method for discovering structure in clustered data*, Pacific Symposium on Biocomputing.

[11] F. Iorio (2006) *ITACA (Integrated Tool for Assessing Clustering Algorithms): Uno strumento integrato per la valutazione del clustering)*, Degree Thesis in Computer Science.

[12] W.B. Johnson, J. Lindenstrauss (1984) *Extensions of Lipshitz mapping into Hilbert space.* In Conference in modern analysis and probability, volume 26 of Contemporary Mathematics.

SUPER-RESOLUTION OF MULTISPECTRAL IMAGES

RAFAEL MOLINA[a*], JAVIER MATEOS[a] and MIGUEL VEGA[b]

a) Dept. Ciencias de la Computación e I. A., Univ. de Granada,
b) Dept. de Lenguajes y Sistemas Informáticos, Univ. de Granada,
18071 Granada, Spain
** E-mail:rms@decsai.ugr.es*

AGGELOS K. KATSAGGELOS

Dept. of Electrical Engineering and Computer Science,
Northwestern University, Evanston, Illinois 60208-3118. USA

In this paper we analyze global and locally adaptive super resolution Bayesian methodology for pansharpening of multispectral images. The discussed methodologies incorporate prior knowledge on the expected characteristics of the multispectral images, uses the sensor characteristics to model the observation process of both panchromatic and multispectral images, and includes information on the unknown parameters in the model in the form of hyperprior distributions. Using real and synthetic data, the pansharpened multispectral images are compared with the images obtained by other parsharpening methods and their quality is assessed both qualitatively and quantitatively.

Keywords: Super-resolution; Bayesian Models; Hyperspectral Images.

1. Introduction

Nowadays most remote sensing systems include sensors able to simultaneously capture several low resolution images of the same area on different wavelengths, thus forming a multispectral image, along with a high resolution panchromatic image. The main characteristics of such remote sensing systems are the number of bands of the multispectral image and the resolution of those bands and the panchromatic image. The main advantage of the multispectral image is to allow for a better land type and use recognition but, due to its lower resolution, information on the objects' shape and texture may be lost. On the other hand, the panchromatic image allows for a better recognition of the objects in the image and their textures but provides no information about their spectral properties.

Throughout this paper the term *multispectral image reconstruction* will refer to the joint processing of the multispectral and panchromatic images in order to obtain a new multispectral image that, ideally, will exhibit the spectral characteristics of the observed multispectral image and the resolution and quality of the panchromatic image.

A few approximations to multispectral image reconstruction have been proposed in the literature (see, for instance, Ref. 1–4) including a few super-resolution based methods [5, 6].

In this paper we follow the hierarchical Bayesian approach to obtain a solution to the super resolution reconstruction of multispectral images problem and discuss the utilization of global and spatially varying image models. Then, applying variational methods to approximate probability distributions, we estimate the unknown parameters, and the high resolution multispectral image.

The paper is organized as follows. In section 2 the Bayesian modeling and inference for super resolution reconstruction of multispectral images is presented. The required probability distributions for the Bayesian modeling of the super resolution problem are formulated in section 3. The Bayesian analysis and posterior probability approximation to obtain the parameters and the super resolution reconstructed image is performed in section 4. Experimental results on a real Landsat 7 ETM+ image are described in section 5 and, finally, section 6 concludes the paper.

2. Bayesian Problem Formulation

Let us assume that $\mathbf{y}$, the multispectral image we would observe under ideal conditions with a high resolution sensor, has B bands $\mathbf{y}_b$, $b = 1, \ldots, B$, that is, $\mathbf{y} = [\mathbf{y}_1^t, \mathbf{y}_2^t, \ldots, \mathbf{y}_B^t]^t$, where each band is of size $p = m \times n$ pixels and t denotes the transpose of a vector or matrix. Each band of this image is expressed above as a column vector by lexicographically ordering its pixels. In real applications, this high resolution image is not available. Instead, we observe a low resolution multispectral image $\mathbf{Y}$ with B bands $\mathbf{Y}_b$, $b = 1, \ldots, B$, that is, $\mathbf{Y} = [\mathbf{Y}_1^t, \mathbf{Y}_2^t, \ldots, \mathbf{Y}_B^t]^t$, where each band is of size $P = M \times N$ pixels with $M < m$ and $N < n$. Each band of this image is also expressed as a column vector by lexicographically ordering its pixels. The sensor also provides us with a panchromatic image $\mathbf{x}$ of size $p = m \times n$, obtained by spectrally averaging the unknown high resolution images $\mathbf{y}_b$.

The objective of the high resolution multispectral image reconstruction problem is to obtain an estimate of the unknown high resolution multispectral image $\mathbf{y}$ given the panchromatic high resolution observation $\mathbf{x}$ and the low resolution multispectral observation $\mathbf{Y}$.

Using the hierarchical Bayesian paradigm (see, for example, Ref. 7) the following joint distribution for $\boldsymbol{\Omega}_M$, $\mathbf{y}$, $\mathbf{Y}$, and $\mathbf{x}$ is defined $\mathrm{p}(\boldsymbol{\Omega}_M, \mathbf{y}, \mathbf{Y}, \mathbf{x}) = \mathrm{p}(\boldsymbol{\Omega}_M)\mathrm{p}(\mathbf{y}|\boldsymbol{\Omega}_M)\mathrm{p}(\mathbf{Y}, \mathbf{x}|\mathbf{y}, \boldsymbol{\Omega}_M)$, where $\boldsymbol{\Omega}_M$ denotes the set of hyperparameters needed to describe the required probability density functions (obviously, depending on the set of hyperparameters the probability models used in the problem will differ).

The Bayesian paradigm dictates that inference on the unknowns $(\boldsymbol{\Omega}_M, \mathbf{y})$, should be based on $\mathrm{p}(\boldsymbol{\Omega}_M, \mathbf{y}|\mathbf{Y}, \mathbf{x}) = \mathrm{p}(\boldsymbol{\Omega}_M, \mathbf{y}, \mathbf{Y}, \mathbf{x})/\mathrm{p}(\mathbf{Y}, \mathbf{x})$.

3. Bayesian Modeling

We assume that $\mathbf{Y}$ and $\mathbf{x}$, for a given $\mathbf{y}$ and a set of parameters $\boldsymbol{\Omega}_M$, are independent and consequently write $\mathrm{p}(\mathbf{Y}, \mathbf{x}|\mathbf{y}, \boldsymbol{\Omega}_M) = \mathrm{p}(\mathbf{Y}|\mathbf{y}, \boldsymbol{\Omega}_M)\mathrm{p}(\mathbf{x}|\mathbf{y}, \boldsymbol{\Omega}_M)$.

Each band, $\mathbf{Y}_b$, is related to its corresponding high resolution image by

$$\mathbf{Y}_b = \mathbf{D}\mathbf{H}\mathbf{y}_b + \mathbf{n}_b, \quad \forall b = 1, \cdots, B, \tag{1}$$

where $\mathbf{H}$ is a $p \times p$ blurring matrix and $\mathbf{D}$ is a $P \times p$ decimation operator and $\mathbf{n}_b$ is the capture noise, assumed to be Gaussian with zero mean and variance $1/\beta_b$.

Given the degradation model for multispectral image super-resolution described by Eq. (1) and assuming independence between the noise observed in the low resolution images, the distribution of the observed $\mathbf{Y}$ given $\mathbf{y}$ and a set of parameters $\boldsymbol{\Omega}_M$ is

$$\mathrm{p}(\mathbf{Y}|\mathbf{y}, \boldsymbol{\Omega}_M) = \prod_{b=1}^{B} \mathrm{p}(\mathbf{Y}_b|\mathbf{y}_b, \beta_b) \propto \prod_{b=1}^{B} {\beta_b}^{P/2} \exp\left\{-\frac{1}{2}\beta_b \parallel \mathbf{Y}_b - \mathbf{H}\mathbf{y}_b \parallel^2\right\}. \tag{2}$$

As already described, the panchromatic image $\mathbf{x}$ is obtained by spectrally averaging the unknown high resolution images $\mathbf{y}_b$, modeled as

$$\mathbf{x} = \sum_{b=1}^{B} \lambda_b \mathbf{y}_b + \mathbf{v}, \tag{3}$$

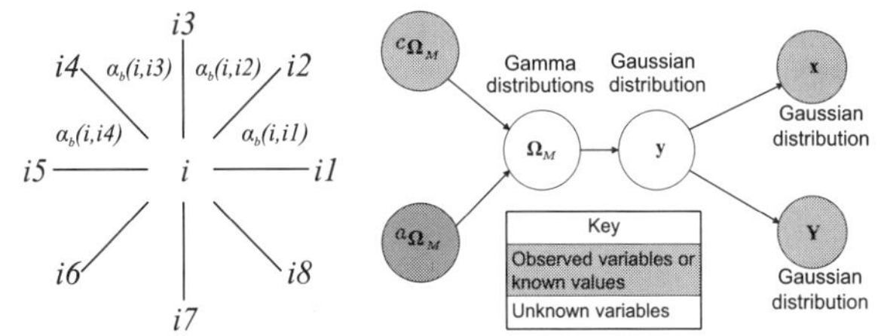

Fig. 1. (a) Pixel and inverse variance notation. (b) Graphical model showing the relationships between the variables.

where $\lambda_b \geq 0$, $b = 1, 2, \cdots, B$, are known quantities that can be obtained, as we will see later, from the sensor spectral characteristics, and $\mathbf{v}$ is the capture noise that is assumed to be Gaussian with zero mean and variance γ^{-1}. Note that, usually, $\mathbf{x}$ does not depend on all the multispectral image bands but on a subset of them, i. e., some of the λ_b's are equal to zero.

Using the degradation model in Eq. (3), the distribution of the panchromatic image $\mathbf{x}$ given $\mathbf{y}$, and a set of parameters $\mathbf{\Omega}$ is given by

$$\mathrm{p}(\mathbf{x}|\mathbf{y}, \mathbf{\Omega}_M) \propto \gamma^{p/2} \exp\left\{-\frac{1}{2}\gamma \parallel \mathbf{x} - \sum_{b=1}^{B} \lambda_b \mathbf{y}_b \parallel^2\right\}. \tag{4}$$

From the above definition the parameter vector $(\gamma, \beta_1, \ldots, \beta_B)$ is a subset of $\mathbf{\Omega}_M$. However, although the estimation of $(\gamma, \beta_1, \ldots, \beta_B)$ can be easily incorporated into the estimation process, we will assume here that these parameters have been estimated in advance and concentrate on gaining insight into the distribution of the prior image parameters, as described next.

3.1. *Global and local image modeling*

In this paper we do not use the correlation among different high resolution bands but concentrate instead on modeling the local variation at each band.

In our global image model we assume a Conditional Auto-Regressive (*CAR*) model [8]. Then we have for the global model $M = G$,

$$\mathrm{p}_G(\mathbf{y}|\mathbf{\Omega}_G) \propto \prod_{b=1}^{B} \bar{\alpha}_b^{\frac{p}{2}} \exp\left\{-\frac{1}{2}\bar{\alpha}_b \left[\mathbf{y}_b^t \mathbf{C} \mathbf{y}_b\right]\right\}, \tag{5}$$

where $\mathbf{C}$ is the laplacian operator. The set of hyperparameters then becomes $\mathbf{\Omega}_G = (\bar{\alpha}_1, \ldots, \bar{\alpha}_B)$.

We now proceed to define a local model $M = L$, for the high resolution multispectral image. In its definition we use the notation $i1, i2, \ldots, i8$ to denote the eight pixels around pixel i (see Fig. 1(a)). Then following the approximation in Ref. 9 which extends Conditional Auto-Regressions to take into account local variability we write (see Ref. 10)

$$\mathrm{p}(\mathbf{y}|\mathbf{\Omega}_L) = \prod_{b=1}^{B} \mathrm{p}(\mathbf{y}_b|\boldsymbol{\alpha}_b) \propto \prod_{b=1}^{B} \prod_{i=1}^{p} \prod_{l=1}^{4} \alpha_b(i,il)^{\frac{1}{8}} \exp\left\{-\frac{1}{2}\alpha_b(i,il)\left[y_b(i) - y_b(il)\right]^2\right\}, \tag{6}$$

where $\alpha_b(i, il)$ controls, for the b-band, the smoothness of the restoration between pixels i and il and $\boldsymbol{\alpha}_b = (\alpha_b(i, il) \mid i = 1, \ldots, p,\ l = 1, \ldots, 4)$.

The set of hyperparameters then becomes $\boldsymbol{\Omega}_L = (\boldsymbol{\alpha}_1, \ldots, \boldsymbol{\alpha}_B)$. Note that if $\alpha_b(i, il) = \bar{\alpha}_b$, $i = 1, \ldots, p$, $l = 1, \ldots, 4$, the local image model becomes the global model defined above.

A large part of the Bayesian literature is devoted to finding hyperprior distributions $\mathrm{p}(\boldsymbol{\Omega}_M)$, $M \in \{G, L\}$, for which $\mathrm{p}(\boldsymbol{\Omega}_M, \mathbf{y}|\mathbf{x}, \mathbf{Y})$ can be calculated in a straightforward way or can be approximated. These are the so called conjugate priors that, as we will see later, have the intuitive feature of allowing one to begin with a certain functional form for the prior and end up with a posterior of the same functional form, but with the parameters updated by the sample information.

Taking the above considerations about conjugate priors into account, we will assume for the hyperparameters of the global model that

$$\mathrm{p}(\boldsymbol{\Omega}_G) = \prod_{b=1}^{B} \mathrm{p}(\bar{\alpha}_b \mid \bar{a}_b^o, \bar{c}_b^o), \tag{7}$$

$\bar{c}_b^o > 0$ and $\bar{a}_b^o > 0$, while for the local model we will use the following distribution on the hyperparameters

$$\mathrm{p}(\boldsymbol{\Omega}_L) = \prod_{b=1}^{B} \prod_{i=1}^{p} \prod_{l=1}^{4} \mathrm{p}(\alpha_b(i, il) \mid a_b^o, c_b^o), \tag{8}$$

where $c_b^o > 0$ and $a_b^o > 0$ (note that the same hyperprior is assumed for all the α's in the same band).

In both, the local and global models, gamma distributions are used to define the hyperpriors of the precision parameters αs, that is, for $\omega \in \boldsymbol{\Omega}_M$ we have

$$\mathrm{p}(\omega \mid u_\omega, v_\omega) \propto \omega^{u_\omega - 1} \exp[-v_\omega \omega], \tag{9}$$

where $u_\omega > 0$ and $v_\omega > 0$. This gamma distribution has the following mean and variance

$$\mathbf{E}[\omega] = u_\omega / v_\omega, \qquad \mathbf{var}[\omega] = u_\omega / v_\omega^2. \tag{10}$$

Finally, combining the first and second stage of the problem modeling we have the global distribution

$$\mathrm{p}(\boldsymbol{\Omega}_M, \mathbf{y}, \mathbf{Y}, \mathbf{x}) = \mathrm{p}(\boldsymbol{\Omega}_M)\mathrm{p}(\mathbf{y}|\boldsymbol{\Omega}_M)\mathrm{p}(\mathbf{Y}|\mathbf{y})\mathrm{p}(\mathbf{x}|\mathbf{y}), \tag{11}$$

for $M \in \{G, L\}$. The joint probability model is shown in Fig. 1(b).

4. Bayesian Inference and Variational Approximation of the Posterior Distribution for Super-Resolution Reconstruction of Multispectral Images

For our selection of hyperparameters in the previous section, the set of all unknowns is $(\boldsymbol{\Omega}_M, \mathbf{y})$.

As already explained, the Bayesian paradigm dictates that inference on $(\boldsymbol{\Omega}_M, \mathbf{y})$ should be based on $\mathrm{p}(\boldsymbol{\Omega}_M, \mathbf{y}|\mathbf{Y}, \mathbf{x})$. Since $\mathrm{p}(\boldsymbol{\Omega}_M, \mathbf{y}|\mathbf{Y}, \mathbf{x})$ can not be found in closed form, we will apply variational methods to approximate this distribution by the distribution $\mathrm{q}(\boldsymbol{\Omega}_M, \mathbf{y})$.

The variational criterion used to find $\mathrm{q}(\Omega_M, \mathbf{y})$ is the minimization of the Kullback-Leibler divergence, given by [11, 12]

$$\begin{aligned} C_{KL}(\mathrm{q}(\boldsymbol{\Omega}_M, \mathbf{y}) || \mathrm{p}(\boldsymbol{\Omega}_M, \mathbf{y}|\mathbf{Y}, \mathbf{x})) &= \int \mathrm{q}(\boldsymbol{\Omega}_M, \mathbf{y}) \log \left(\frac{\mathrm{q}(\boldsymbol{\Omega}_M, \mathbf{y})}{\mathrm{p}(\boldsymbol{\Omega}_M, \mathbf{y}|\mathbf{Y}, \mathbf{x})} \right) d\Omega_M d\mathbf{y} \\ &= \int \mathrm{q}(\boldsymbol{\Omega}_M, \mathbf{y}) \log \left(\frac{\mathrm{q}(\boldsymbol{\Omega}_M, \mathbf{y})}{\mathrm{p}(\boldsymbol{\Omega}_M, \mathbf{y}, \mathbf{Y}, \mathbf{x})} \right) d\boldsymbol{\Omega}_M d\mathbf{y} + \mathrm{const}, \end{aligned} \tag{12}$$

which is always non negative and equal to zero only when $\mathrm{q}(\mathbf{\Omega}_M, \mathbf{y}) = \mathrm{p}(\mathbf{\Omega}_M, \mathbf{y}|\mathbf{Y}, \mathbf{x})$.

We choose to approximate the posterior distribution $\mathrm{p}(\mathbf{\Omega}_M, \mathbf{y}|\mathbf{Y}, \mathbf{x})$ by the distribution

$$\mathrm{q}(\mathbf{\Omega}_M, \mathbf{y}) = \mathrm{q}(\mathbf{\Omega}_M)\mathrm{q}_D(\mathbf{y}), \tag{13}$$

where $\mathrm{q}(\mathbf{\Omega}_M)$ denotes a distribution on $\mathbf{\Omega}_M$ and $\mathrm{q}_D(\mathbf{y})$ denotes a degenerate distribution on $\mathbf{y}$.

Note that other distribution approximations are also possible. However, as we will see later the one used here alleviates the problem of having to estimate an enormous amount of hyperparameters. We now proceed to find the best of these distributions in the divergence sense. Let us assume that $\mathbf{y}^k$ is the current estimate of the multispectral image where $\mathrm{q}_D(\mathbf{y})$ is degenerate.

Given $\mathrm{q}_D^k(\mathbf{y})$, we can obtain an estimate of $\mathrm{q}(\mathbf{\Omega})$ which reduces the KL-divergence by solving

$$\mathrm{q}^{k+1}(\mathbf{\Omega}_M) = \arg \min_{\mathrm{q}(\mathbf{\Omega}_M)} C_{KL}(\mathrm{q}(\mathbf{\Omega}_M), \mathrm{q}_D^k(\mathbf{y}) \parallel \mathrm{p}(\mathbf{\Omega}_M, \mathbf{y}|\mathbf{Y}, \mathbf{x})). \tag{14}$$

Differentiating the integral in the right hand side of Eq. (14) with respect to $\mathrm{q}(\mathbf{\Omega}_M)$ and setting it equal to zero we have that if $M = G$ then $\mathrm{q}^{k+1}(\mathbf{\Omega}_G)$ satisfies $\mathrm{q}^{k+1}(\mathbf{\Omega}_G) = \prod_{b=1}^{B} \mathrm{q}^{k+1}(\bar{\alpha}_b)$, where $\mathrm{q}^{k+1}(\bar{\alpha}_b) = \mathrm{p}\left(\bar{\alpha}_b | \bar{a}_b^o + \frac{p}{2}, \bar{c}_b^o + \frac{1}{2}\left[\mathbf{y}_b^{k^t}\mathbf{C}\mathbf{y}_b^k\right]\right)$. These distributions have the following means

$$\mathbf{E}[\bar{\alpha}_b]_{q^{k+1}(\mathbf{\Omega}_G)} = \frac{\bar{a}_b^o + \frac{p}{2}}{\bar{c}_b^o + \frac{1}{2}\left[\mathbf{y}_b^{k^t}\mathbf{C}\mathbf{y}_b^k\right]}, \quad b = 1, \ldots, B, \tag{15}$$

which can be rewritten as

$$\frac{1}{\mathbf{E}[\bar{\alpha}_b]_{q^{k+1}(\mathbf{\Omega}_G)}} = \bar{\mu}_b \frac{\bar{c}_b^o}{\bar{a}_b^o} + (1 - \bar{\mu}_b)\frac{\mathbf{y}_b^{k^t}\mathbf{C}\mathbf{y}_b^k}{p}, \quad b = 1, \ldots, B, \tag{16}$$

where $\bar{\mu}_b = \frac{\bar{a}_b^o}{p/2 + \bar{a}_b^o}$, $b = 1 \ldots, B$.

The above equations indicate that $\bar{\mu}_b$, $b = 1, \ldots, B$, can be understood as normalized confidence parameters taking values in the interval $[0, 1)$. That is, when they are zero no confidence is placed on the given hyperparameters, while when the corresponding normalized confidence parameter is asymptotically equal to one it fully enforces the prior knowledge of the mean (no estimation of the hyperparameters is performed). Furthermore, for each hyperparameter, the inverse of the mean of its posterior distribution approximation is a weighted sum of the inverse of the mean of its hyperprior distribution (see Eq. (10)) and its maximum likelihood estimate.

If we use the local image model, that is, $M = L$, we have

$$\mathrm{q}^{k+1}(\mathbf{\Omega}_L) = \prod_{b=1}^{B}\prod_{i=1}^{p}\prod_{l=1}^{4} \mathrm{q}^{k+1}(\alpha_b(i, il)),$$

where $\mathrm{q}^{k+1}(\alpha_b(i, il)) = \mathrm{p}\left(\alpha_b(i, il) \mid a_b^o + \frac{1}{8}, \frac{1}{2}[y_b^k(i) - y_b^k(il)]^2 + c_b^o\right)$. These distributions have the following means

$$\mathbf{E}[\alpha_b(i, il)]_{q^{k+1}(\mathbf{\Omega}_L)} = \frac{a_b^o + \frac{1}{8}}{c_b^o + \frac{1}{2}[y_b^k(i) - y_b^k(il)]^2} = \alpha_b^{k+1}(i, il). \tag{17}$$

Note that Eq. (17) can be rewritten as

$$\frac{1}{\mathbf{E}[\alpha_b(i, il)]_{q^{k+1}(\mathbf{\Omega}_L)}} = \mu_b \frac{c_b^o}{a_b^o} + (1 - \mu_b)4[y_b^k(i) - y_b^k(il)]^2, \tag{18}$$

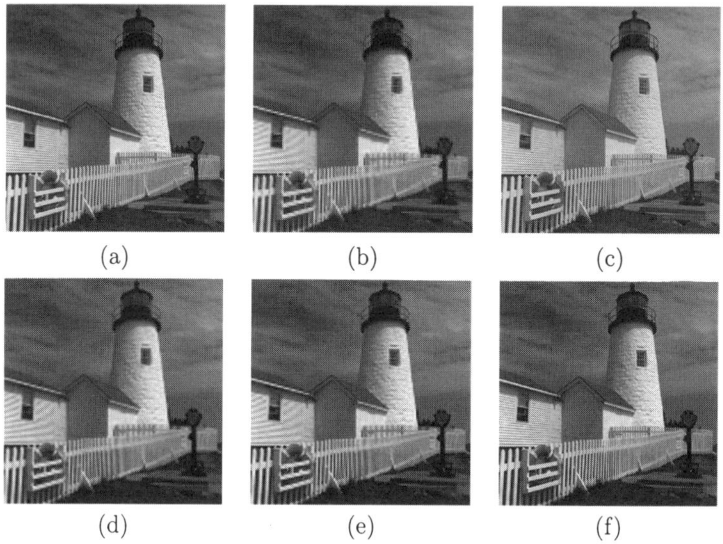

Fig. 2. (a) Original HR color image; (b) Observed LR color image (the image has been resized by zero-order hold to the size of the high resolution image for displaying purposes); (c) Panchromatic HR image; (d) Bicubic interpolation of (b); (e) Reconstruction using the global image model proposed in Ref. 8; (f) Reconstruction using the local image model proposed in Ref. 10.

where $\mu_b = \frac{a_b^o}{a_b^o + \frac{1}{8}}$. These equations indicate, as for the global model, that μ_b can be understood as a normalized confidence parameter taking values in the interval $[0, 1)$.

Given now $\mathrm{q}^{k+1}(\mathbf{\Omega}_M)$ we can obtain an estimate of $\mathbf{y}_M^{k+1}$ (the value where $\mathrm{q}_D^{k+1}(\mathbf{y})$ is degenerate, which obviously will depend on the image model used) which reduces the KL-divergence by solving

$$\mathbf{y}_M^{k+1} = \arg\min_{\mathbf{y}} \left\{ -\mathbf{E}[\log \mathrm{p}(\mathbf{\Omega}_M, \mathbf{y}, \mathbf{Y}, \mathbf{x})]_{\mathrm{q}^{k+1}(\mathbf{\Omega}_M)} \right\}.$$

The convergence of the parameters defining the distributions $\mathrm{q}^{k+1}(\mathbf{\Omega}_M)$ and $\mathbf{y}_M^{k+1}$ can be used as stopping criterion for the iterative procedure that alternates between the estimation of both distributions.

5. Experimental Results

Let us now compare the use of the described local and global image models in the reconstruction of synthetic color images and real Landsat ETM+ images.

Following Eq. (2), the color image in Fig. 2(a) was convolved with the mask $0.25 \times \mathbf{1}_{2\times 2}$ to simulate the sensor integration and then downsampled by a factor of two in each direction. Zero mean Gaussian noise with variance 4 was then added to obtain the observed LR image in Fig. 2(b). The panchromatic image, depicted in Fig. 2(c) was obtained from the original HR color image using the model in Eq. (3) with $\lambda_b = 1/3$, for $b = 1, 2, 3$, and Gaussian noise with variance 6.25.

The reconstruction provided by the global model proposed in Ref. 8 is shown in Fig. 2(e). The method in Ref. 8 was also used to estimate the parameters, β_b, $b = 1, 2, 3$, and γ. This method also provides values for the parameters $\bar{\alpha}_b$, $b = 1, 2, 3$ of the global

Table 1. PSNR and COR values for the color image reconstructions.

	PSNR			COR		
Band	1	2	3	1	2	3
Bicubic interpolation	12.7	12.7	12.7	0.50	0.50	0.51
Using the global image model	13.5	13.4	13.4	0.68	0.68	0.68
Using the local image model	18.9	19.0	18.9	0.99	0.99	0.99

image model. Several local model reconstructions were then obtained using the method in Ref. 10, they correspond to using as u_b^o/c_b^o values ranging from 10^{-2} to 10^2 times $\bar{\alpha}_b$ and in Eq. (18) values of μ_b ranging from 0 to 1 (note that knowing u_b^o/c_b^o and μ_b, the values of u_b^o and c_b^o can be calculated easily).

The spatial improvement of the reconstructed image has been assessed by means of the correlation of the high frequency components (COR) which measures the spatial similarity between each reconstructed multispectral image band and the panchromatic image, and spectral fidelity by means of the peak signal-to-noise ratio (PSNR) between the reconstructed and original multispectral image bands. Bicubic interpolation of each band was used as a reference method for comparison.

Table 1 depicts the resulting PSNR and COR values for all the reconstructed images. The table clearly shows that the proposed methods performs better than bicubic interpolation and that using a local image model provides considerably better results than using a global image one. Visual inspection of the results shows that with the use of a global image model we obtain improved spatial resolution but the details in the image are still oversmoothed. Using the local image model, however, we are able to incorporate the high frequency information from the panchromatic image into the reconstruction while preserving the spectral properties of the multispectral image.

Global and local image models are also compared on a real Landsat ETM+ image. Figure 3(a) depicts a 64×64 pixels false RGB color region of interest composed of bands 4, 3, and 2 of the Landsat ETM+ multispectral image, and Fig. 3(b) its corresponding 128×128 panchromatic image. The multispectral image was resized to the size of the panchromatic image for displaying purposes. The contribution of each multispectral image band to the panchromatic, that is, the values of λ_b, $b = 1, 2, \ldots, 4$, were calculated from the spectral response of the ETM+ sensor. The obtained values were equal to $0.0078, 0.2420, 0.2239$, and 0.5263, respectively. Reconstructions using the global and local image models are shown in Fig. 3(c) and 3(d), respectively. From these results it is clear that the local image model based method preserves the spectral properties of the multispectral image while successfully incorporating the high frequencies from the panchromatic image.

6. Conclusions

In this paper the reconstruction of multispectral images has been formulated from a super-resolution point of view. A hierarchical Bayesian framework has been presented to incorporate global and local prior knowledge on the expected characteristics of the multispectral images, model the observation process of both panchromatic and low resolution multispectral images, and also include information on the unknown parameters in the model in the form of hyperprior distributions. The methods have been tested experimentally.

References

[1] W.J. Carper, T.M. Lillesand and R.W. Kiefer,*Phot. Eng. & Rem. Sens.*,**56**, 459 (1990)
[2] J. Nuñez, X. Otazu, O. Fors, A. Prades, V. Pala, R. Arbiol, *IEEE Trans on Geosc. & Rem. Sens.*,**37**, 1204 (1999)

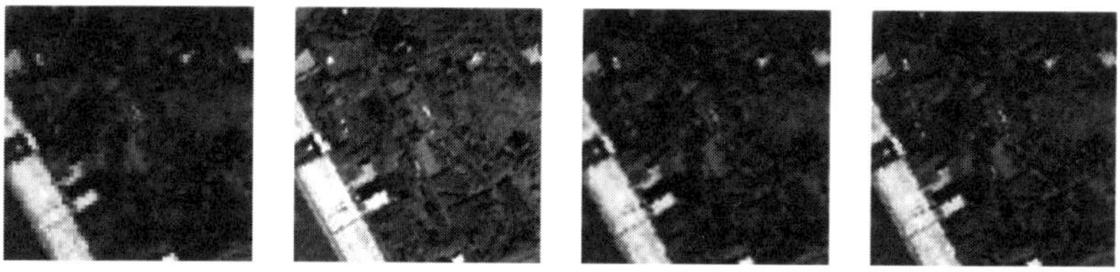

(a) (b) (c) (d)

Fig. 3. (a) Observed multispectral image; (b) Panchromatic image; (c) Reconstruction using the global image model; (d) Reconstruction using the local image model.

[3] V. Vijayaraj, "A Quantitative Analysis of Pansharpened Images", Master's Thesis, Mississippi St. Univ., (2004)

[4] J.C. Price, *IEEE Trans. on Geosc. & Rem. Sens.*, **37**, 1199 (1999)

[5] M.T. Eismann and R.C. Hardie, *IEEE Trans. on Geosc. & Rem. Sens.*, **43**, 455 (2005)

[6] T. Akgun, Y. Altunbasak and R.M. Mersereau, *IEEE Trans. on Img. Proc.*, **14**, 1860 (2005)

[7] R. Molina, A. K. Katsaggelos and J. Mateos, *IEEE Trans. on Img. Proc.*, **8**, 231 (1999)

[8] R. Molina, M. Vega, J. Mateos and A.K. Katsaggelos, *Applied and Computational Harmonic Analysis (accepted for publication)* (2007)

[9] J. Chandas, N. P. Galatsanos and A. Likas, *IEEE Trans. on Image Processing*, **15**, 2987 (2006)

[10] R. Molina, J. Mateos and A.K. Katsaggelos, "Super Resolution of Multispectral Images using Locally Adaptive Models", in SUPER *2007 European Sig. Proc. Conf. (EUSIPCO 2007), submitted,* (2007)

[11] S. Kullback and R.A. Leibler, *Annals of Math. Stat.*, **22**, 79 (1951)

[12] S. Kullback, *Information Theory and Statistics*, Dover Publications, (1959)

FROM THE QUBIT TO THE QUANTUM SEARCH ALGORITHMS

GIANFRANCO CARIOLARO and TOMMASO OCCHIPINTI*

Department of Information Engineering, University of Padova, Padova, 35140, Italy

** E-mail: tommaso.occhipinti@dei.unipd.it*

The main topic of this paper is to introduce the Quantum Information theory, emphasizing its good properties, but also clarifying common misunderstandings. Quantum Information theories and technologies have indeed many positive and interesting features, and hold promises of great developments. However, there are also problems. At present time, we can count a very small number of quantum algorithms, for example Quantum Fourier Transform, Shor's and Grover's. Grover's Algorithm in particular is capable of accelerating the search of elements in a large database. In our opinion, future evolutions of such algorithms will be successfully applied to the world of data handling, for example to satisfy astronomical needs when its time resolution capabilities will be pushed toward Heisenberg quantum limit.

Keywords: Quantum Information; Qubit; Coherent States; Quantum Algorithms; Quantum Communications; Quantum Astronomy.

1. Introduction

Quantum mechanics made a great revolution in physics, and today it imposes intriguing and very innovative implications also in the world of information area. The fundamental idea inside the so called Quantum Information (quantum mechanics applied to information theory) is to use a quantum system for the storage, management and transmission of the information. Quantum system, in contrast with "classical system", means a physical system whose behavior is strongly dominated by the rules of quantum mechanics. Hence, according to the formulation of Quantum Mechanics (P.A.M. Dirac [1] and J. von Neumann [2]), any property of a quantum system can be derived by four postulates:

- **Postulate 1** gives us the universal mathematical model of any physical system: a vector Hilbert space on the complex numbers;
- **Postulate 2** describes the temporal evolution of a closed physical system;
- **Postulate 3** is about "quantum measurements" and indicates the way to extract information from a quantum system in a precise instant;
- **Postulate 4** formalizes the interaction of many physical systems with the combination of different Hilbert spaces coming to a unique Hilbert space.

In particular the first postulate asserts that to each quantum system is always associated a Hilbert space $\mathcal{H}$, called "states space", and that at each instant of its evolution, a quantum state is *completely* described by its "state" $|\psi\rangle$[a], witch is a unitary vector of the Hilbert space $\mathcal{H}$. A very interesting implication of the first postulate is the **superposition** principle, derived by the linearity property of Hilbert spaces: if $|\psi_1\rangle, |\psi_2\rangle, \ldots, |\psi_n\rangle$ are n different states of quantum systems, also the linear combination

$$|\psi\rangle = a_1 |\psi_1\rangle + a_2 |\psi_2\rangle + \cdots + a_n |\psi_n\rangle, \tag{1}$$

with a_i complex numbers, is a valid state and the system is said to be in the superposition of the n states.

1.1. *Quantum systems for information*

Quantum mechanics has many fundamental implications on every aspect of physics. The relatively new idea, which is dealt with in this paper, is the application of quantum concepts to information area. Trying to implement a memory cell or a signal into a quantum system is not only an "interesting" idea but a truly innovative one, which could revolutionize the applications of the theory to our "information society". By moving from a classical bit of information to a bit stored into a quantum system we make a big step toward quantum information. Hence, the passage from the realm of classical to quantum information implies also a big change in the implementation technologies. Any physical system governed by the above mentioned four postulates can become a quantum system useful as a medium for information, for example a single electron, a set of nuclei, a single photon, a coherent state of light.

As the number of useful candidates for quantum states of information is high, a general theory can be developed only starting from the definition of an abstract quantum system capable of staying into two different base states: **the qubit**. Another very general tool encountered in quantum information is the concepts of **coherent states** representing electromagnetic radiation produced by physical devices such as lasers. In the next sections we will outline both the qubit and the coherent states in order to describe some applications, with the further aim to express the great innovation that these ideas imply in the computation (qubit) and transmission (coherent states) of the information.

We conclude the paper with reference to quantum astronomy applications (see C. Barbieri [3], this conference). We think indeed that quantum astronomy is a perfect example of a scientific field where quantum mechanics is the fundamental inspiring principle, and where quantum computing and quantum information theory could become excellent tools to satisfy its peculiar and huge computational needs.

2. The Qubit

The qubit is an elementary example of quantum system that can be described by a bi-dimensional Hilbert space $\mathcal{H}$. A base of this Hilbert space consists of two orthonormal vectors of $\mathcal{H}$, $|0\rangle$ and $|1\rangle$. Then a general state of a qubit is:

$$|\psi\rangle = a\,|0\rangle + b\,|1\rangle \tag{2}$$

with a and b complex numbers and where $|a|^2 + |b|^2 = 1$. One interesting point is that implementing a qubit is not particularly simple, but it can be done in different ways. For example a qubit can be a single polarized photon, an electromagnetically confined ion, a superconductive devices like a Josephson junction or a set of nuclei in a nuclear magnetic

[a]This notation is called Dirac notation [1] or "bracket notation" and the symbol $|\psi\rangle$ is called *ket* while the transposed conjugate of $|\psi\rangle$ is $\langle\psi|$ and it is called *bra*.

resonance apparatus. Each different technology is characterized by the fact that the qubit is "written" on a property of the quantum system itself. Taking the single polarized photon as representative of the qubit technologies, one can write the information 0 into the base state $|0\rangle$ of horizontal polarization and the value 1 into the other base state $|1\rangle$ representing the vertical polarization.

Considering now an ensemble of qubits, the superposition of the states can be a good property of a *quantum memory* where two or more qubits are described by the overall quantum state, according to Postulate 4. The resulting state of the system composed by the two qubits in $\mathcal{H}$ is described by a Hilbert space that is the tensor product $\mathcal{H} \otimes \mathcal{H}$. For example, for two qubits we have:

$$|\phi\rangle = a\,|00\rangle + b\,|01\rangle + c\,|10\rangle + d\,|11\rangle \tag{3}$$

where $|00\rangle$ is a symbol that represents the tensor product: $|0\rangle \otimes |1\rangle$.

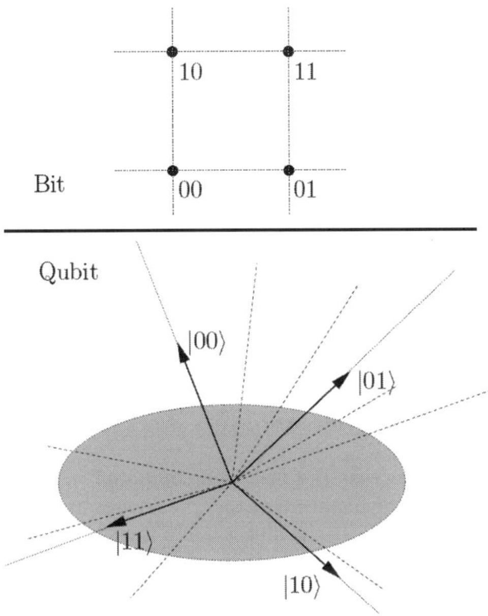

Fig. 1. These figures represent the states of a classical memory register in a conventional computer, and the states of a quantum memory formed by two qubits.

We can now observe that a collection of qubits may contain an enormous quantity of information in comparison with a collection of classical bits. Unfortunately, this is not completely true, as in the process of measurement the quantum state of the qubit is definitively one of the two base state. The superposition only tell us that the probability of obtaining a particular outcome is determined by the parameters a and b [1].

2.1. *Qubit applications*

If we consider the qubits as the elementary units in computation, we can think to use them to realize a quantum computer, inside which it is possible to write information on the qubits, to read it from them by applying some measurements operations (Postulate 3), and obviously to calculate functions on them. These functions, derived directly from Postulate 2, are unitary transformations called **quantum gates**.

A quantum logic gate is a device which performs a unitary operation on selected qubits in a fixed period of time. A quantum computer can be thought of as a device consisting of quantum logic gates whose computational steps are synchronized in time. The outputs of the gates are connected to the inputs of the others. There are many types of quantum gates and they can be divided into passive or conditional gates.

The passive are the single port gates like the NOT or the very important Hadamard transformation H. If we represent the qubit as $|\psi\rangle = a\,|0\rangle + b\,|1\rangle$ with the matrix expression $\begin{pmatrix} a \\ b \end{pmatrix}$ one can write the quantum gates as matrix of finite dimension. For example, the NOT and the Hadamard gates are:

$$\mathrm{NOT} = \begin{pmatrix} 0 & 1 \\ 1 & 0 \end{pmatrix}, \quad \mathrm{H} = \frac{1}{\sqrt{2}} \begin{pmatrix} 1 & 1 \\ 1 & -1 \end{pmatrix} \tag{4}$$

The controlled gates are intuitively quantum multiple qubits gates where one or more qubits have the capability of controlling the functionalities of the gate itself (ex. CNOT quantum gate). The most famous controlled gate is Toffoli's gate, together with the Hadamard gate it forms a *universal set of quantum gates.* Any n-qubit unitary operation can be performed by a combination of Toffoli's and Hadamard's gates.

2.2. *Quantum algorithms*

Consider now a quantum memory formed by two qubits, the state of this memory can be express like in equation (equation 3). Consider also to apply a generic function $f(\cdot)$ to this memory in order to compute something on the qubits stored in the memory. For the linearity, the output of the computation of $f(\cdot)$ on $|\psi\rangle$ is then:

$$f\left(|\psi\rangle\right) = a\,|f(00)\rangle + b\,|f(01)\rangle + c\,|f(10)\rangle + d\,|f(11)\rangle \tag{5}$$

We can conclude that the function $f(\cdot)$ has been computed *in parallel* in one single step on each of the four values of the quantum memory. All cases are present, the first qubit of the memory is 0 and the second is 0, the first is 0 and the second is 1 and so on. This fact, generalized to multiple qubits inputs on quantum gates, is called **quantum parallelism.**

We must point out a limitation of such quantum parallelism. The output of the quantum gates is in the superposition of all possible computed results (Postulates 1 and 2). If we are interested in some particular value, we have to measure the output obtaining only one quantum state with a fixed probability, for example $|f(00)\rangle$ with measurement probability $|a|^2$. This is due to Postulate 3.

Table 1. Measurement probabilities of the outcomes from a dual qubit quantum gate.

Result	Probability of the result
$\|f(00)\rangle$	$\|a\|^2$
$\|f(01)\rangle$	$\|b\|^2$
$\|f(10)\rangle$	$\|c\|^2$
$\|f(11)\rangle$	$\|d\|^2$

This property brought to design a new class of algorithms for quantum computers. As all the results of a particular computation are present in a superposition of the states, we must manipulate the evolution of the quantum system toward the solution that we are looking for. We must be confident that, at the end of this type of computation the target result has a very high probability of been read in a process of quantum measurement (Table 1). This process is a sort of probability amplification of the results, and for this reason the **quantum algorithms** are similar to classical probabilistic algorithms. The great difference between quantum and classical ones is that the operations that are possible inside a quantum computer are ruled by the above four postulates.

2.3. *Quantum search on unstructured databases*

In the literature several quantum algorithms [4] are present. They are fundamentally divided into two families: those based on the Quantum Fourier Transform (QFT) and those based on Grover's [5] quantum search algorithm.

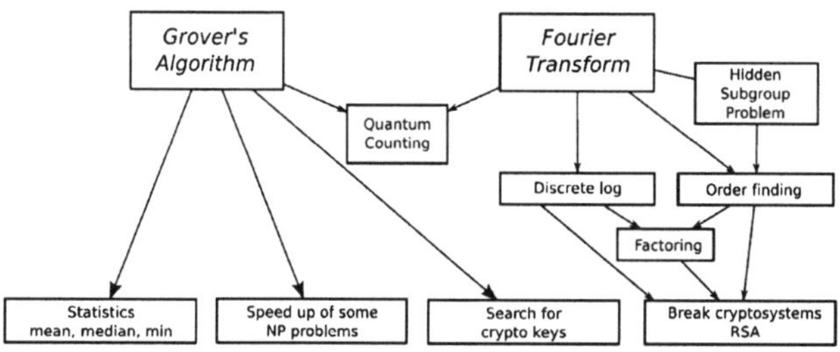

Fig. 2. The variety of quantum algorithms [4].

Looking to Fig. 2, it is possible to appreciate that another large class of quantum algorithms is derived from Grover's quantum search. For example it implies speeding up the computation of some NP problems[b]. Grover's algorithm searches for an object inside a set of N elements. If we define computational complexity the number of steps to perform a cascade of quantum gates, we notice that Grover's algorithm needs only $\sqrt{N}$ steps which is essentially a $O(\sqrt{N})$ complexity. This complexity is lower in comparison to the same parameter of a classical search algorithm, which is $O(N)$.

Considering only the computational complexity, Grover's algorithm is truly powerful. However, there is an often forgotten problem, namely how to introduce classical information on a quantum processor capable of performing Grover's algorithm. We could store the N elements of the database inside a classical memory, and use a quantum computer to search for the element.

We would consequently need a hybrid apparatus in order to read (*load*) and to write (*store*[c]) the data from/to the memory. This apparatus is hybrid in the sense that it will use several quantum gates for accessing a classical memory. It has been shown [4, 5] that this apparatus has complexity $O(N \log N)$, actually deteriorating the performances of a practical implementation of the quantum search algorithm.

[b] NP problems are one of most famous computational class in computer science. The meaning of this class is out of the scope of this paper

[c] Load and store are more appropriate terms in database applications

We conclude this brief overview of quantum computation with a question: with the help of a quantum computer and the application of QFT or Grover's algorithm, will it be possible to accelerate and optimize the data analysis, when the storage becomes very large and the classical computational power is not sufficient? This is really a still open question, but some considerations can be derived from what we said before.

We think that Grover's algorithm has to become an intermediate step inside a larger quantum algorithm, where finding an element in a collection of data is not the final target. This future quantum algorithm will certain benefit from the speed of Grover's one, but only in a particular step of its overall elaboration. Our impression is motivated by an important example: one of the most powerful quantum algorithms, namely Shor's factorization, uses the QFT as a crucial intermediate step in determining the factors of a large integer number, however the calculation of the Fourier Transform of the data is not the final scope of the algorithm.

3. The Coherent States

Coming back to the discussion of quantum states useful for information theory, we want to outline now the coherent states representing electromagnetic radiation produced by physical devices such as lasers. Such coherent states are as important for optical communications as for a number of optics experiments.

The mathematical description of a coherent state of a single mode of radiation has been formalized by Glauber [6] and it is summarized by the expression:

$$|\alpha\rangle = e^{-\frac{1}{2}|\alpha|^2} \sum_{n=0}^{\infty} \frac{\alpha^n}{\sqrt{n!}} |n\rangle \tag{6}$$

The quantum state of the coherent radiation is $|\alpha\rangle$ and it is expressed in the form of superposition of orthonormal states $|n\rangle$ called *number states* or *number eigenstates*. We can think to the number eigenstates as the basis states that "contain" exactly n photons.

3.1. *Coherent states applications*

Coherent states are the ideal tool for the so called quantum communications. In particular these states permit to improve the performances of two particular areas of optical telecommunications: the reception and the security[d] of the signal.

A communication system is generally composed by a transmitter, a communication channel and a receiver. At the transmitter, if we take a laser source that produces a train of quantum states $|\alpha\rangle$ (equation 6), we can modulate the phase of these pulses performing the so called *Phase Shift Keying* (PSK) modulation. For example it is possible to transmit a binary message $a \in \{0, 1\}$ producing the coherent states $|\alpha\rangle$ for the bit 0 and $-|\alpha\rangle$ for 1. Moreover at the receiver, as the bits are transmitted with coherent states, it is necessary to perform the quantum measurements based on Postulate 3. In particular the detector, that converts the optical pulses to electrical signals, must have the ability to detect one single photon[e].

---

[d]Quantum Key Distribution based on coherent states is now a very interesting topic in telecommunication system security.
[e]Usually in quantum communication, the detector is a SPAD, Single Photon Avalanche Diode. Hence, the SPADs are very useful in Quantum Astronomy application (see C. Barbieri [3], this conference)

In telecommunications, a very important performance parameter is the error probability P_e in the communication system, which actually gives the probability of having an error in the message demodulated at the receiver. It has been shown [7, 8] that the error probability of an optical transmission system based on the rules of quantum mechanics is much lower than the classical optical communication system that uses the same modulation scheme.

It is interesting that quantum communications have been implemented for the first time at the Jet Propulsion Laboratory [9] (May, 2006), trying to improve the performances (and consequently the communication data rates) of the Deep Space Network[f] that is progressively moving to the optical wavelength domain.

We want to conclude this work with an observation. We think that the study of coherent states from a telecommunication engineering perspective will be very useful also for quantum astronomy. Both in optical communication and in quantum astronomy the tools to acquire the data are based on the same principles (the four Postulates) and instruments. In quantum astronomy [3, 10] we mainly want to find particular signatures of the astronomical sources, looking to the statistics of the photons coming to the observing telescope. This can be done only if it is possible to discriminate the coherent states from the incoherent light of other sources and from the electronic noise of the acquisition system. This is essentially the same purpose of the engineering of a good communication system.

For example, the design of a very performing receiving electronics for quantum communication will be very helpful for astronomy applications. Both quantum communication and quantum astronomy share the acquisition front-end: we refer to the previously mentioned SPADs, but also to the optical wavelength filters, to the beam splitters, to the polarization analyzer and so on. The result of the acquisition in both quantum applications is a stream of time tags[g], which has to be processed and stored. For this reason, we think that a set of Field Programmable Gate Array (FPGA) will be the right solution for the needs of reconfigurability and computational power. We must remember that one of the main scientific targets of quantum astronomy is the searching of some particular temporal correlation inside the stream of photons; this operation is essential also in a quantum communication system, for example in the time synchronization of the transmitter and the receiver.

We are then convinced that quantum astronomy can be improved not only by the realization of the quantum computer, but also by the development of innovative quantum telecommunication systems.

4. Conclusions

This paper is essentially a short tutorial of Quantum Information Technology and Computing and, in particular, it has the specific purpose of discussing its applications to *quantum astronomy*.

From the same initial point, namely of quantum physical systems, entirely described by the *quantum state* $|\psi\rangle$, two different concepts useful to quantum astronomy have been derived.

First, we considered the quantum states for the elaboration of information in the so called quantum computers: the qubits. Then, we recalled the definition of coherent quantum states $|\alpha\rangle$, which can bring us to overcome the limits of optical telecommunications.

---

[f]The Deep Space Network is a receiving system global network, that permits to acquire the signals, usually based on radio frequencies, from the solar system exploring probes.

[g]A time tag is the time reference of a generic event, for example the arrival of a single photon.

Qubits and coherent states have both great relevance to quantum astronomy. Glauber's theory connected to the coherent states, and finally to the detection of one, two-, or multiple-photon processes in the light from celestial sources, is one of the pillars of quantum astronomy.

References

[1] P. A. M. Dirac: *The Principles Of Quantum Mechanics*, Oxford University Press, 1958.

[2] J. von Neumann, *Mathematical Foundation of Quantum Mechanics*, Princeton Univ. Press, Princeton 1955.

[3] C. Barbieri, *Quantum Astronomy and Information*, this conference.

[4] M. A. Nielsen, I. L. Chuang: *Quantum Computation and Quantum Information*, Cambridge: University Press, 2000.

[5] L. K. Grover, *A fast quantum mechanical algorithm for database search*, Proceedings, 28th Annual ACM Symposium on the Theory of Computing, (May 1996) p. 212

[6] R. J. Glauber, *Coherent and Incoherent States of the Radiation Field*, The Physical Review, vol. 131, no. 6, pp. 2766-2788, September 15, 1963.

[7] C. W. Helstrom, J. W. S. Liu, and J. P. Gordon, *Quantum Mechanical Communications Theory*, Proceedings of the IEEE, vol. 58, no. 10, pp. 1578-1598, October 1970.

[8] V. Vilnrotter and C. W. Lau, *Detection Theory for the Free-Space Channel*, IPN Progress Report 42-146, April-June 2001, Jet Propulsion Laboratory, Pasadena, California, August 15, 2001.

[9] C. W. Lau, V. A. Vilnrotter, S. Dolinar, J. M. Geremia and H. Mabuchi, *Binary Quantum Receiver Concept Demonstration*, IPN Progress Report 42-165, May 15, 2006.

[10] Naletto G, Barbieri C, Dravins D, Occhipinti T, Tamburini F, Da Deppo V, Fornasier S, D'Onofrio M, Fosbury RAE, Nilsson R, Uthas H, Zampieri L, (2006), *QuantEYE: A Quantum Optics Instrument for Extremely Large Telescopes*, Ground-Based and Airborne Instrumentation For Astronomy, SPIE Proc. Vol. 6269.

VISUALIZATION AND DATA MINING IN THE VIRTUAL OBSERVATORY FRAMEWORK

MARCO COMPARATO *, UGO BECCIANI, BJORN LARSSON and ALESSANDRO COSTA

INAF - Catania Astrophysical Observatory
via S. Sofia 78, I-95123 Catania, Italy
** E-mail: marco.comparato@oact.inaf.it*
www.oact.inaf.it

CLAUDIO GHELLER

CINECA, v. Magnanelli 6/3,
I-40033 Casalecchio di Reno (BO), Italy

VisIVO is a package for supporting the visualization and analysis of astrophysical multi-dimensional data and has several built-in tools which allow the user an efficient manipulation and analysis of data. We are integrating VisIVO with VO services: connection to VO web services, retrieval and dealing with data in the VOTable format, interoperation with other VO compliant tools.

Keywords: Astronomical data; Virtual Observatory; Visualization; Numerics

1. Introduction

Data represent a critical issue for scientists, and in particular, for astronomers. Observational instruments (telescopes, satellites etc.) produce enormous quantities of images and information. Computers and numerical simulations generate huge amount of data and all these data must be stored, managed and analysed. These steps require a great human effort, large scale facilities and efficient, powerful tools. In the last years the Astrophysical community has started a joint effort, in order to create an infrastructure which could provide the maximum availability of data and common tools to deal with them. This infrastructure is the Virtual Observatory (VO). The development of the VO is coordinated by the International Virtual Observatory Alliance (IVOA). The IVOA work encompasses all the aspects related to data management, access and manipulation. Most of the work of IVOA has focused on observational data. However, the interest toward theoretical data, produced by numerical simulations, is rapidly growing. Broadly speaking, the goal of the Theoretical VO (TVO) is to create a database of simulated data, accessible from anywhere by web based tools in a easy and transparent way. Using the same metaphor of the VO (and of the web), its just like as if the researcher has all the simulation data in his/her pc. Data mining, analysis and visualisation of these data require tools capable to interact with the TVO and data from catalogue, accessing and exploiting its resources. VisIVO, the software that we present in this paper, is one of these tools. It as an effective instrument for the visualization and analysis of astrophysical data. It is VO standards compliant and it supports the most important and popular astronomical data formats such as ASCII,

FITS, HDF5 and VOTables. Data can be retrieved directly connecting to an available VO service (like, for example, Vizier [3]) and loaded in the local computer memory where they can be further selected, visualized and manipulated. It can deal with both observational and simulated data and it focuses on multidimensional datasets (e.g. catalogues, computational meshes etc.). It is completely open source and the binaries and sources can be downloaded from the web site `http://visivo.cineca.it`. At present the software is fully tested on Microsoft Windows XP and GNU/Linux systems. This work is only one of many different applications that the Italian astronomical community is developing in the VO framework. This effort is led and coordinated by the National Institute for Astrophysics (INAF), in collaboration with other institutions like, CINECA, the largest Italian academic supercomputing centre.

2. VisIVO: A Tool for 3D Visualization

VisIVO is specifically designed to deal with multidimensional data. Catalogues or numerical simulations, rather than 2D images, represent the basic target of VisIVO. Different quantities can be visualised and treated at the same time. The architecture of VisIVO strictly reflects the structure of a typical scientific application built on the Multimod Application Framework [4]. On this framework VisIVO implements all the elements that are specific to the visualization and analysis of astronomical data. VisIVO embodies a *"select object-apply operation"* utilization metaphor which is somehow similar to that of many graphical commercial applications (such as Adobe PhotoShop). The usage of VisIVO generally begins with an operation of data loading. Data types that are currently supported by the application are VOTables, FITS, HDF5 as well as ASCII and binary raw data (dump of memory). The result of a loading/importing/modifying any type of data is represented in a data tree as a node. In order to display data loaded in tree, it is necessary to instantiate a View and associate it with the chosen tree node. A View is a rendering window that gives a particular representation of the data tree. VisIVO currently supports the following Views: Points, Vectors, Volume Rendering, Isosurface, Glyph, Stereo, Histogram. Each view displays data according to a specialised visualization pipeline. It is possible to instantiate several Views and each View can display a different subset of the data tree. This is a point of strength of VisIVO as many specialised views, displaying different (or same) data, can be instantiated and data can be manipulated and analysed independently of the views displayed. The properties of the selected view (e.g. camera perspective, axes, logarithmic scale) can be adjusted via the contextual menu. In order to analyse and modify data loaded in the tree, a set of operations are available to the user. There are operations that simply modify the data, operations that perform some statistical analyses on data and operations that do both things. All of these generate output nodes that can be displayed according to the type of output (e.g. An output that represents statistical 2D data can be displayed in a Histogram View).

VisIVO supports several visualization techniques for data. Unstructured points are visualized as pixels. The user can set the transparency of the pixels and their colour. It is possible to colour each pixel differently, according to an associated quantity. For example, gas particles can be coloured by their temperature (e.g. blue for cold particles, red for hot particles). Points can also be described with glyphs (solid geometrical shapes), whose shapes and size can be parametrized to some physical quantities associated with data. Mesh-based data can be visualized with two different techniques, isosurfaces and volume rendering. Isosurfaces are surfaces characterized by a given fixed value of the plotted quantity. They separate regions with higher values from regions with lower values. Different isosurfaces can be calculated and visualized at once. They can be characterized by different colours, according to the contour value. Using the volume rendering technique, different

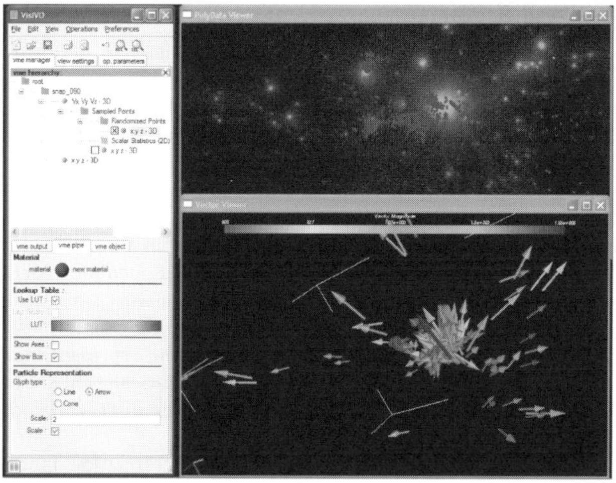

Fig. 1. VisIVO visualizing an HDF5 dataset using the points viewer to show the points position and the vector viewer to show the velocity field associated with the points. Here Glyph visualization is used to highlight points of interest. On the left, VisIVO's control panel.

values of the quantity are represented by different colours and different transparencies. The overall effect is a cloud appearance.

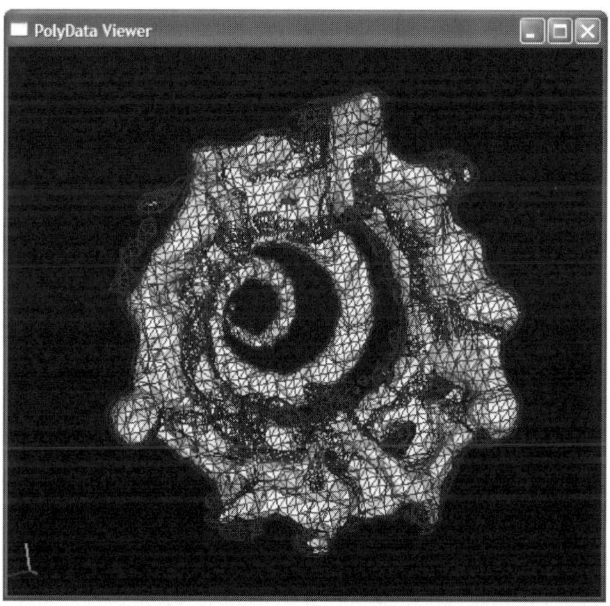

Fig. 2. An example of isosurface view, two isosurfaces are represented with two different materials.

3. VisIVO for Data Analysis

VisIVO provides various built-in utilities that allow the user to perform mathematical operations and to analyze data. It is possible to apply algebraic and mathematical operators to the loaded data. Basic arithmetic operations (addition, subtraction, multiplication division) as well as logarithm, power law, absolute value and many others are supported. Scalar product, magnitude and norm of vector quantities are available too. In this way, new physical quantities can be calculated. For example the X-ray emission due to thermal Bremsstrahlung is proportional to the product between the square of the mass density and the square root of the temperature of the gas distribution in a simulated galaxy cluster. If these two quantities are available, the emission can be immediately derived. It is also possible to merge two different Table Data structures to create a new one. Data in the resulting table can be treated as a single dataset. The merging capabilities and the mathematical operations give great flexibility in data analysis and representation.
Several built-in functions allow the user to perform a statistical analysis of a points distribution. The **Scalar Distribution** function calculates the distribution of any quantity loaded in the Table Data and plots it as a histogram.
The **Correlation Filter** calculates the linear two-point correlation function of a point set. The Fourier transform of the correlation function is represented by the **Power Spectrum**, which can be estimated by VisIVO as well. The last available analysis tool is for **Minkowski Functionals** (*MFs*), they describe the geometry, the curvature and the topology of a point-set [5].

4. Interacting with the VO

In a first phase of development, the only possible interaction of VisIVO with the VO was off-line. Data could be downloaded from usual on-line services in a standard format like VOTables or FITS and visualized by the software. Now, VisIVO can get data directly from a remote database. The software implements a web-service based approach to interact with remote resources. At present it supports access to the Strasbourg VizieR data service. This is an observational data archive, but the same architecture can be extended to the future TVO infrastructure. Therefore, VizieR represents a meaningful case study. It provides the developers a Web Service interface, which has been exploited by VisIVO to access the remote resource. The result of the VizieR query is the list of catalogues matching the request. At this point, the user can choose a specific catalogue and gather the associated data from the web service. At this point, the data can be treated as local data, using all the functionalities of the software. VisIVO is able to interact with many other VO compliant tools using PLASTIC. PLASTIC (PLatform for AStronomical Tool InterConnection) is a collaboration between the teams behind Aladin, Topcat, VisIVO, AstroGrid and others to develop interoperability standards for client-side virtual observatory tools. VisIVO's interface to the PLASTIC hub allow the user to have many points of view of the data, not only the ones provided by VisIVO, the user can get images making VisIVO to interop with Aladin, or directly see the loaded table data using TOPCAT etc.

5. Conclusion

VisIVO represents the first experience of an immersive Visualisation and Data Analysis Tool in astrophysics. It is one of the first visualization tools that operates in the VO framework and its development will follow the IVOA recommendations and issues.

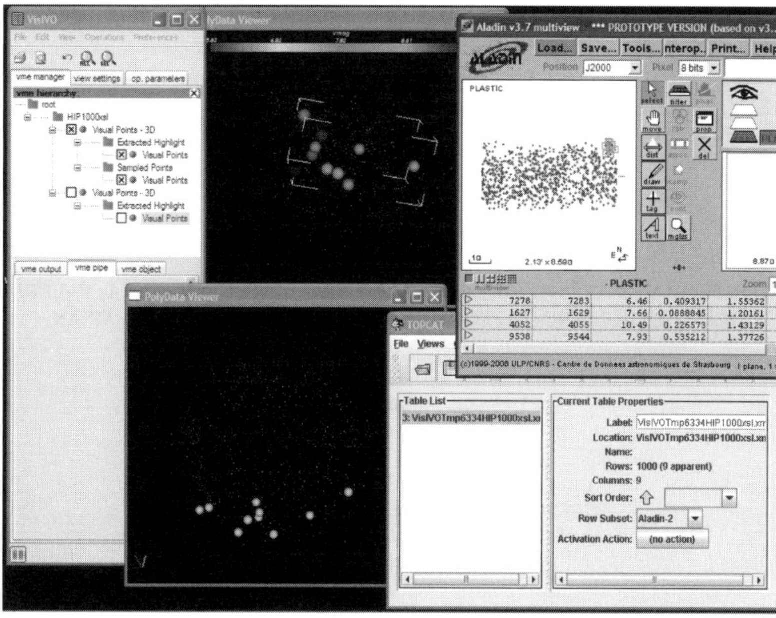

Fig. 3. An example of interoperability between VisIVO, Aladin and Topcat

References

[1] C. Gheller , E. Rossi, U. Becciani et al, 2004, AstroMD 3.1 User Guide ISBN 88-86037-33-3

[2] J. Huchra and M. Geller, Groups of Galaxies I. Nearby Groups., 1982, ApJ 257 423.

[3] F. Ochsenbein, P. Bauer and J. Marcout, 2000, A&AS 143, 221

[4] M. Viceconti, L. Astolfi, A. Leardini et al. 2004, The Multimod Application Framework IEEE Proceedings of IV2004, pp. 15-20

[5] Platzoder M, and Buchert T. *Applications of Minkowski-functionals to the Statistical Analysis of Dark Matter Models.* 1995, A. Weiss, G. Raelt, W. Hillebrandt, and F. von Feilitzsch, Proc. of 1st SFB workshop on Astro-particle physics, Ringberg, Tegernsee, pages 251

AN ARCHIVE OF COSMOLOGICAL SIMULATIONS AND THE ITVO MULTI-LEVEL DATABASE

PATRIZIA MANZATO*, RICCARDO SMAREGLIA, LUCA MARSEGLIA, VALERIA MANNA, GIULIANO TAFFONI, FEDERICO GASPARO, FABIO PASIAN

INAF-SI and O.A.Trieste,
v. G.B.Tiepolo 11, 34143 Italy
* *E-mail:manzato@oats.inaf.it*
http://wwwas.oat.ts.astro.it/

CLAUDIO GHELLER

CINECA High Performance System Division,
v. Magnanelli 6/3, 40033 Casalecchio di Reno (Bologna), Italy
E-mail: c.gheller@cineca.it

UGO BECCIANI

INAF - O.A.Catania,
v. S.Sofia 78, 95123 Catania, Italy
E-mail: ube@oact.inaf.it

As of today many cosmological simulations exist spread throughout the world and it is difficult for an astronomer to find the one he/she is interested in to compare observational data or to compare the data of different types of simulations. The aim of this work is to follow the Virtual Observatory idea of simplifying the work to astronomers and begin unifying the world of simulations just as it has been made by IVOA with the observational data. The Italian Theoretical Virtual Observatory (ITVO) database (DB) was born with the idea of drawing a DB structure for the cosmological simulations general enough to ingest not only the metadata of one specific simulation but of many different types (N-body, N-body+SPH, Mesh, etc.). The goal is to be able to provide a web service to the astrophysical community allowing to search with one query in many kinds of simulations archives and also in data obtained through many levels of post-processing and to use appropriate IVOA tools to analyze the data of every levels. So we present the first DB structure in which two type of metadata coexist, one coming from an N-body+SPH simulation and another from an N-body+Mesh simulation, and some examples of using tools that permit searching and immediately comparing theoretical and observational data. This project is being developed as part of VO-Tech/DS4, ITVO and VObs.it assets.

Keywords: Astronomical simulation, cosmology, Virtual Observatory, GRID

1. Introduction

The ITVO DB is a multi-level database where the metadata of the cosmological simulations archive are stored. We focus our work on a set of numerical simulations of galaxy clusters identified at various red-shifts generated by the GADGET-2 code [12] and halos simulation generated by the Enzo code [5]. In particular, for the clusters at red-shift zero, we offer the possibility of visualizing the temperature and density profiles and downloading the maps via a new web interface which is connected to the ITVO DB. Such an interface gives the opportunity to search easily in the simulated data by selecting many parameters that permit a simple comparison with the observational data. The first level of DB concerns the cosmological parameters used in the simulation, the type of algorithm used and all the parameters used on that algorithm, the physical quantities calculated for the particles or grid points and the output data of the simulation. The second and the upper levels concern post-processing data (in which for example some type of objects have been identified), the code used to obtain post-processed data and the output files of this post-processing. The upper levels of the DB are all the metadata of the files obtained from other levels of post-processing on the previous data level. An example could be the data of simulated observations obtained by running a code that implements a virtual telescope. A further step will be to implement within the DB the usage of a protocol, SNAP [6] (Simple Numerical Access Protocol), which produces on-the-fly the dataset request via web, for example making a cut-out of the initial simulated box. This can be done using GRID technologies like the ones described in G.Taffoni et al. [13] and U.Becciani et al. [2] .

2. The Stored Simulations

Nowadays many types of simulation exist: N-body, Mesh, AMR (Adaptive Mesh Refinement), N-body+SPH (Smoothed Particles Hydrodinamics), N-body+Mesh, N-body+AMR, etc. The scope of these methods is to simulate many different kinds of objects, for example: dark matter haloes, globular clusters, etc. We start to deal with two different simulations. The first one is a large cosmological hydrodynamic simulation [4] , which used the massively parallel tree N-body/SPH code GADGET-2.0 to simulate a concordance ΛCDM cosmological model within a box of 192 h^-1 Mpc. The cosmological parameters assumed were $\Omega m = 0.3$, $\Omega b = 0.04$, $H_0 = 70$ km s^-1 Mpc^-1 and $\sigma_8 = 0.8$. This simulation took care of star formation, radiative cooling, metal production and galactic wind. It produced 102 snapshots for a total amount of approximately 1.2 TB of raw data. All these parameters and results have been included in the first level of the DB. Then, subsequently, from the raw data of the initial big box, a first level of post processing has been implemented to identify the groups and clusters of galaxies, first applying a friends-to-friends halo finder to the distribution of dark matter particles, then running a spherical overdensity algorithm: 400 haloes, 72 clusters were identified. The total number of files stored in the second level of the DB are 1986 boxes that refer to the clusters at different red-shifts. Then, another step of post-processing has been implemented on the clusters at red-shifts equal to zero: these are 117 clusters for which we can extract the 2-D maps for the density, the spectroscopic-like temperature, the mass-weighted temperature, the Sunyaev-Zel'dovich emission and the X luminosity in 0.5-2(10^{44} erg/s) band and the corresponding profiles for the density, the spectroscopic-like temperature, the mass-weighted temperature and emission weighted temperature. These more refined data are stored in the third level of the DB. In the table (Tab. 1) below the minimum and maximum of the more important quantities can be found.

The second simulation is an AMR [10], grid-based hybrid code (N-Body + hydrodynamic) which is designed to simulate cosmological structure formation; this used the Enzo code. We stored two simulations made with this code, both of them simulate a ΛCDM universe with the following cosmological parameters: $\Omega m = 0.27$, $\Omega b = 0.044$, $H_0 = 71$ km s^-1 Mpc^-1 and $\sigma_8 = 0.94$. The results are two boxes, one at red-shift zero at another at red-shift 0.5.

Table 1. Minimum and maximum of quantities of 117 simulated galaxy groups and clusters at redshift zero.

QUANTITY	MIN	MAX
T-M (keV)	0.9277	5.4657
Rvir (kpc/h)	879.64	2233.69
Mtot(Msol/h)	0.795 E+14	13 E+14
Mstar(Msol/h)	0.0179 E+14	0.284 E+14
Lx bol. (10^{44}erg/s)	0.1705	13.4355
Vdisp. (km/s)	433.069	1081.25
T-em (keV)	1.1813	7.0955
Mgas/Mtot	0.08828	0.10262

3. The ITVO Database and Archive

The variety of the data that we want to store persuaded us to use a relational DB, whose query engines allow the user to made complex queries in a standard language, SQL, which is also the standard query language used by many tools developed under the IVOA standards [7]. The DB has been designed to store all kinds of cosmological simulations that one can make as today: N-body, Mesh, SPH, etc. The DB is a multilevel one, as can be seen in the figure 1, every level holds the data of one step of the whole data process. The first level

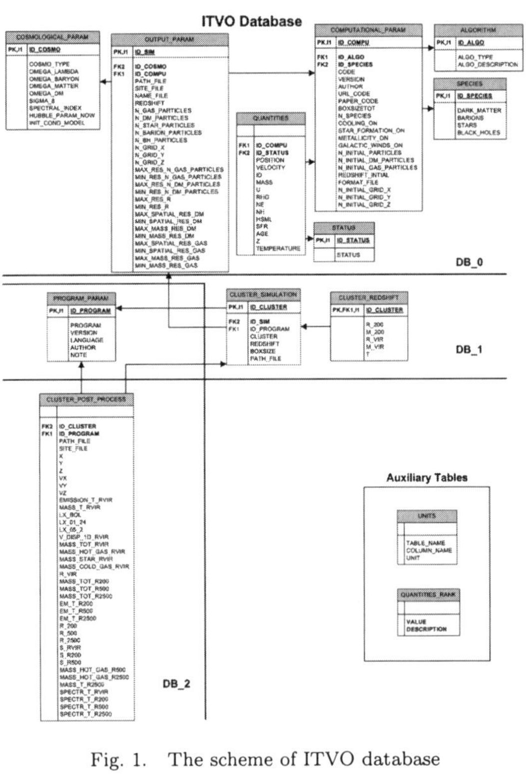

Fig. 1. The scheme of ITVO database

contains the description of the algorithm, the computational and cosmological parameters used to make the computational run, the species of matter inserted in the simulation, the physics quantities linked to the particles or grid points, and also the format (HDF5 in our case), resolution and redshift of the output file. The next level holds the data of the code used for the post-processing, for example the code used to extract the clusters and group of galaxies from the initial box and all the metadata of these new output file, like virial radius, virial mass, temperature, etc. Another level refers to more refined post-processing: it stores all the metadata of the FITS files concerning the bidimensional maps.

All these data are stored in tables which are linked each other by foreign keys on primary keys to avoid any duplication of the information at all levels. There are also two auxiliary tables, one to take into account the units used for the different quantities and another to take into account the rank value of the quantities. The figure (Fig. 1) depicts the pattern of the ITVO multilevel DB. Another DB for theoretical data was made by G. Lemson et al. [9] for the Millenium simulation that produced halos and galaxies, but it contains only the metadata of post-processing products of the simulation.

The first level of the simulations archive contains raw simulated data: the entire simulation outcome, the particles positions and velocities within the simulation box or the physical quantities at each grid point, known as a snapshot at a specific instant. The second level of archive contains the boxes where the code has identified the objects of interest; up to now there are the HDF5 files of the boxes that include the clusters or groups of galaxies. Now also the FITS files of all the maps of the clusters at red-shift zero and the related jpg files of the profiles have been saved, but these could be created on-the-fly and removed whenever appropriate.

4. The ITVO Web Interface and Scientific Use

At the moment there are three β versions of the web interface accessible from the IA2 (Italian Astronomical Archives Center), http://wwwas.oats.inaf.it/IA2/ITVO/. One to find raw data box of the simulation, another to find data of the simulated clusters at several red-shift values and the last one to search and download the maps of the clusters of galaxies, to visualize and print the graphics, to visualize the preview of the maps and the header of the FITS file and to save as a VOTable the catalogue obtained as a result of the query to the DB. In the web interface, is also possible to personalize the SQL query, by making a more complex one and/or filtering on different fields. One example of scientific query for every level of DB is:

(1) discover a snapshot in an assigned red-shift range and/or a specific cosmological parameters (Ω_m, Ω_Λ, H_0, etc.) range and/or originating with a specific algorithm;

(2) find all galaxy clusters in a given red-shift range and/or in a specific boxsize range;

(3) find all maps of galaxy clusters in a selected mass-weighted Temperature range and/or a virial radius range and/or X luminosity and/or on the ratio of two selected quantities.

This functionality permits a search on a large amount of scientific parameters so to allow a simple and direct query to find the data that better satisfy the characteristics of a research typology.

Another archive for theoretical data, with similar characteristics of the above-mentioned archive in Trieste, is in progress at the INAF Astrophysical Observatory of Catania. The preliminary version is accessible at the following address: http://woac.oact.inaf.it/itvo

5. Tools and Scientific Use

As entry level for processing data using VO tools to recover and analyze the theoretical data and compare them with the observational is appropriate. To find and visualize the N-D boxes data we are working with the VisIVO [1] application, that is able to open HDF5 files, Gadget files, fits files and other formats, to yield it able to retrieve a cut-out box from the big initial one via SNAP.

To study the 2-D maps we modified the Aladin tool to enable the search of simulated galaxy clusters by virial mass, virial radius, X luminosity and mass-weighted temperature within the virial radius to be immediately compared with an observational one (Fig. 2). Our preference to recover and visualize the 1-D temperature and density profiles of clusters falls on TOPCAT which is platform independent and can open both fits tables and VOTables and can easy create plots on the selected data. All of these tools can be connected to each other using the PLASTIC [3] hub, a software designed for interoperability between astronomical VO applications.

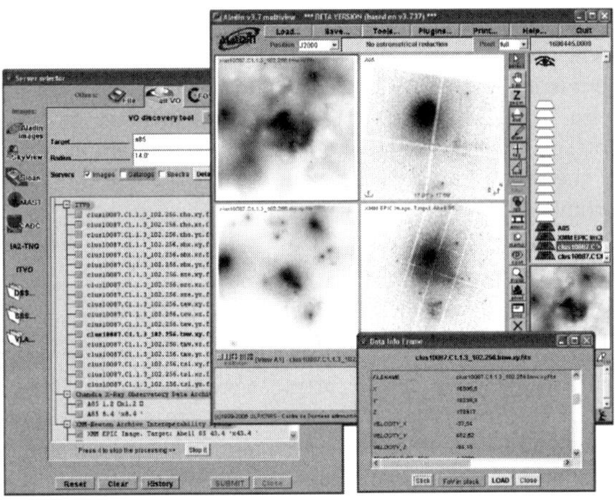

Fig. 2. The Aladin VO tool showing the comparison of simulated and observed data

6. The SNAP and Theory Data Model

SNAP is a mechanism being currently defined within IVOA to allow accessing numerical simulations of astrophysical relevance just as if they were observational data. SNAP is expected to deal with simulations of "objects evolving in 3D space". This will allow the concept of spatial subset to be predefined, including various geometries; any simulation that can be expressed as a N-D box could/should also be included. The multilevel DB permits to access via SNAP the simulation data at every level and not only [11] (Fig. 3). In the future someone could create the post-processed data of the specific desired sub-box coming from the raw simulated data or for example create on-the-fly the cluster map or profile and if the request involves a lot of time we can think to use parallel computation through the GRID.

Also the theory Data Model [8] is under development by a group of interest within IVOA in which we take part, to define a model for the metadata required to be associated with the publication of numerical simulations and to create a logical model for the SNAP.

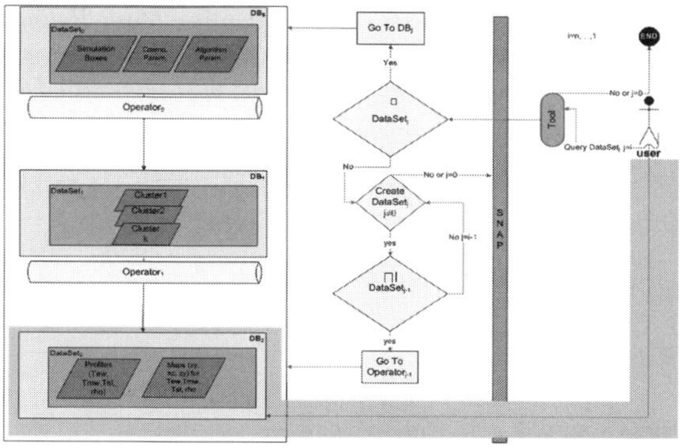

Fig. 3. The diagram of db level

7. Conclusions

This are initial steps of a work to compare in an easy way theoretical and observational data using an appropriate Virtual Observatory tools for every levels of data: VisIVO for the N-D boxes, Aladin for 2-D maps and TOPCAT for 1-D profile. This work also provides a way to reuse the simulation data, that take a long CPU time to be produced, for different scientific scopes and targets than the initial one. We want to contribute to promote the research on exploring how the galactic and dark matter structures evolve in our universe, while following the new VO standards. SNAP is indeed a new protocol that we are participating in writing and contributing to the IVOA together with the new Data Model for theoretical data.

We are grateful to S. Borgani and S. Ameglio for having provided us with the theoretical data.

References

[1] U. Becciani et al., *VisIVO: a Visualization Toolkit towards Grid Environment* (ASP 2006, 351-445 C)

[2] U. Becciani et al., *Theoretical Virtual Observatory and Grid Web Services: VisIVO and new Capabilities* (2006, IAUSS 3E 69C)

[3] T. Boch et al., *PLASTIC- a protocol for desktop application interoperability*, (http://www.ivoa.net/Documents/latest/PlasticDesktopInterop.html)

[4] S. Borgani et al.,*X-ray properties of galaxy clusters and groups from a cosmological hydrodinamical simulation* (MNRAS 348, 1078-1096, 2004).

[5] Enzo Code, (http://www.cosmos.ucsd.edu/enzo/)

[6] C. Gheller et al., *Simple Numerical Access Protocol (SNAP) for theoretical data*, (http://www.ivoa.net/twiki/bin/view/IVOA/IVOATheorySNAP 2006)

[7] R.J. Hanisch & P.J. Quinn, *International Virtual Observatory Alliance*, (http://www.ivoa.net/pub/info/ 2003).

[8] G. Lemson et al., *Data model for theoretical(meta-)data*, (http://www.ivoa.net/twiki/bin/view/IVOA/IVOATheorySimulationDatamodel 2007)

[9] G. Lemson et al., *Halo and Galaxy Formation Histories from the Millenium Run: Public realise of a VO-oriented and SQL-queryable database for studying the evolution of galaxies in the ΛCDM cosmology*(Astro-Ph/0608019, 2006)

[10] M.L. Norman & G.L. Bryan, *Cosmological Adaptive Mesh Refinement* (Numerical Astrophysics, eds. S. Miyama & K. Tomisaka, 1999, pp. 19-28)
[11] F. Pasian et al., *Interoperability and Integration of Theoretical Data in the Virtual Observatory* (long abstract, Highlights of Astronomy, Vol. 14 XXVIth IAU, 2006 K.A. van der Hucht, ed.)
[12] V. Springel et al.,*GADGET: a code for collisionless and gasdynamical cosmological simulations* (New Astronomy, 2001, 79-117)
[13] G. Taffoni et al., *Bridging the Virtual Observatory and the GRID with the query element* (Astro-Ph, 2006).

STUDYING COMPLEX STELLAR DYNAMICS USING A HIERARCHICAL MULTI-AGENT MODEL

JEAN-CLAUDE TORREL*, CLAUDE LATTAUD

Lab. d'Intelligence Artificielle de Paris V, Université René Descartes,
45 rue des Saints Pères, 75006 Paris, France
** E-mail: jean-claude.torrel@devinci.fr*

JEAN-CLAUDE HEUDIN

International Institute of Multimedia, Pôle Universitaire Léonard de Vinci,
92916 La Défense Cedex - France

This paper defines a new approach for cosmological simulation based on complex systems theory: a hierarchical multi-agent system is used to study stellar dynamics. At each level of the model, global behavior emerges from agent interactions. The presented model uses physically-based laws and agent-interactions to present stellar structures has the result of self-organization. Nevertheless a strong bond with cosmology is kept by showing the capacity of the model to exhibit structures close to those of the observable universe.

Keywords: complexity, hierarchical multi-agents systems, cosmology

1. Introduction

For years numerical simulation in cosmology has tried to reproduce and explain the wide variety of patterns (from globular clusters to spiral galaxies) and complex behaviors that cosmological evolution shows. Algorithms used are based on strictly reductionnist models (such as [1], [2]). Even if this approach has carried out to successes, some problems remain unsolved: observed dynamics highly depends on the number of point-mass particles used in simulation [3, 4](strict application of physical laws requires the use of point-mass particles in simulation models). Using this approach, some complex patterns do not appear in the cosmological models till a high number of particles (over 10^9). In addition, a realistic number of point-mass particles should be around 10^{41} [5] for a typical spiral galaxy what is a calculative impossibility. The current approach to solve these problems is to define increasingly precise models taking in account more and more parameters [6]: each physical phenomenon is calculated by a dedicated algorithm (such as gravity [7, 8]) and the results are combined according to the goal of the experiment. The average number of particles currently used is between $2 \cdot 10^6$ and 10^8. In the same time, the study of complex systems obtained a certain success, in particular by the use of models like cellular automata [9–11]. Even if results get by theses models are structurally close to the observable, they are often too abstract and too far away from physical reality to be easily accepted by the cosmologists community. Based on previous works [12], our approach is to set up of a multi-agent model using a multi-layers system to mitigate the calculative problem. The model is using physical laws as intra-levels interactions and agents at a given level are aggregated to form structures of higher level, subjected to another physical law. The aim of this paper is to present and investigate a new kind of model to show that such models need a lower number of agents than what is necessary for cosmological models (from a calculative point of view) to exhibit observable structures.

2. TreeCode

Gravitation is the main law of all celestial mechanics in the majority of theories and thus takes a significant part in the models which intend to simulate this dynamics:

$$\vec{F}_{ij} = -G \cdot \frac{m_i m_j}{r_{ij}^2} \tag{1}$$

with: G the gravitational constant, m_k the mass of the agent k and r_{ij} the distance between agent i and agent j. Few algorithms exist to calculate the impact of the gravitation force (1) on the various elements of a system. Presented by Barnes & Hut [7] the TreeCode is one of these algorithms designed to calculate the effect of gravity. It was proposed to obtain a significant number of particles and a correct resolution. This algorithm allows to specify a parameter θ which determines from which distance the action of n close particles is not dissociable from the action of a particle having the added mass of these n particles. Its use in a wide number of cosmological algorithms and its adaptability justifies its use in our model.

3. Hierarchical Multi-Agent Model

The obvious complexity of the problem and wide variety of forces and scales [6] led us to use of a hierarchical multi-agent system. The aim is to define a high-level model, highly parametrizable to get a picture of universe and to study how cosmological patterns can emerge from interaction within and between complexity levels.

3.1. *Overview*

Our model is composed by four kinds of agents and complexity levels (these levels have been detailed in [13]):

- Level 1 agent : internal interactions, this is an abstraction of mass matter subjected to gravitational attraction.
- Level 2 agent : local interactions and interactions applying on a short distance (approximated as a vicinity) such as accretion.
- Level 3 agent : long range interactions and forces which apply without any distance constraints such as gravity.
- Level 4 agent : environmental actions: all the forces which apply to the whole agents in the universe, such as expansion for instance.

3.2. *Inter-levels exchange*

Each level N agent transmits, to all level N-1 to level N-3 agents (if they exist), the whole variations it was subjected to. For instance, if the position (X_3, Y_3, Z_3) of a level 3 agent is modified (δ_x, δ_y, δ_z), the position (X_2, Y_2, Z_2) of the whole level 2 agents, contents in this agent, evolves by (δ_x,δ_y, δ_z), just as the position (X_1, Y_1, Z_1) of level 1 agents.

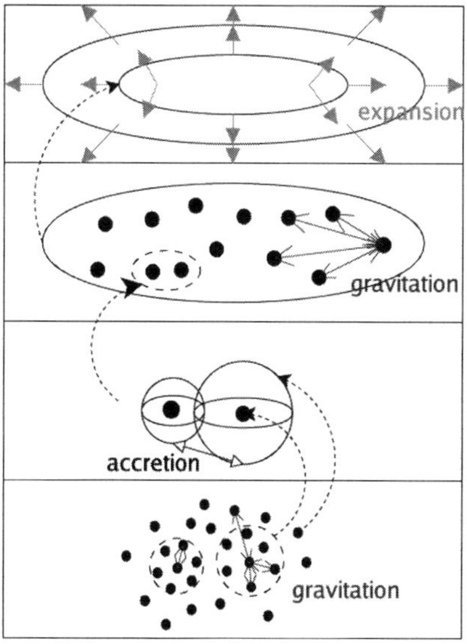

Fig. 1. Schematic representation of our hierarchical multi-agent model

3.3. *Static modifiers*

In addition to the intra-levels and inter-levels elements, the model contains static parameters, applied to the results of the equations above: level 1 : gravity modifier g_1 / level 2 : global accretion power μ / level 3 : gravity modifier g_3 / level 4 : expansion power e. These modifiers are used to widen and facilitate the system parametrization. They are hand-fixed at the beginning of an experiment and do not evolve till the end. Gravity modifier (g_1 and g_3) are used to take into account some physical theories using different kind of gravitation according to the studied celestial object. Accretion power parameter allows to modify, even remove, the viscosity of the whole universe. Expansion power varies the force of expansion to take into account various assumptions: an expanding universe accelerated, a fixed universe or a universe breaking down on itself.

4. Experimental Validation

4.1. *Model validation*

In order to validate the model defined above, we show that the emergent structures are coherent with the observable ones. In this experiment we simulate the collision of two particular galaxies (identified as G1 and G2) and compare it with other simulation and real pictures from Hubble space telescope. The structure, called "The Mice", is often studied in cosmology [14–16]. The experiment is undertaken with the following parameters [14] : *mass* $(G1) = 3.95 \cdot 10^{41}$ kg / *mass* $(G2) = 4.05 \cdot 10^{41}$ kg, *radius* $(G1) = 9.40$ kpc[a] / *radius* $(G2) = 11.0$ kpc, 1*s* "*simulation time*" $= 1.10^{14}s$ "*cosmological time*". The

[a] $1kpc = 3.08568025 \cdot 10^{19}$ m

observational data indicate that the collision tooks place 160 Myr[b] ago. To get the image of simulation corresponding to the current state of these galaxies, it is necessary to let the simulation evolves during 220 Myr (which is equivalent to 69s in simulation time with our system). Distribution of elements is as follow [14] : 10% of gas (represented in light gray), viscous and subjected to gravitation forces, and 90% of dark matter (represented in dark gray), only affected by gravity. Fig.2 shows the result of this evolution and a comparison with other data. These results are obtained after the same evolution time and show that the dynamics and the resulting emergent patterns are the same.

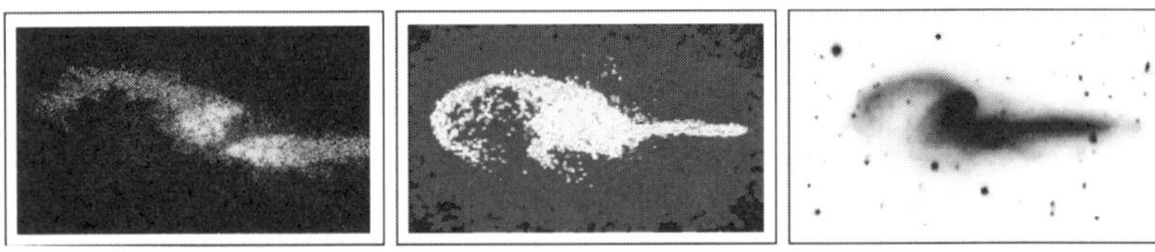

Fig. 2. *Left, the evolution of our model with 300 000 agents. Center, result of a simulation using dedicated cosmological algorithm by J. Hibbard. Right, a picture get by the Hubble Space Telescope (http://hubblesite.org/).*

4.2. *Experimental protocol*

To check if our model is less dependent of the number of elements used than a classical model, a series of experiments aiming at repeating the same simulation, with a decreasing number of agents/particles, has been carried out. Seven simulations have been led using: 100, 1 500, 3 000, 10 000, 30 000, 100 000 and 300 000 agents with our hierarchical model and a classical Treecode. Same experimental conditions are used in each simulation (as describe in 4.1). All the static modifier are set to 1 (neutral value) in these experiments.

4.3. *Results*

Fig.3 presents the same kinds of experiments conducted with a small number of agents and the various generated dynamic. With such few agents (compared to classical cosmo-

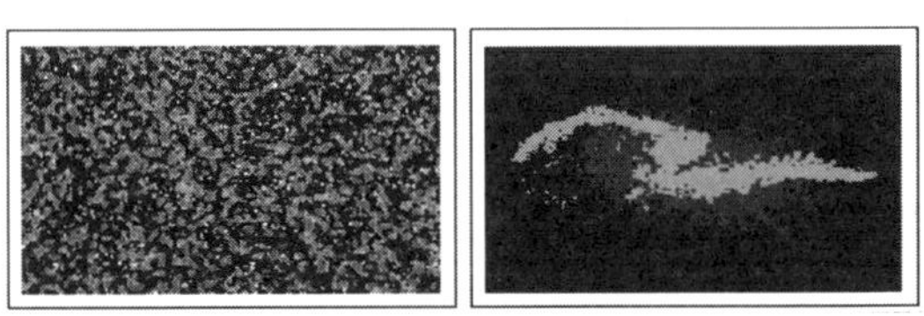

Fig. 3. Left, the result of a simulation using dedicated cosmological algorithm (TreeCode) with 3 000 particles. Right, the same simulation time with our model (still 3 000 agents). Dark matter in dark gray. Gas in light gray.

logical simulations), a structure, qualitatively similar to "The Mice", emerges from our

[b] 1 Myr = 1 million years = $3.1536.10^{13}$s

model. Same experiment with a TreeCode does not lead to any stable structure and all the agents are scattered in the universe. Fig. 4 shows the difference of behavior according to the model used: the top row shows the final step of evolution with various densities of agents (1 500, 3 000 and 100 000) and the bottom row shows same experiments using a TreeCode. The experiment with 1 500 agents does not lead to the formation of any stable

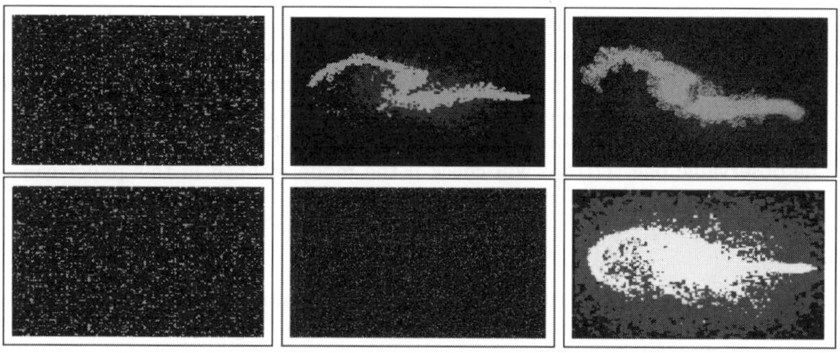

Fig. 4. Top row, simulations with 1500, 3000 and 100 000 agents, using the hierarchical multi-agent model. Bottom row, same simulations using a Treecode.

structure: dynamics is linear and the agents scatter progressively. Dark matter and gas form aggregates of matter, visible on the picture. Experiments with 3 000 and 100 000 show the same dynamic: after G1 and G2 collision both structures remain compact and it appears a "tail" of agents because of their inertia, but they are still subjected to the gravitation of the central mass. In these experiments, dark matter remains distinct from gas and forms a halo around the central structure. The size of this halo is close to the gas structure one, even if, in the experiment with 100 000, as we can see, many dark matter particles escape from it. In the experiment with 1 500 and 3 000 agents we find the same spatial distribution of the agents as with the hierarchical model in the experiment with 1 500 agents: gas and dark matter aggregate and are scattered in all the universe. It is only with 100 000 agents that dynamics, carrying out to the formation of a structure similar to "The Mice", is found.

5. Discussion

Results presented above qualitatively show that, in contrast with classical cosmological models, beyond a threshold, the number of elements does not influence dynamics any more. Such a threshold seems to be around 2 000 elements in these experiments but further studies need to be done to specify the conditions and the reasons of appearance of this threshold value. In contrast to the physical models where a reduction of the number of particles leads to a less homogeneous application of forces, the evolution of the agents at each level on our model compensates this drift: in both models point-mass particles (TreeCode) / level 1 agents (our model) are subjected to less important gravity forces. In a traditional model that leads to an expansion dynamics of all the elements. In the hierarchical model this leads to an increase of the size of the level 2 agents and the size of the vicinity of these agents. Such an increase modifies the importance and the range of the transition functions applied by this level (accretion). These transition functions increase the total cohesion between agents, compensating gravity and improving formation and survival of complex structures.

6. Conclusion

In this paper a new hierarchical multi-agent model aiming to solve the inherent problems of the point-mass particle approach, used by the cosmologists, have been introduced. Qualitative evidences have been presented that a hierarchical multi-agent model is less sensitive to the number of elements used for a simulation than traditional models: beyond some threshold, emergent dynamics remain the same. Future works include a quantitative analysis of the results presented here as well as a study of the complexity classes described by a wide parametrization of this model: by replacing the universe *as-we-know-it* in a larger picture of universe *as-it-could-be*, we will try to understand patterns formation and dynamical evolution. As it exists a difference between observation and numerical simulation on spheroidal galactic formation (radial velocity of peripheral elements does not match with the observable one [17]) we will check if the model defined in this paper can bring some answers.

References

[1] J.M.Alimi, A.Serna, C.Pastor and G.Bernabeu, "Smooth particle hydrodynamics : importance of correction terms in adaptive resolution algorithms", *Journal of Computational Physics*, 2003

[2] S.Gelato, D.F.Chernoff and I.Wasserman, "An Adaptative Hierachical Particle Mesh code with Isolated Boundary Conditions", *The Astronomical Journal*, 1997

[3] V.N.Snytnikov et al, "Space chemical reactor of protoplanetary disk", *Adv. Space Res.*, 30 6, 1461-1467, 2002

[4] E.A.Kuksheva et al, "Numerical Simulation of Self Organisation in Gravitationally Unstable", *Media on Supercomputers*, PaCT 2003, LNCS 2763, 354-368, 2003

[5] J.M.Dawson, "Gravitational N-Body Problem", edited by M.Lecar (Reidel, Dordrecht), 315, 1972

[6] J.M.Alimi, J.P.Chièze, R.Teyssier, A.Serma and E.Audit, "Simulations numériques en cosmologie", *Calculateurs parallèles*, 11, 255-273, 1999

[7] J.Barnes and P.Hut, "A hierarchical O(N log N) force calculation algorithm", *Nature* 324. 446-449, 1986

[8] Hocney and Eastwood, "Simulations using Particles", 1980

[9] L.S.Schulman and P.E.Seiden, "Percolation and Galaxies", *Science*, 233, 425-430, 1986

[10] S.Wolfram, "Cellular Automation Fluids: Basic Theory", *Journal of Statistical Physics*, 45, 471-526, 1986

[11] U.Frisch, B.Hasslacher and Y.Pomeau, "Lattice gas automata for the Navier Stokes equation", *Physical Review Letters*, 56, 1505-1508, 1996

[12] J.C.Heudin, "Complexity Classes in Three-dimensional Gravitational Agents",*Artificial Life VIII*, 9-13, 2003

[13] J.C.Heudin, "Modeling Compexity using Hierarchical Multi-Agent Systems", *these proceedings*

[14] J.C.Mihos, G.D.Bothun and D.O.Richstone, "Modeling the Spatial Distribution of Star Formation in Interacting Disk Galaxies", *Astrophysical Journal*, 418, 82-99, 1993

[15] J.E.Barnes and J.E.Hibbard", "Model of Interacting Galaxies", http://www.ifa.hawaii.edu/˜barnes/research/

[16] J.E.Hibbard and J.E.Barnes, "Observations and Simulation of N4676", http://www.cv.nrao.edu/˜jhibbard/n4676/

[17] A.Burkert and T.Naab, "The Formation of Spheroidal Stellar Systems", *Carnegie Observatories Astrophysics Series, Vol. 1 - Coevolution of Black Holes and Galaxies*, 1-16, 2004

AIDA: ASTRONOMICAL IMAGE DECOMPOSITION AND ANALYSIS

MICHELA USLENGHI

INAF/IASF–Milano,
Via Bassini 15, I–20133 Milano, Italy
E-mail: uslenghi@iasf-milano.inaf.it

RENATO FALOMO

INAF – Osservatorio Astronomico di Padova,
Vicolo dell'Osservatorio 5, I-35122 Padova, Italy
E-mail: falomo@pd.astro.it

We present a package (Astronomical Image Decomposition and Analysis, AIDA), developed in IDL, specifically designed to perform 2D model fitting of quasar images. The software provides simultaneous decomposition into nuclear and host components. AIDA is a "user-friendly" software, which can work both in "interactive" and "batch" mode and manage complex PSF models (including HST and AO psf), and allows one to characterize its spatial variability in the field.

Keywords: Software; Model fitting

1. Introduction

Quasars are one of the most energetic phenomena in the universe, and can be traced out to very large redshifts. The determination of the properties of the galaxies hosting quasars at high z is an important way to investigate not only the link between host properties and nuclear activity, but also to study the history of formation of massive spheroids.

In this context it is important to push as far as possible the direct detection and characterization of QSO host galaxies. In particular, a key point is to probe the host properties at epochs close to (and possibly beyond) the peak of quasar activity ($z \sim 2.5$).

However, because of the faintness of the host galaxies (affected by the rapid cosmological dimming of the surface brightness $\propto (1+z)^4$) and the presence of the strong unresolved nuclear component, the characterization of the properties of high z QSO hosts requires to combine excellent capability of spatial resolution, i.e. narrow Point Spread Function (PSF) with very good throughput to detect and measure the faint nebulosity surrounding the bright QSO nuclei. Recently, the introduction of Adaptive Optics (AO) systems at 8m class telescopes for the first time provided the spatial resolution and the adequate sensitivity for pushing the detection of QSO hosts at z>2. On the other hand, these instruments often shows complex PSF shape, generally variable in the field of view. Detection of the faint extended emission surrounding a bright point source obviously requires careful characterization of the PSF. The most critical part of the analysis is thus to build a detailed PSF model, taking in account possible dependence on the location. Once obtained it, the quasar host galaxies information can be retrieved with simultaneous decomposition into nuclear and host galaxy components by 2D model fitting (correct decoupling of the shape of the host galaxy and shape of the PSF requires 2D modelling, see for example Ref.1–3), considering that the QSO image formed in the focal plane will be:

$$(galaxy + nucleus) \otimes PSF \tag{1}$$

2. AIDA: Software Overview

Several groups of researchers, involved in this kind of study, developed 2D model fitting codes in various languages, but these programs are usually intended for personal use only, they are not available for other users or not documented (see, for example Ref. 3). The only "public" software largely used by the community is GALFIT [2], which has been written and optimized for galaxy dominated objects. Moreover, this software provides 2D model fitting of the object but does not perform PSF model extraction (the PSF model must be provided by the user). Also it requires external software to prepare the image to be analyzed.

In general, no graphical, interactive software is available to support analysis of complex cases.

AIDA (Astronomical Image Decomposition & Analysis) has been developed in IDL (Interactive Data Language) 6.0 and tested under Linux SuSE, Linux Red Hat and Windows XP. The software can run also under Virtual Machine, thus not requiring an IDL license. It makes use of widget based Graphical User Interfaces to provide a user friendly interaction with the analysis procedure. However, when a large number of objects is to be analyzed a fully interactive analysis is unpractical. For this reason, we implemented also a "non-widget" mode which bypass the graphical interfaces and allows to use AIDA as an environment under which running batch procedures.

When running AIDA, a bar including the main menu is always at the top of the screen. During operations, system messages are displayed in the area below the menu. Depending on the task, different secondary GUI are also displayed. At any moment, the overall session of analysis can be saved and restored later.

The objective of the software is to perform simultaneous decomposition of the object image into nuclear and host components. This is done in two main steps:

(1) PSF model extraction (basing on stars images)

(2) Target model fitting

2.1. *PSF Modeling*

The original image can be visualized with a version of ATV [4] which has been modified to be integrated in AIDA. When a new fits image is loaded, the software try to recognize the relevant information by parsing the header. Missing values can be provided by filling a form.

A first selection of sources to be used in the analysis (both stars, for PSF modelling, and QSOs) can be done by the software itself, on the basis of FWHM, sharpness, roundness and signal to noise ratio, or by mouse-selecting them. Then a GUI (Fig. 1) assists the user in choosing sources and preparing them for subsequent analysis. For each selected source, information about its characteristics (FWHM, apparent magnitude, ellipticity, position angle, local background) are displayed, along with a 4-panels visualization of the object (image, contour plot, 3D surface and radial profile). Several interactive controls are available in the right side of the GUI, allowing changing the visualization appearance. Each plot can be exported in eps files and a text report of the characteristics of all the sources can be generated.

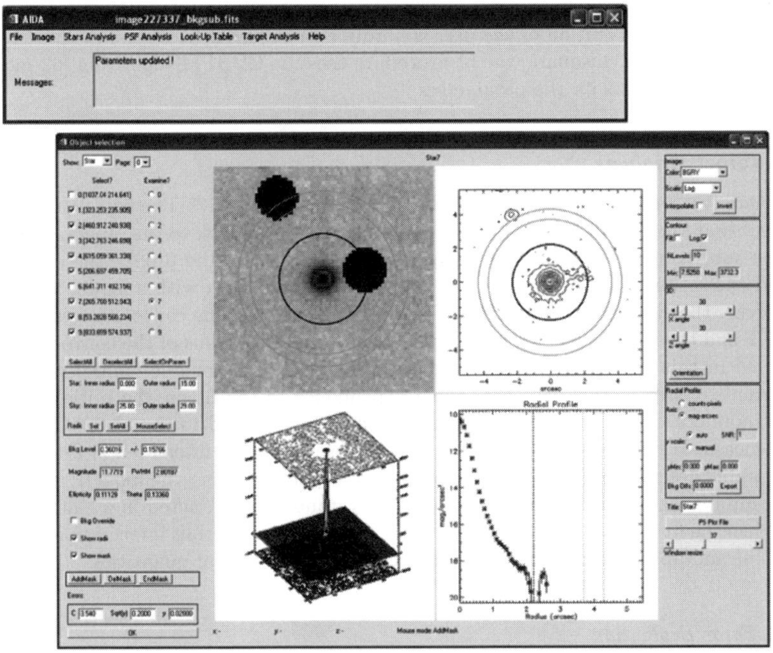

Fig. 1. Main menu bar and "Object Selection" GUI.

As stated before, the most critical part of the analysis is to perform a detailed PSF model. AIDA can manage both analytical and empirical PSF (in the form of Look-Up table), even mixing them. At the moment, AIDA does not assist in building empirical look-up table, that should be provided by the user (in FITS format or as IDL variable). Support is specifically given for HST TinyTim [5] generated PSF (allowing oversampled PSF and convolution with blurring kernel).

The PSF modelling is done by fitting reference stars with a parametrized model including any combination of the available 2-d functions (user-defined functions can be added to the default ones) and/or empirical look-up table. It is also possible to define 2 regions with different PSF models (see the case in section 3.1).

If the PSF is supposed to be invariant in the field of view (see the case of section 3.2), multiple star images can be fitted simultaneously with the same model parameters. This allows, for example, to combine information on the core shape from even moderately faint stars and on the wings shape from bright, saturated stars. Alternatively, individual stars can be fitted allowing parameters changes. Using analytical models with a limited number of parameters, dependence of the PSF parameters on the position can be modeled.

One of the critical points is providing initial guesses. This can be done by filling the related form (also constraints can be user-provided for each parameter); however, if a PSF with multiple 2-D components is chosen, the number of parameters rises quickly making unpractical this approach. Thus the software can provide a reasonable set of initial guesses and constraints automatically, with a multiple step procedure, mainly based on 1-D fitting of the radial profile of the best reference star (preferably not saturated and with high S/N ratio); ellipticity and position angle are instead evaluated by fitting ellipses on isophotes. In this way the intervention of the user is hardly ever required.

To minimize the chisquare, the fit procedure uses the CURFIT algorithm [6], modified to manage constraints on the parameters.

2.2. *Target Analysis*

There are two kind of analysis that can be performed on the QSO. The first one is a simple PSF subtraction: this kind of analysis is useful to give a first look to the object. The target will be subtracted by a PSF normalized to some fraction of the flux of the object. The second one is the true model fitting, providing a fit of the object with a model obtained by convolving the galaxy with the PSF and adding the nucleus. The convolution is computed by FFT and the user can select the sampling factor and the radius of the convolution area.

Even in this case, the software can provide a set of (reasonable) initial guesses and constraints. Since in this case a possible dependence of the fit on the initial guesses is much more critical (because the parameters are strictly related to the physical properties to be measured, whereas, in the case of the PSF, we were only interested in finding one of the possible description of the PSF shape, even overparametrizing the model), a procedure can compute the fit with different initial guesses, randomly extracted in a suitable range.

When the fit is done, a text report is generated with the relevant information, including apparent and absolute (if a cosmology is provided) astronomical quantities.

2.2.1. *Error Evaluation*

Evaluation of the confidence limits on the estimated model parameters is a complex task. In fact, even being trustful on the knowledge of the errors model associated with the original image, and on the exact error propagation associated to the data reduction process (that, for example, in the case of IR images obtained with tip-tilt technique is not obvious), the impact of the PSF model uncertainty on the computed parameters should be also taken in account. Two tools are implemented in AIDA to assist the error evaluation on the parameters: Monte Carlo simulation of synthetic data sets and constant chi-square boundaries on chi-square map.

3. Examples of astronomical data

AIDA has been extensively tested with simulated images, then has been used to analyze several astronomical data sets generated by different instruments, with different PSF features. In the following sections a very short description is given.

3.1. *Low redshift quasars observed by HST-WFPC2*

We analyzed QSOs images obtained with the second Wide Field and Planetary Camera (HST/WFPC2) of the Hubble Space Telescope (HST). The software TinyTim [5] produces tabulated PSFs which correctly model the PSF in the core but underestimate it in the wings [7] (where it is most important to be able to discriminate the light of the galaxy from the PSF).

We used a mixed PSF model: empirical in the inner part (TinyTim generated); empirical +analytical (3 exponential components) in the wings. Figure 2 shows the PSF generated by TinyTim and the model produced by AIDA, compared with the radial profile of a bright star. Five objects have been analyzed and their results are reported in Ref. 8.

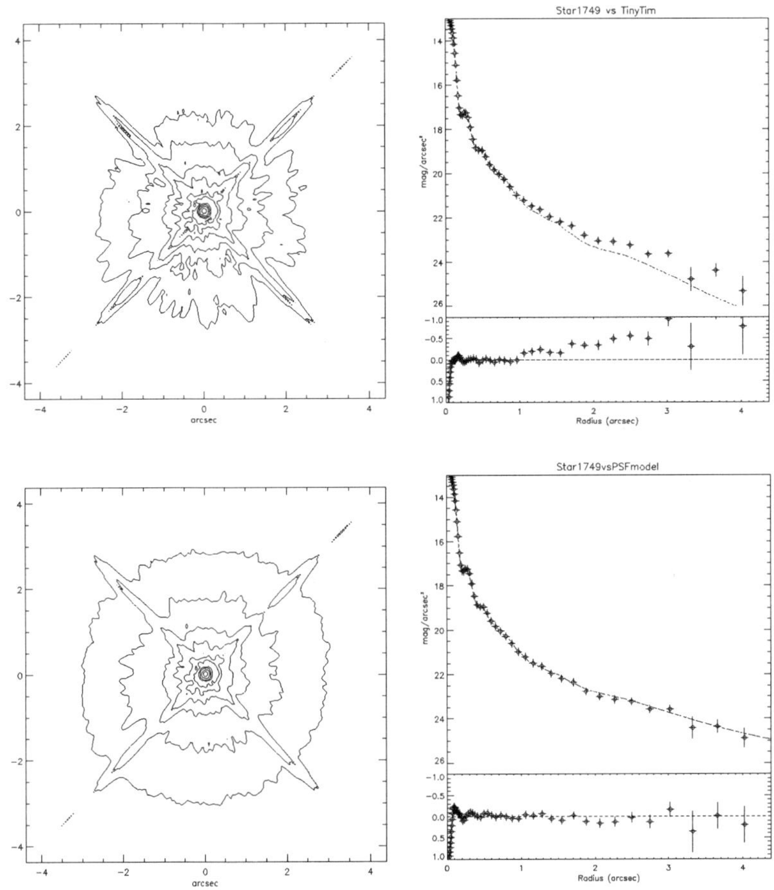

Fig. 2. HST PSF. Upper panel: TinyTim PSF; Lower panel: AIDA mixed PSF model.

3.2. *NIR images of quasar at $1 < z < 2$ observed with VLT+ISAAC*

Deep images of the quasars in the H- or K-band were obtained using the near-infrared (NIR) ISAAC camera, mounted on UT1 (Antu) of VLT at the European Southern Observatory (ESO). The Short Wavelength (SW) arm of ISAAC is equipped with a 1024 x 1024 px array, with pixel scale of 0.147 arcsec/pixel, giving a field of view of 150 x 150 arcsec. We analyzed a sample of 15 objects with $1.221 \leq z \leq 1.895$, 13 resolved, 1 marginally resolved [9]. For this configuration of instruments, the PSF appeared to be invariant in (a large part of) the field of view, then, for each image, all the stars have been fitted with the same model. We adopted a fully analytical PSF, formed by 5 gaussians and 3 exponentials. In figure 3 a typical result is given, showing the radial profile of one object with its nucleus+galaxy modelization.

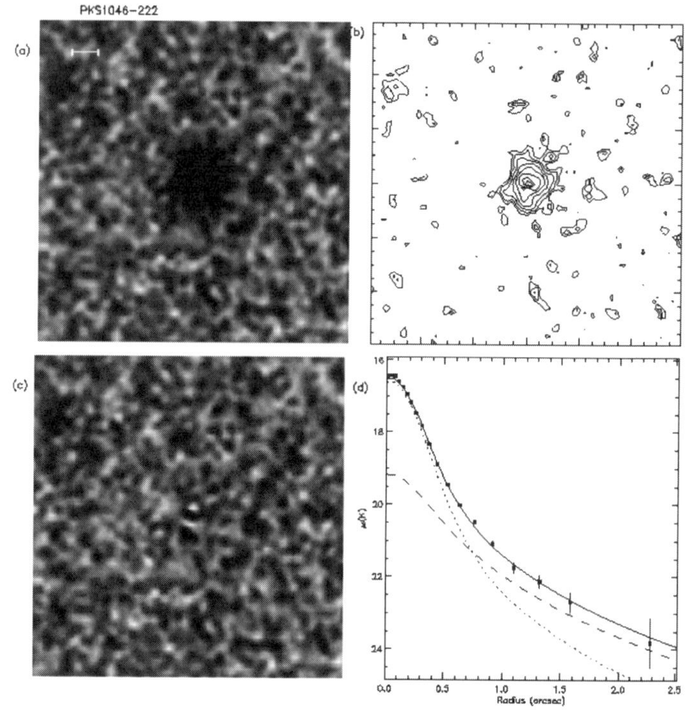

Fig. 3. Example of VLT-ISAAC results. H-band image of PKS 1046-222, from top to bottom a) the original image, b) the contour plot of the image after subtracting the PSF, c) residuals after the model fitting, d) the observed radial surface brightness profiles (solid points with error bars, overlaid with the scaled PSF model (dotted line), the de Vaucouleurs $r^{1/4}$ model convolved with the PSF (long-dashed line), and the fitted PSF + host galaxy model profile (solid line). The Y-axis is in mag·arcsec^{-2}.

References

[1] B. Kuhlbrodt, L. Wisotzki, K. Jahnke, MNRAS **349**, 1027 (2004)
[2] C.Y. Peng, L.C. Ho, C.D. Impey, H.W. Rix, AJ **124**, 266 (2002)
[3] G.L. Taylor et al., MNRAS **283**, 930 (1996)
[4] Barth, A. J., in ASP Conf. Ser., Vol. **238**, Astronomical Data Analysis Software and Systems X, eds. F. R. Harnden, Jr., F. A. Primini, H. E. Payne (San Francisco: ASP), 385 (2001)
[5] J.E. Krist, R.N. Hook, *The Tiny Tim Users Manual*, STScI (1997)
[6] P. Bevington, *Data Reduction and Error Analysis for the Physical Sciences*, McGraw-Hill (1991)
[7] R. Scarpa, C.M. Urry, R. Falomo, J.E. Pesce, A. Treves, ApJ **532**, 740 (2000)
[8] M. Labita, A. Treves, R. Falomo, M. Uslenghi, MNRAS **373**, 551 (2006)
[9] J.K. Kotilainen, R. Falomo, M. Labita, A. Treves, M. Uslenghi, to appear in ApJ, astro-ph/0701417 (2007)

COMPARISON OF STEREO VISION TECHNIQUES FOR CLOUD-TOP HEIGHT RETRIEVAL

ANNA ANZALONE

INAF - Istituto di Astrofisica Spaziale e Fisica Cosmica di Palermo
Palermo, Italy
E-mail:Anna.Anzalone@ifc.inaf.it

FRANCESCO ISGRÓ

Dipartimento di Scienze Fisiche, Università degli Studi Federico II,
Napoli, Italy

DOMENICO TEGOLO

Dipartimento di Matematica ed Applicazioni, Università di Palermo
Palermo, Italy

This paper presents an ongoing study for the estimation of the cloud-top height by using only geometrical methods. In agreement with some recent studies showing that it is possible to achieve reliable height estimations not only with the classical methods based on radiative transfer, this article includes a comparison of performances of a selected set of vision algorithms devoted to extract dense disparity maps or motion fields from Infra Red stereo image pairs. This collection includes both area-based techniques and an optical flow-based method and the comparison is accomplished by using a set of cloudy scenes selected from the Along-Track Scanning Radiometer (ATSR2) database. The first group of algorithms yields results generally comparable in terms of the measures used for the inter-comparison, while the maps from the optical flow-based method are slightly better.

Keywords: Cloud-top height, multi view, stereo matching algorithms, satellite infra red images, optical flow

1. Introduction

Observations from space of atmosphere and in particular of clouds and their related features are crucial for climate and weather forecasting studies. Earth's warming or cooling are strictly dependent on the cloud composition and height, therefore the process for reliable cloud parameter estimations play an important role. It is also important in the study of cosmic ray radiation at ultra high energies ($E > 5 * 10^9 eV$). Current and future ground and space-based observatories [1, 2] monitor the Earth atmosphere by means of large field-of-view telescopes to detect the (300-400 nm) UV fluorescence track produced by a relativistic cosmic ray interacting with the atmosphere. From the direction of the track and its light content, it is then possible to reconstruct the kinematical and dynamical features of the primary cosmic ray. Certainly the monitoring of the atmospheric parameters including clouds, within the field-of-view of the specific telescope, is crucial for the proper reconstruction of the event [3].

Current operational methods for indirect measurements from space, in particular for cloud-top height retrieval are mostly based on radiative transfer. Brightness Temperature ($11\mu m$), $CO_2-slicing$ ($15\mu m$) and O_2-A band methods ($76\mu m$) [4–6] need extra information such as ambient temperature-pressure profiles, surface reflectivity data coming from other sources that can affect the accuracy of the estimates. Some recent studies [7, 8] show that it is possible to obtain reliable height estimations also using stereo vision techniques whose relevant advantage is being dependent only on the geometry of the observations Fig.4.

Naud et al. [4–6] discussed inter-comparisons between height maps derived by the different methods, matched up them to the one from ground instrument and finally compared them with heights derived from a back-scattering lidar. In all papers a sufficient variety of cloud characteristic are taken into account and a good agreement of the stereo method with the reference values has been observed.

Taking into account these outcomes, in this paper we present some preliminary results of a comparison of performances of a selected set of well known vision algorithms devoted to extract dense disparity maps or motion fields and having as input IR stereo image pairs.

Starting from the study described by Anzalone et al. [9], classical area based techniques for stereo matching are firstly inter-compared in order to establish a good set of parameters, then are compared with the algorithm based on a differential approach for the computation of the optical flow [10]. After having applied the matchers to a collection of cloudy scenes, the outcoming maps of disparity can be transformed in height maps according to the geometry of the reference data system Fig.4. Similarly the optical flow extractor is performed on the same scenes and the resulting motion fields can also be converted in depth map. However this contribution focuses attention only on the assessment of disparity maps for pairs of satellite image instead of height maps as explained in section 4.

In section 2 a brief introduction on the most diffuse approaches used to deal with the problem of correspondence between images, is given. Moreover the same section summarizes the matching algorithm set included in the experiments and briefly presents the method developed to compute the optical flow. The following section 3 reports some notes about the input data, while in section 4 the experiments and the results are discussed.

2. Disparity Map Computation

Passive stereo remains one of the fundamental technologies for estimating 3-D information. It is desirable in applications because it requires no modifications to the scene, and because dense information (that is, at each image pixel) can nowadays be achieved at video rate on standard processors for medium-resolution images.

There are two broad classes of correspondence algorithms, seeking to achieve, respectively, a sparse set of corresponding points (yielding a sparse disparity map) or a dense set (yielding a dense disparity map). The algorithms in the first category select feature points independently in the two images, then attempt to match them. Algorithms in the second category select templates in one image (usually patches with some texture information), then look for corresponding points in the other image using some similarity measure [11–13]. The output of a dense matching algorithm is a correspondence or disparity map, which associates to each pixel in one image its displacement (disparity) vector with respect one of the other images. For cloud images it is a challenge detecting cloud features suitable for the algorithms of the first category, as they are generally smoothed and change their appearance from different viewing angles. Therefore at the moment we focus our attention on area-based approaches.

Dense stereo matching is a well-studied topic in image analysis [11]. Scharstein gave an excellent review including suggestions for comparative evaluation [14]. We refer the reader to this paper for an exhaustive list of known algorithms.

The optical flow [11], first introduced by Horn and Schunk in 1980, is usually computed with differential algorithms as it is defined as the velocity field which warps one image into another (generally very similar) image.

Set of matching algorithms For the purpose of the experimental evaluation we adopted a few algorithms well known among the computer vision community. Following four steps for a general stereo matching algorithm [14] can be considered: a) matching cost computation; b) cost aggregation; c) disparity compuations/optimisation; d) disparity refinement. Needless to say that the step sequence can be different, depending on the particular algorithm considered. For the experiment carried out in this work, we implement only the first three steps.

The pixel based matching cost functions that we adopted are the squared intensity differences and the absolute intensity differences. Usually, for local methods, the cost function is computed (still pixelwise) for each image point over a neighbourhood, that is the aggregation window. The cost function can be aggregated following several rules, and different aggregation windows can be chosen: some algorithms, for instance, adopt multiple aggregation windows. Here, for simplicity, we decided to adopt a single aggregation window, centred in the image point for which we are computing the disparity and in the candidate match. We used aggregation windows of different sizes in the range of 9-17 pixels.

For the third step we must distinguish between local and global methods. Local methods usually adopt the winner takes all strategy, as the core of these algorithms is in the first two steps. Therefore they simply choose as best match, the one with the highest matching score.

Conversely global methods may sometimes skip the aggregation step at all. Most of the algorithms in this class are formalised as an energy minimisation problem, i.e. they aim to find a disparity function minimising some energy function. Among the various methods proposed in literature for locating the minimum for the energy function, we adopted the simulated annealing [15]. Different global methods are based on dynamic programming, where the optimisation problem is solved for each scan-line independently. More recent work focuses on the scan-line optimisation problem, where disparities are obtained by considering the matrix of all pairwise matching costs between two scan-lines, and computing the minimum cost path through this matrix. We considered both dynamic programming and scan-line optimisation in our work.

Optical flow based algorithm The algorithm presented by Anzalone et al. [10] approaches the problem of establishing correspondences between the images of the same scene, by computing the apparent motion in the image plane. This motion cannot be very large, under the assumption that the images can be acquired at relatively high frame rate; if the temporal and spatial differences are small, the retrieved motion field coincides with the stereo disparity map.

The algorithm we implemented is a non-iterative version of the Lucas-Kanade optical flow algorithm [16]. Taking into account that the final motion field can be affected by corrupted pixels and that the kind of images we generally deal with can contain a significant amount of these pixels, the optical flow is robustly computed [10].

Furthermore the multi-resolution approach is used to cope with large vector motion likely present if the time interval between two consecutive frames is large: the motion is computed for coarsest images and then used to predict the solution for the finest level.

The algorithm could be speeded up, by filtering out from the optical flow estimation the pixels marked as cloud free (land point) by a cloudiness mask.

As schematically depicted in Fig.1, firstly the Multiresolution Optical Flow Estimation Module establishes dense correspondences between the input image pair. It estimates the 2D motion field from IRimage1(I_1) to IRimage2(I_2) and viceversa. Then the Consistency Check and Interpolation Module, based on a straightforward and widely used method, evaluates the difference according to a fixed metrics, between the two computed motion fields M_{12} and M_{21} by warping forward I_1 through M_{12} and warping back this new image utilizing M_{21}. If this process validates the motion fields they are merged into a single field. In other words for each pixel p a new motion field M is produced via the following rule:

$$M(p) = \begin{cases} M_{12}(p) \; if ||M_{12}(p) - M_{21}(p + M_{12}(p))|| < \lambda \\ \infty \qquad\qquad otherwise \end{cases}$$

where λ is a threshold value meaning that the error we can tolerate in the motion field is not larger than λ pixels. We decided to use a relatively large value (i.e., greater than 1) for the parameter λ in order to compensate the effects of the wind, and the value is set accordingly to the resolution of the images. For the experiments shown in this paper we set this value to 3.

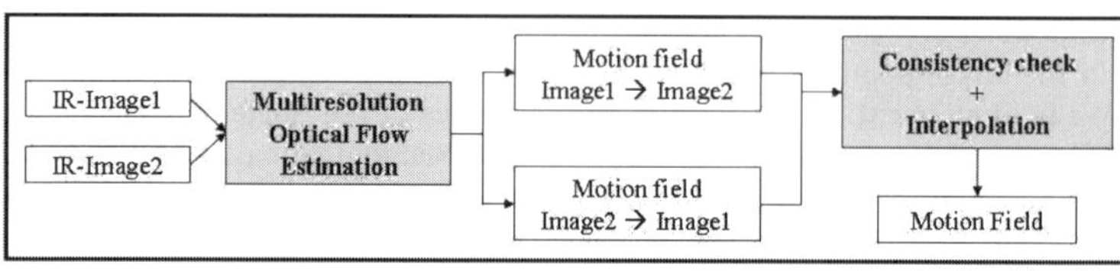

Fig. 1. Schematic representation of the optical flow algorithm

3. ATSR2 Data Set

Along Track Scanning Radiometer (ATSR2) [17] is a multiangle instrument onboard the ERS2 satellite flying at the altitude of about 780km. Two different views in seven spectral bands ranging from infra red to visible wavelengths, are recorded along the flight direction, within a temporal gap of 120 sec. The forward view is taken by the sensor while the satellite flies towards the scene; the nadir one is acquired when the satellite sees the scene at the nadir position, such that the same scene is mostly in the field of view of the sensor at the two different times.

Forward and nadir view pairs are given as input (see Fig. 2) to the set of stereo matching algorithms and selected only from the IR bands (11 or 12μm), where clouds radiate the energy maximum behaving as a blackbody. Additional ancillary information provided by the satellite data base, such as the forward/nadir masks of flags resulting from cloud clearing tests, were exploited to build masks of pixel cloudiness to carefully filter the data to eliminate images not including a meaningful number of mostly cloudy pixels.

4. Experiments and Results

In our experiment we preferred to compare the disparity maps instead of the height maps, to test the goodness of the results independently of the reconstruction phase. This step in fact, can introduce artefacts due to the geometry of the acquisition system and for ATSR2 data, depends on the camera model chosen to calculate the forward and nadir zenith angles used in the Prata and Turner height formula [18]. Furthermore a preliminary evaluation of the accuracy of the retrieved heights has been already reported [9] where Digital Terrain Elevation Data were used as ground truth for nearly cloud free scenes and the maps were found in good agreement with the reference ones. A last consideration is that more reliable height values could be obtained if the contribution of the along-track cloud motion component, likely present for ATSR2 data, is not neglected. Here attention is focused on the assessment and comparison of the disparity values. The algorithms compute the disparity map along the direction of the satellite motion so that for the optical flow vector we consider only the component in the along-track direction. Fig.3 schematizes the applied methodology architecture and highlights the main operative modules: after an exhaustive analysis obtained by the set of combinations of the algorithms and parameters proposed in section 2, the Matching Module estimates dense disparity maps with the selected set of parameters necessary for each sequence step. In the meanwhile the Optical flow Module calculates the optical flow as described in section 2 and output a dense disparity map for the same input image pair. Afterwards all maps are assessed by the PSNR Module, that yields a final selection of the best performer.

As measure of goodness for the estimation of the disparity maps we used the PSNR between the original image and the reconstructed one as follows: given the disparity component between image I_1 and I_2, it is possible to warp back I_2 using d and obtaining an image I'. Note that in the noiseless case and assuming a perfect disparity map, I' should be an exact copy of I_1. PSNR is defined as: $PSNR(I, I') = 20 \log_{10} \left(\frac{255}{\sum_p (I(p) - I'(p))^2} \right)$.

We also consider the percentage of pixels ($PGaps$) where the consistency check step (applied to all retrieved maps) did not validate the computed disparity value, to obtain a weighted measurement of the PSNR (WPSNR) values as $WPSNR(I, I') = (1 - PGaps)PSNR(I, I')$. These Gaps pixels obtained a final disparity value through an interpolation phase.

We tested the different algorithms on several ATSR2 images, neglecting to examine the influence of different cloud configurations on the quality of the final retrieved disparities. In fact at this stage, we need to know how the chosen algorithms perform and understand how the different method combinations influence the results. For this reason only their basic versions have been tested without considering any optimisation step. Finally we were interested in comparing the results of the optical flow based method with the others.

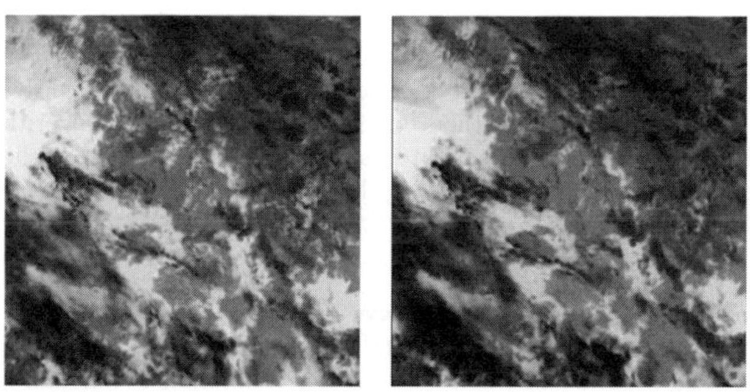

Fig. 2. An example of stereo $11\mu m$ ATSR2 input image pair. On the left the *forward* view and on the right the *nadir* view.

Here we show results relative to a pool of the input image pairs: each plot shown in Fig.6.a and Fig.6.b, report respectively the PSNR and the WPSNR values for 80+1 disparity maps (80 are the different combinations of matching algorithms and parameters; 1 is the optical flow) provided as result of the methods, where the first value of the x axis corresponds to the optical flow based method. It can be noted that this value is slightly better in all cases and improves a bit more when the consistency in both directions is taken into account (WPSNR). In fact the second measure considers the number of points where the disparity values from I_1 to I_2 and from I_2 to I_1, do not satisfy the consistency check test illustrated in 2 and then the final disparity value is assigned by interpolation. So the fluctuation in the WPSNR graphs mean that some combination of parameters and matching steps provide more consistent disparity maps than others. On the contrary this result it is not highlighted by the nearly regular behaviour of the PSNR values. Similar graphs have been obtained for the other input image pairs. Whereas, the different sequences of algorithms were found to yield very similar results in terms both of PSNR and of disparity values as shown in Fig.5, where relatively to an input image pair, separate difference distributions between pairs of disparity maps from different matcher combinations, are displayed. The plot shows a blow up of the central peak for all distributions and small difference values.

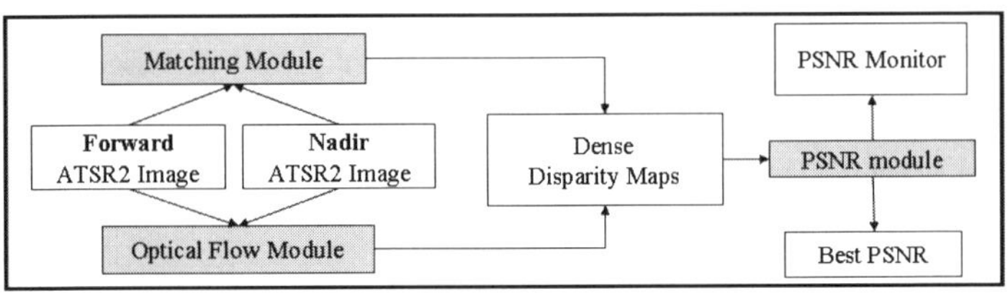

Fig. 3. Architecture of the applied methodology.

5. Conclusions

The paper presented a study on classical stereo vision methodologies applied to the remote sensing field, in particular to the problem of estimating the cloud-top height from IR images. The performance of different matching algorithms were tested on several ATSR2 IR image pairs, focussing attention on the assessment of the goodness of the disparity maps and on the inter comparison of the results. The experiments show a general agreement between the maps retrieved by the selected methods and highlight a slightly better performance of the optical flow based method. In the future these sets of algorithms should be tested on data coming from other satellite data sources and the maps validated through maps provided by other type of sensors (e.g. lidar, radar, etc).

References

[1] Pierre Auger Observatory (AUGER), `www.auger.org`
[2] Extreme Universe Space Observatory (EUSO), `www.euso-mission.org`
[3] Anzalone, A., Comella, M., D'Alì Staiti, G., Raimondi, F. M. and Rinella, V., *Il Nuovo Cimento* 29, 519 (2006)
[4] Naud, C. et al., Inter-comparison of MERIS, MODIS and MISR cloud top heights, *Proc. MERIS User Workshop (ESA SP-549) ESA-ESRIN*, (2003)

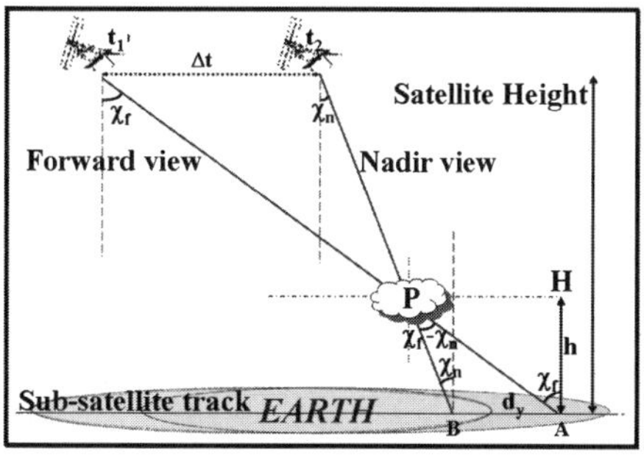

Fig. 4. Geometry of the system. χ_n and χ_f are satellite zenith angles projected on the sub-satellite track plane; d_y is the pixel shift in the along track direction.

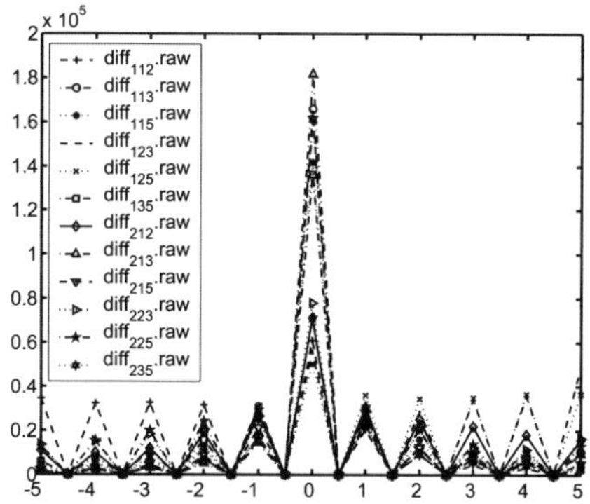

Fig. 5. Pixel by pixel comparison of disparities retrieved by the 80 different matcher method combination.

[5] Naud, C. et al., *Annales Geophysicae*, 23, 2415 (2005)
[6] Naud, C. et al., *Geophysical Research Letters*, 31 (2004)
[7] Muller, J. P. et al., *IEEE Trans. Geosci. & Remote Sensing*, 40, 1547 (2002)
[8] Moroney, C. et al., *IEEE Trans. Geosci. & Remote Sensing*, 40, 15320 (2002)
[9] Anzalone, A., Isgrò, F., and Tegolo, D., Stereo Matching Techniques for Cloud-Top Retrieval, *Proceedings of SPIE Image and Signal Processing for Remote Sensing XII*, ed. L.Bruzzone (2006)
[10] Anzalone, A., Isgrò, F., and Tegolo, D., A study on recovering the cloud top height from infra-red video sequences, *Proceedings of MDIC* (2004)
[11] Trucco, E. and Verri, A., *Introductory techniques for 3D computer vision*, Prentice Hall (1998)
[12] Goshtasby, A., Gage, S.H., and Bartholic, J.F., *IEEE PAMI*, 6, 374 (1984)
[13] Chou, C.H. and Chen, Y.C., *Pattern Recognition*, 23, 461 (1990)
[14] Scharstein, D. and Szeliski, R., *Int. Journal of Computer Vision*, 37, 7 (2002)
[15] Barnard, T., *Int. Journal of Computer Vision*, 3, 17 (1989)
[16] Lucas, B.D. and Kanade, T., An iterative image registration technique with an application to stereo vision, *Proc. International Joint Conference on Artificial Intelligence*, 674 (1981)
[17] ATSR2 project, `www.atsr.rl.ac.uk/index.shtml`
[18] Prata, A.J. and Turner, P.J., *Remote Sensing Environment*, 59, 1 (1997)

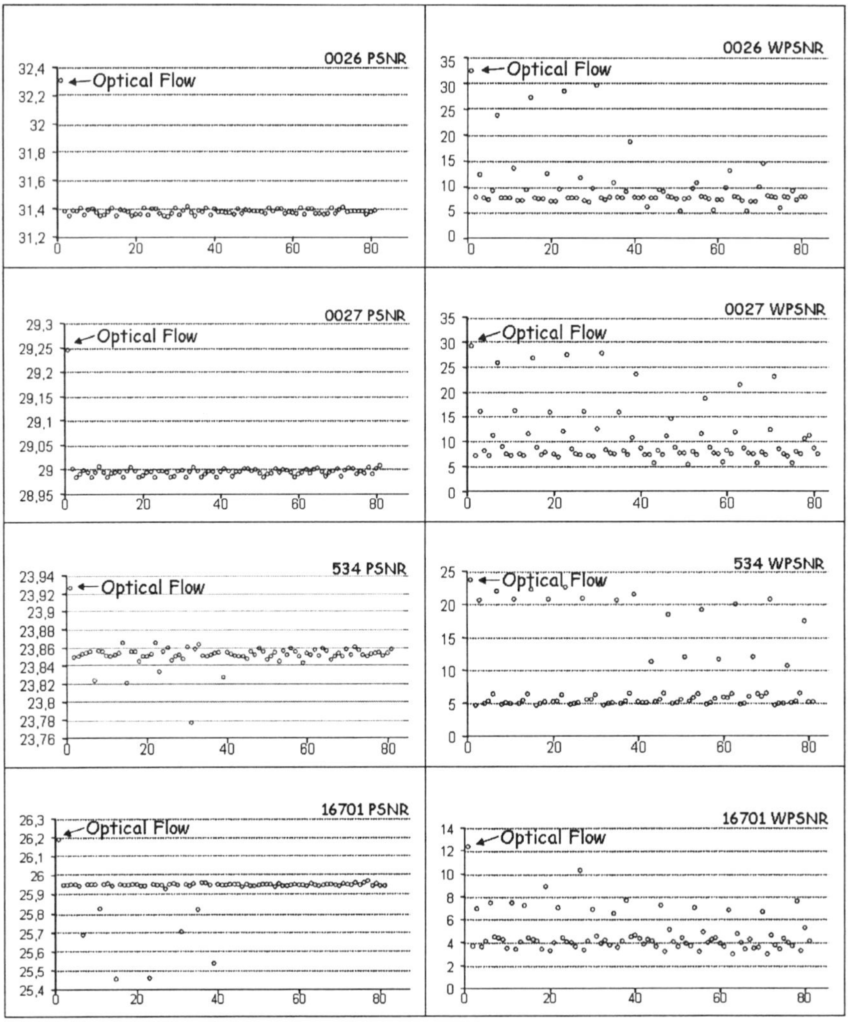

Fig. 6. a) PSNR and b) WPSNR values for 80+1 disparity maps from different methods obtained as combinations of different steps of the matching algorithms and parameters. The first value on the x axis concerns the optical flow based method.

AUTHOR INDEX

PARTICIPANTS

ANZALONE, Anna
Ist. Astrofisica Spaziale e Fisica Cosmica
IASF-Palermo/INAF
Palermo, Italy
anna.anzalone@iasf-palermo.inaf.it

BARBIERI, Cesare
Department of Astronomy
University of Padua
Padova, Italy
cesare.barbieri@unipd.it

BELLAVIA, Fabio
DMA, Dept. Mathematics & Applications
University of Palermo
Palermo, Italy
fbellavia@math.unipa.it

BOI, Luciano
EHESS, Center of Mathematics
Paris, France
boi@ehess.fr

CHAKRABARTY, Dalia
School of Physics & Astronomy
University of Nottingham
Nottingham, UK
dalia.chakrabarty@notthingham.ac.uk

CIPOLLA, Marco
DMA, Dept. Mathematics & Applications
University of Palermo
Palermo, Italy
mcipolla@math.unipa.it

COLOMBO, Paolo
Inst. of Biomedicine and Molecular Immunology
IBIM/CNR
Palermo, Italy
paolo.colombo@ibim.cnr.it

COMPARATO, Marco
Catania Astrophysical Observatory
OACT/INAF
Catania, Italy
marco.comparato@oact.inaf.it

CONTRERAS ALBORNOZ, Pedro
Dept. of Computer Science
Royal Holloway, University of London
Egham, Surrey, UK
pedro@cs.rhul.ac.uk

CORONA, Davide
Inst. Telethon Dulbecco
University of Palermo
Palermo, Italy
dcorona@unipa.it

DE SABBATA, Venzo
University of Bologna
Bologna, Italy

DI GESÚ, Vito
CITC and DMA, Dept. Mathematics & Applications
University of Palermo
Palermo, Italy
digesu@math.unipa.it

GALATI, Vita
Universitá di Modena e Reggio Emilia
Modena, Italy
vitagalati@virgilio.it

HEUDIN, Jean-Claude
Int. Inst. of Multimedia
Pôle Universitaire Leonard de Vinci, La Défense
Paris, France
jean-claude.heudin@devinci.fr

GAUTARD, Valerie
Dapnia, Astrophysics Dept.
CEA/Saclay
Gif sur Yvette, France
valerie.gautard@cea.fr

GHOSH, Joydip
George Mason University
Fairfax VA, USA
jghosh1@gmu.edu

KAFATOS, Menas
CEOSR, Center for Earth Observ. & Space Res.
George Mason University
Fairfax VA, USA
mkafatos@gmu.edu

KNAPP, Johannes
School of Physics and Astronomy
University of Leeds
Leeds, UK
j.knapp@leeds.ac.uk

LO BOSCO, Giosué
DMA, Dept. Mathematics & Applications
University of Palermo
Palermo, Italy
lobosco@math.unipa.it

LONGO, Giuseppe
Department of Physics
University Federico II
Napoli, Italy
longo@na.infn.it

LOUCHET, Jean
COMPLEX Team
INRIA
Rocquencourt, France
jean.louchet@gmail.com

MACCARONE, Aurora
Department of Physics
University of Palermo
Palermo, Italy
aurora@maccarone.it

MACCARONE, Maria Concetta
Ist. Astrofisica Spaziale e Fisica Cosmica
IASF-Palermo/INAF
Palermo, Italy
cettina.maccarone@iasf-palermo.inaf.it

MAKOWIECKI, Wojciech
Computer Science Dept.
AGH Univ. of Science and Technology
Cracow, Poland
wojtech1982@gmail.com

MANZATO, Patrizia
Astronomical Observatory of Trieste
OATS/INAF
Trieste, Italy
manzato@oats.inaf.it

MASSONE, Anna Maria
Lab. of Innovative and Artificial Materials
National Inst. for the Physics of Matter
INFM-LAMIA/CNR
Genova, Italy
massone@ge.infm.it

MASULLI, Francesco
DISI, Computer and Information Science Dept.
University of Genoa
Genova, Italy
masulli@ disi.unige.it

MILANESI, Luciano
Institute of Biomedical Technologies
ITB/CNR
Segrate, Milano, Italy
luciano.milanesi@itb.cnr.it

MILLONZI, Filippo
DMA, Dept. Mathematics & Applications
University of Palermo
Palermo, Italy
millonzi@math.unipa.it

MOLINA, Rafael
Dept. of Comp. Science & Artif. Intelligence
University of Granada
Granada, Spain
rms@ decsai.ugr.es

MURTAGH, Fionn
Dept. of Computer Science
Royal Holloway, University of London
Egham, Surrey, UK
fmurtagh@acm.org

NAPOLITANO, Francesco
DMI, Mathematics and Informatics Dept.
University of Salerno
Fisciano (SA), Italy
franapoli@gmail.com

NEČESAL, Petr
Institute of Physics, AS CR
Prague, Czech Republic
necesal@ fzu.cz

OCCHIPINTI, Tommaso
DEI, Information Engineering Dept.
University of Padua
Padova, Italy
tommaso.occhipinti@dei.unipd.it

PAL, Sankar K.
Indian Statistical Institute
Kolkata, India
sankar@isical.ac.in

PATITUCCI, Davide
Centro Enrico Fermi
Roma, Italy

PERES, Giovanni
Physics & Astronomical Science Dept.
University of Palermo, Palermo, Italy
peres@astropa.unipa.it

PETRERA, Sergio
Dept. of Physics and INFN
University l'Aquila
L'Aquila, Italy
sergio.petrera@aquila.infn.it

PINELLO, Luca
DMA, Dept. Mathematics & Applications
University of Palermo
Palermo, Italy
lucapinello@gmail.com

RÖPKE, Friedrick
Max-Planck-Institut fr Astrophysik
Garching, Germany
fritz@mpa-garching.mpg.de

ROY, Malabika
Indian Statistical Institute, Calcutta, India
George Mason University, Fairfax VA, USA
mroy@science.gmu.edu

ROY, Sisir
Indian Statistical Institute, Calcutta, India
George Mason University, Fairfax VA, USA
sisir@isical.ac.in

RUBINI, Francesco
Dept. Of Astronomy and Space Science
University of Florence
Firenze, Italy
rubini@arcetri.astro.it

SACCO, Bruno
Ist. Astrofisica Spaziale e Fisica Cosmica
IASF-Palermo/INAF
Palermo, Italy
bruno.sacco@iasf-palermo.inaf.it

SANCHEZ UBEDA, Eugenio F.
Inst. of Technological Investigation
Universidad Pontificia Comillas
Madrid, Spain
eugenio.sanchez@iit.upcomillas.es

SCARSI, Marco
Biozentrum
University of Basel
Basel, Switzerland
marco.scarsi@unibas.ch

SEDMAK, Giorgio
Physics Department
University of Trieste
Trieste, Italy
sedmak@ts.infn.it

SONWALKAR, Vikas
Electrical and Comp. Engin. Dept.
University of Alaska Fairbanks
Fairbanks, Alaska, USA
ffvss@uaf.edu

SUTERA, Alfonso
Department of Physics
University of Rome La Sapienza
Roma, Italy
alfonso.sutera@roma1.infn.it

TAGLIAFERRI, Roberto
DMI, Mathematics and Informatics Dept.
University of Salerno
Fisciano (SA), Italy
robtag@unisa.it

TORREL, Jean-Claude
Lab. Artficial Intelligence, Paris V,
University Ren Descartes
Paris, France
jean-claude.torrel@devinci.fr

TRAMONTANO, Anna
Dept. of Biochemical Science
University La Sapienza
Roma, Italy
anna.tramontano@uniroma1.it

USLENGHI, Michela
Ist. Astrofisica Spaziale e Fisica Cosmica
IASF-Milano/INAF
Milano, Italy
uslenghi@iasf-milano.inaf.it

WADE, Rebecca C.
EML Research GmbH
Heidelberg, Germany
rebecca.wade@eml-r.villa-bosch.de

ZAVIDOVIQUE, Bertrand
AXIS Dept.
Inst. of Fundamental Electronics
University Paris Sud
Orsay, France
zavido@ief.u-psud.fr